Jaime Delgado   George D.
Al Mullery   Didoe Prevedc
Keith Start (Eds.)

Telecommunications and IT Convergence Towards Service E-volution

7th International Conference
on Intelligence in Services and Networks, IS&N 2000
Athens, Greece, February 23-25, 2000
Proceedings

Springer

Series Editors

Gerhard Goos, Karlsruhe University, Germany
Juris Hartmanis, Cornell University, NY, USA
Jan van Leeuwen, Utrecht University, The Netherlands

Volume Editors

Jaime Delgado
Pompeu Fabra University, Department of Technology
La Rambla, 30-32, 08002 Barcelona, Spain
E-mail: jaime.delgado@tecn.upf.es

George D. Stamoulis
University of Crete, Department of Computer Science
Knossos Ave., P.O. Box 2208, 7144 09 Heraklion, Crete, Greece
E-mail: gstamoul@csd.uoc.gr

Al Mullery
I C Europe
06450 Belvédère, France
E-mail: al.mullery@bigfoot.com

Didoe Prevedourou
INTRACOM S.A., Hellenic Telecommunications and Electronics Industry
Markopoulou Ave., 19 002 Peania, Attica, Greece
E-mail: dpre@intracom.gr

Keith Start
ORCA Research Associates
6 Balally Hill, Dublin 16, Ireland
E-mail: keiths@eircom.net

Cataloging-in-Publication Data

Die Deutsche Bibliothek - CIP-Einheitsaufnahme
Telecommunications and IT convergence : towards service evolution : proceedings /
7th International Conference on Intelligence in Services and
Networks. IS&N 2000. Athens, Greece, February 23 - 25. 2000. Jaime
Delgado ... (ed.). - Berlin ; Heidelberg ; New York ; Barcelona ;
Hong Kong ; London ; Milan ; Paris ; Singapore ; Tokyo : Springer. 2000
(Lecture notes in computer science ; Vol. 1774)
ISBN 3-540-67152-8

CR Subject Classification (1998): C.2, I.2.11, D.2, H.4.3, H.5, B.4.1, K.4, K.6

ISSN 0302-9743
ISBN 3-540-67152-8 Springer-Verlag Berlin Heidelberg New York

Springer-Verlag is a company in the specialist publishing group BertelsmannSpringer
© Springer-Verlag Berlin Heidelberg 2000
Printed in Germany

Typesetting: Camera-ready by author
Printed on acid-free paper   SPIN: 10719902   06/3142   5 4 3 2 1 0

Telecommunications and IT Convergence
Towards Service E-volution

Telecommunications is the enabler of the emerging Information Society, whose members will greatly benefit from the provision of affordable, high-quality, ubiquitous services. A revolution in information technology and telecommunications is already in progress, and will escalate during the beginning of the 21st century. The two worlds have already begun to converge; this convergence path is marked by the continuously expanding penetration and scope of telecommunications services, thus broadening the opportunities available to the users.

However, the forthcoming integration of information technology and telecommunications implies neither that computers will disappear behind an "active network" nor that networks will disappear behind the "all-pervasive" computers; on the contrary, both will appear within a single system. In this combined system, active computation power will be pervasive (in every room, in every appliance, for every person), while all this "computation" will be effectively totally interconnected; more importantly, its scope will not just be the home or the business, it will be the entire globe.

One of the driving forces of convergence is the increasing competition among the different service-supplier chains in the field. Successful and innovative combination of computation and telecommunication capabilities and features tends to help the players involved to acquire competitive advantage. However, there is still a considerable way to go, and several key questions have yet to be answered. Who will be taking the market-lead in this race and what appear to be the most likely evolution paths? Which technologies are or will be playing a major role? What will the role and impact of the Internet be in this convergence process and how is it expected to evolve through this? What will be the impact of other factors such as regulatory and pricing policies?

The IS&N conferences have been devoted to techniques and tools for development and management of services and communications. With the theme of IS&N'2000 being *Telecommunications and IT Convergence Enabling Service Evolution*, the conference especially relates to trends and enabling technologies that can be brought to bear on this new, integrated world and on the breakthrough it promises in everyday life in the new millennium.

The IS&N conferences have their origins in the RACE programme of the European Union co-funded research. What was initiated as a set of project workshops addressing communications management and programming infrastructure, became a series of truly international conferences, given the general interest of the subject matter. These conferences offered a unique opportunity to discover and discuss the overall technical area of telecommunications services and the relevant software, covering subjects such as communications management, Intelligent Networks, mobility, architecture, brokerage, security, service modelling and creation, etc. The Conference has become, over the years, a major forum for the discussion of issues and the exchange of outstanding technical ideas and results related to the areas above.

For IS&N'2000, seven chapters, with a total of 23 papers, have been selected for this book; namely, eCommerce, Developing Communications Efficiency through Management and Control, Architectures for Multimedia Communications, Service

Creation Techniques for Software Development and Deployment, Agent-Based Management, Virtual Home Environment, Integrated and Scalable Solutions for Telecommunications Management. This shows that the issues related to communications management, architectures, and service creation are still of great interest, while the virtual home environment is emerging as a new key topic in IS&N.

In summary, this book reflects the state of the art in research on IS&N topics, with the focus mentioned above, not only from European Union co-funded projects (mainly in the ACTS programme), but also from research organisations around the globe.

February 2000   Jaime Delgado
                                                     George D. Stamoulis
                                                          Alvin Mullery
                                                       Didoe Prevedourou
                                                            Keith Start

Previous IS&N Conferences and Proceedings

The first IS&N conference was organised in 1992 in Paris, France. Since then, the IS&N conferences have been held almost every year, with proceedings published as part of the Lecture Notes in Computer Science (LNCS) series of Springer-Verlag. These are as follows.

"Towards a Pan-European Telecommunication Service Infrastructure – IS&N'94", Hans-Jürgen Kugler, Al Mullery, Norbert Niebert (Eds.), Aachen, Germany, September 1994, LNCS 851, ISBN 3-540-58420-X.

"Bringing Telecommunication Services to the People – IS&N'95", Anne Clarke, Mario Campolargo, Nikos Karatzas (Eds.), Heraklion, Greece, October 1995, LNCS 998, ISBN 3-540-60479-0.

"Intelligence in Services and Networks: Technology for Cooperative Competition – IS&N'97", Al Mullery, Michel Besson, Mario Campolargo, Roberta Gobbi, Rick Reed (Eds.), Cernobbio, Italy, May 1997, LNCS 1238, ISBN 3-540-63135-6.

"Intelligence in Services and Networks: Technology for Ubiquitous Telecom Services – IS&N'98", Sebastiano Trigila, Al Mullery, Mario Campolargo, Hans Vanderstraeten, Marcel Mampaey (Eds.), Antwerp, Belgium, May 1998, LNCS 1430, ISBN 3-540-64598-5.

"Intelligence in Services and Networks: Paving the Way for an Open Service Market – IS&N'99", Han Zuidweg, Mario Campolargo, Jaime Delgado, Al Mullery (Eds.), Barcelona, Spain, April 1999, LNCS 1597, ISBN 3-540-65895-5.

Acknowledgements

The authors of the papers in this volume are warmly thanked for their contribution. The authors of the many submitted papers that could not be accepted are also thanked for their efforts.

The editors would like to thank the members of the Technical Program Committee and the numerous reviewers, listed below, for their effort in assuring the quality of this volume, as well as the section editors, who have written the introductions to each section. We would also like to thank Fofy Setaki for her most valuable contribution to the organisation of the conference and Raymond Llavador, who assisted in the review process. Finally, special thanks go to Mario Campolargo, Kyriakos Baxevanidis and Alessandro Barbagli of the European Commission for their support and encouragement.

Steering Committee

Alvin Mullery (Chairman), Jaime Delgado, Didoe Prevedourou, George D. Stamoulis

Technical Programme Committee

Jaime Delgado   (Co-Chairman)
George Stamoulis (Co-Chairman)
Chelo Abarca
Miltos Anagnostou
Alessandro Barbagli
Speros Batistatos
Kyriakos Baxevanidis
Nuno Beires
Hendrik Berndt
Jose Maria Bonnet
Steffen Bretzke
Stefan Covaci
Petre Dini
Sofoklis Efremidis
Dieter Gantenbein
Anastasius Gavras

Brendan Jennings
Henryka Jormaka
Andreas Kind
Thomas Magedanz
Reinhard Posch
Kimmo Raatikainen
Stelios Sartzetakis
Peter Schoo
Paul Spirakis
Keith Start
Sebastiano Trigila
Eric Veldkamp
Mike Wooldridge
Fabrizio Zizza
Johan Zuidweg

Organising Committee

Didoe Prevedourou, Fofy Setaki

List of Reviewers

Chelo Abarca
Miltos Anagnostou
Stefano Antoniazzi
Alessandro Barbagli
Speros Batistatos
Kyriakos Baxevanidis
Nuno Beires
Johan E. Bengtsson
Hendrik Berndt
Lennart H. Bjerring
Jose Maria Bonnet
Indranil Bose
Steffen Bretzke
Heinz Brüggemann
Gaetano Bruno
Magdaleine Chatzaki
Jaime Delgado
Luca Deri
Bruno Dillenseger
Manos Dramitinos
Sofoklis Efremidis
Pier Luigi Emiliani
Takeyuki Endoh
Motohisa Funabashi
Isabel Gallego
Dieter Gantenbein

Takeo Hamada
Harri Hansen
Keith Howker
Brendan Jennings
Mikael Jørgensen
Henryka Jormaka
Jorma Jormakka
Toni Jussila
Dimitrios Kalopsikakis
Fotis Karayannis
Andreas Kind
Anna Kyrikoglou
Maria Lambrou
Herbert Leitold
Beat Liver
Ferdinando Lucidi
Thomas Magedanz
Yannis Manolessos
Claudio Maristany
Ramon Martí
Al Mullery
Norbet Niebert
Vincent Olive
Thanassis Papaioannou
Symeon Papavasiliou
George Pavlou

Reinhard Posch
Martin Potts
Didoe Prevedourou
Kimmo Raatikainen
Tony Richardson
Jose Luis Ruiz
Stelios Sartzetakis
Peter Schoo
Hans-Detlef Schulz
Fofy Setaki
Vjekoslav Sinkovic
Paul Spirakis
George D. Stamoulis
Keith Start
Peter Stollenmayer
Thanassis Tiropanis
Sebastiano Trigila
Kirsi Valtari
Eric Veldkamp
Iakovos Venieris
Paul Wallace
Frans Westerhuis
Mike Wooldridge
Fabrizio Zizza
Han Zuidweg
Kostas Zygourakis

Table of Contents

eCommerce

Miltiades Anagnostou

National Technical University of Athens,
Department of Electrical and Computer Engineering,
Athens 15780, Greece.
miltos@cs.ntua.gr

Electronic commerce provides the capability of buying and selling products and information on the Internet and other online services [1]. eCommerce is about building better relationships among customers, products, and suppliers. Among its enablers are traditional tools and techniques, like e-mail, electronic funds transfer and WWW, while at the same time designers are eager to take advantage of new technologies, like mobile and intelligent agents, and service engineering. Online auctions, networked catalogs, portals, brokers, data mining, are concepts, practices and tools under constant development in the eCommerce arsenal [2].

Usability and trust are key issues. Any function of an eCommerce application has to be measured against its overall usability by the targeted user group [3]. Trust does not only involve progress in security and the existence of trusted third parties (TTPs), but also the proper usage of multimedia, which are challenged to somehow replace direct human contact. Let us remember that the electronic technology can easily shake a user's trust, since it is particularly well suited to produce illusory impressions of the attributes of a company by using interfaces and virtual spaces, and thus hiding its overall size and structure.

Quality is still another key issue, which can hardly be overestimated [4]. Everybody has seen poor Web page designs and defective payment facilities even in the largest companies and organisations. Since numerous businesses are now in their first attempt to enter the eCommerce realm, quality is a particularly important requirement, which can easily be sacrificed to cost cuts.

Life around the eCommerce business can be fast and interesting. In 1999 the EITO (European Information Technology Observatory) has surveyed 570 European businesses. By the end of this year 47% of this sample expected to be using eCommerce applications [5]. This figure was down to 6% in 1997.

While the proper exploitation of the analogy with traditional commerce concepts and the usage of proper metaphors will appeal to the customer and will facilitate the establishment of reliable procedures, one should never forget that eCommerce involves a deep change in the technological infrastructure of commerce itself. Therefore the deepening of the knowledge of the enabling technologies should gradually produce new concepts and subsequently new tools and practices. Thus we shall be led to a second generation eCommerce, which will efficiently rely on the new technologies.

Both papers in this session are perhaps unusual. While communications and legal issues are important in e-commerce, these papers rather deal with e-commerce *of* communications and legal issues.

Trading often not only involves easy to mechanise transactions, like carrying out a payment and delivering the goods, but also more demanding ones, like negotiations and comparison of similar products. In fact the negotiation process has always been

Trading often not only involves easy to mechanise transactions, like carrying out a payment and delivering the goods, but also more demanding ones, like negotiations and comparison of similar products. In fact the negotiation process has always been considered to consume high-level resources of the human brain. In some cases it relies on various capabilities of the negotiating parties, including cheating and other unorthodox behavioural patterns. However, as soon as a particular human activity is invaded by computer technology, there is an immediate demand for its rationalisation, and commerce seems to be no exception to this rule.

The first paper of this session describes how mobile agent technology can facilitate brokering activities for network resources and services [6]. The authors have shown that the use of mobile agents can give a new impetus to the provision of network services. Mobile agents are used to speed up most transactions and negotiation tasks. Separate processing of different services facilitates competition.

The second paper examines e-commerce of administrative and legal issues. The authors use as a starting point the fact that not only are documents distributed electronically but also we are able to automatically control the workflow of legal and administrative services. Subsequently they propose a distributed system that integrates all of these functionalities using electronic commerce.

References

1. R. Kalakota, A.B.Whinston, "Electronic Commerce", Addison Wesley. 1997.
2. R.Aaron, M.Decina, R.Skillen, "Electronic Commerce: Enablers and Applications", *IEEE Communications Magazine,* Vol. 37, No. 9, Sept. 1999, pp. 47-52.
3. M. Boscolo, M.Mastretti, E.Paollilo, "Technology as the catalyst of users' acceptance in Electronic Commerce", *Flexible Working: New Network Technologies*, IOS Press, Amsterdam, 1999, ISBN 1 58603 028 0, pp. 121-132.
4. C.K.Prahalad, M.S.Krishnan, "The New Meaning of Quality in the Information Age", *Harvard Business Review,* Sept.-Oct. 1999, pp 109-122.
5. European Commission, "Information Society Technologies: 2000 Workprogramme", draft version, IST Conference-Helsinki, 22-24 Nov. 1999.
6. "Agents technology in Europe -ACTS activities", ACTS project InfoWin AC113, Berlin 1999, ISBN 3-00-005267-4.

A Mobile Agent Brokering Environment for the Future Open Network Marketplace

David Chieng[1], Ivan Ho[2], Alan Marshall[1], and Gerard Parr[2]

[1] Advanced Telecommunications Systems Laboratory,
School of Electrical and Electronic Engineering, Ashby Bld, Stranmillis Road,
The Queen's University of Belfast, BT9 5AH Belfast, Northern Ireland, UK
{d.chieng, a.marshall}@ee.qub.ac.uk
[2] Telecommunication and Distributed Systems Group,
School of Information and Software Engineering,
University of Ulster at Coleraine, BT52 1SA Northern Ireland, UK
{wk.Ho, gp.Parr}@Ulst.ac.uk

Abstract. The growth of commercial activities across networks has led to the network itself becoming a competitive marketplace with a multitude of vendors, operators and customers. In such an environment, users will 'shop around' for the best deals in terms of network financial costs and services. Intelligent, autonomous and mobile software agents introduce an alternative approach that facilitates expertise-brokering activities on behalf of the users. In this paper we present such an agent supported brokering scenario. The scenario involves interactions between Java-based mobile agents and an interactive model of the network. The paper describes how this prototyping environment can be used to examine the impact of mobile agents for brokering network resources in the future open network marketplace.

1   Introduction

Over the past few years, a diverse variety of services have been rapidly introduced into the network. These have allowed a tremendous growth of trading activities across the networks. For example, in 1997, the global E-Commerce market was estimated at $10bn and this is predicted to rise to $200-300bn, by 2002. Additionally, over the same period, the global multimedia market will quadruple, from $150bn to $600bn [1].

This growth has led to the network itself becoming a competitive marketplace with a multitude of vendors, operators and customers. In this environment, users will be able to choose from a wide range of network services, which provide different bandwidth requirements and various Quality of Service (QoS), and which will operate under different pricing schemes [2]. It is anticipated that marketable resources will not be limited to end-user services. According to [3], technical and market forces are driving evolution towards the Traded Resource Service Architecture (TRSA) model whereby network resources such as bandwidth, buffer space and computational power will be traded in the same manner as existing commodities. To cope with these fast changing and complicated trading environments, users, whether buyers or sellers, need tools that facilitate expertise brokering activities such as buying or selling the right products, at the right price, and at the right time. This paper describes the use of intelligent mobile agent technology as a brokering tool for the various parties involved in a future network marketplace.

In this paper, the trading of resources such as network bandwidth and end-user services is examined. Similar studies have been conducted by Gibney [4], where a market-based approach was used to allocate network resources. Here, cost is imposed on each link and traded by software agents. Sun Microsystems Laboratories and Agorics Inc., have also demonstrated a system whereby ATM network bandwidth can be traded in real time between a bandwidth provider and a video server, and between the video server and a client, with the assistance of the bidding agents [5]. Both studies have demonstrated the benefits of employing software agents for the brokering of the network resources.

Section 2 gives a brief introduction to mobile agent technology and highlights its advantages over traditional client/server approach. A prototyping framework consisting of a multi-operator network model and a number of associated real-time agents, has been developed, and this is described in section 3. Section 4 explains the agent components' interactions and management. A number of case studies were carried out using this prototype system and their results are discussed in section 5. Section 6 presents the conclusions and some comments on the future directions for this work.

2   Mobile Agent Technology

Mobile Agent technology has been widely proposed as a key technology in the creation of an open, active, programmable and heterogeneous network environment. Many network systems employ client-server architectures which usually requires multiple transactions before a given task can be accomplished. This results in increased signalling throughout the network which can rapidly escalate, and reduce network utilisation as well as increasing the operating costs. An alternative approach is to dispatch a mobile agent to a host, where the interactions can take place locally, thus reducing the level of remote interactions required. Additionally, mobile agent executions are independent of both their runtime environments (i.e. operating systems) and the transmission system they transverse.

Software agents can be delegated to complete specific tasks on their own, providing that a certain set of constraints or rules have been defined. With the ability to move across the network, a software agent is no longer bound to the system where it is created and executed. When an agent starts executing in a server and tries to request a service that is not provided by that server, the agent then saves its current state and process and moves itself to a particular network node which provides that service. Upon arrival, the agent unwraps its process and state and resumes execution there.

The attributes of a mobile agent can be classified as follows:

Table 1. Mobile agent attributes and descriptions.

Attribute	Description
Autonomous	Agents can be proactive and able make autonomous decisions, based on their environment.
Communication	Agents can communicate with one another.
Cooperate	Agents can cooperate among themselves to achieve a task.
Mobility	The ability to move itself across any networks and platforms (Java-based).

There have been numerous works that have investigated the application of Mobile Agent technology to Network Management [6], [7] and [8]. Agent Technology offers an alternative solution to Network Management issues such as automatic network discovery, fault detection and more efficient bandwidth utilisation. This involves the remote execution and transmission of executable code between clients and servers in a distributed network environment. Hence, it becomes unnecessary to transfer intermediate data across the network.

With the exponential growth on the usage of Internet, E-Commerce is another area in which agent technology can be adopted [9]. Using agent technology, buying or selling for a product on-line becomes more convenient. Users can predefine the constraints for a product or service by which the agent would buy or sell for before it is launched into the network.

In order for mobile agents to work in a heterogeneous network, a platform independent programming language is necessary. Java has become the *de facto* language for creating mobile agents. Java offers portability, automated garbage collection and memory management required by agents. An agent environment framework is also required for the implementation, execution and management of agents. Several commercial agent frameworks have been developed such as ObjectSpace Voyager [10], IBM Aglet [11], General Magic Odyssey [12] and Meitca Concordia [13]. The ObjectSpace™ voyager ORB version 3.0 has been adopted for the implementation of our Agent prototype system solely because of the detailed documentation available on the framework, the ability to construct remote objects in the remote host and a set of control mechanisms that offer more flexible instructions on how the agent should terminate itself. There are two types of communication mechanism, namely Method Calling and ObjectSpace™ by which the Voyager agents interact with each other. The former mechanism enables an agent to call methods of another agent. This requires that the *calling* agent knows *a-priori* the method interface of the *called* agent. The latter mechanism enables the voyager agent to multi-cast an event message to other agents.

The level of intelligence in agents can range from hard-coded procedures to sophisticated reasoning and learning capabilities. The specific functions and requirements of an intelligent agent are the prime determinant of which techniques should be used. The amount of intelligence required by an agent is related to the degree of autonomy and the mobility that is required. If an agent must deal with a wide range of situations then it needs a broad knowledge base and a flexible inference engine. If it is mobile, there may be a premium on having a small, compact knowledge representation and a lightweight reasoning system in terms of code size.

For a software agent to perform actions, it must perceive events occurring around it, and have some concept of the state of its world by using *sensors*. A fundamental component of perception is the ability to recognise and filter out the expected events and attend to the unexpected. A software agent must be able to gather information about its environment actively or passively. The former involves sending messages to other agents or systems and the latter receives a stream of event messages from the system, the user or other agents. For example, in the context of an Interface Agent, it must be able to distinguish normal events (mouse movements) from significant events (double click on an action icon). In a modern GUI environment such as Windows or Xwindows, the user generates a constant stream of events to the underlying windowing system. The agents can monitor this stream and recognise sequences of basic use action as signalling some larger scale semantic event or user action.

Once an agent has perceptively recognised that an event has occurred, the next step is to take some action. This action could be to realise there is no action to take or it could be to send a message to another agent to take an action. Intelligent agents use *effectors* to take actions either by sending messages to other agents or by calling application libraries or system services directly.

3   The Prototype Multi-Operators Network Model

In order to carry out research on the agent brokering system, a prototype network model has been developed using the Block Oriented Network Simulator (BONeS) [15] as an agent testbed. A scenario, which involves three competing networks and a system of associated agents, is illustrated in Figure 1.

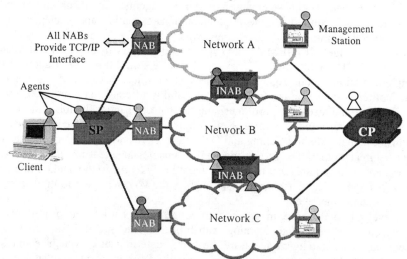

Fig. 1. Prototype Network Model for Agent Brokering Environment

The scenario presents three individual networks owned by different operators; each is identical in terms of the number of devices, topology and available resources. Currently each network consists of nodes connected in a mesh topology. As a means of creating competition, all three networks offer the same access to a remote *Content Provider (CP)* which provides multimedia services such as voice, data and video. Network operators A, B and C will then be competing to sell their network links to clients through a representative agent host, the *Network Access Broker (NAB)*. The *Service Provider (SP)* acts as a single point of contact between a *Client* and multiple *NABs*, rather than the client having their own negotiating agents hoping between multiple network hosts. Hence several different customers' resource requirements can be aggregated together in order to achieve a lower cost to the individual [Greenwood, 2]. The *NAB* provides TCP/IP socket interfaces for software agents to communicate with the model. There is a *Management Station (MS)* for each network, and this maintains routing tables, network pricing, resource updates, other management information and network statistics. The **Inter Network Access Broker (INAB)** serves as an intermediary between two neighbouring networks. This allows negotiation for access to one another's resources. This is useful if, for example, when Network A is unable to supply the requested bandwidth asked by a *SP* agent. Rather than turning down the request and hence losing potential revenue, Network A can buy supplemental bandwidth from Network B to compensate for the shortfall [Greenwood, 2]. Background traffic loads are supplied to the networks in order to create a dynamic, realistic scenario of a network system.

4   Agent Interactions and Simulator Interface

Figure 2 illustrates the system components and its agent interactions. Our agents are IPC (Inter Process Communication) enabled, and this allows them to interact with the Network Simulator using a predefined protocol. At this stage, the security issues have not been considered as essential for demonstrating the system. However, it is anticipated that future development of the agent system will need to address these issues.

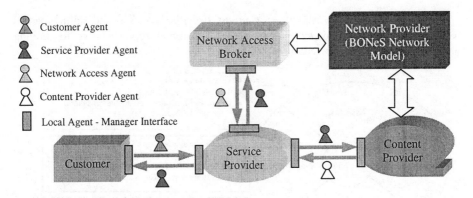

Fig. 2. System Components and Agent Interactions

There are five component parties in this architecture: the *Customer, Service Provider, Content Provider, Network Access Broker* and the *Network Provider*, which constitute the agent system. Each component consists of its own database, agent(s) and a manager. The manager's job is to provide service for any arriving agent, handle data transactions, storage retrieval, agent creation and task assignment. Figure 2 shows that a customer agent requests a service by moving itself to the central service point (*Service Provider*) having being launched by the *Customer* and interacts with the local object (*Manager*), which contacts its server to create an agent and delegates tasks to it. The event diagram of general inter-agent interactions is illustrated in figure 3.

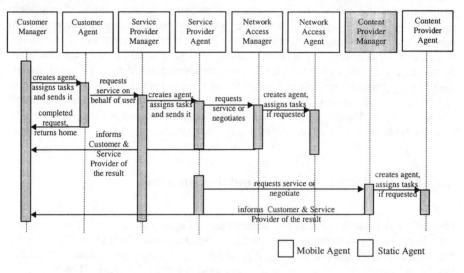

Fig. 3. Event Diagram for General Inter-Agent Interactions

The *Network Access Broker* and *Content Provider* serve as the selling parties whereas the *Service Provider* agent acts as a customer representative for buying network resources and services.

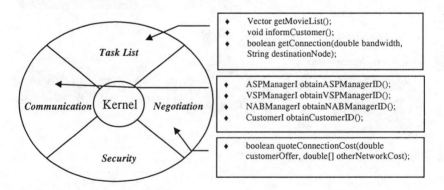

Fig. 4. Agent Model

An agent must follow its task list in order to fulfil the assignment given by its manager as shown in figure 4. This list contains a set of operations that need to be carried out and the locations to be visited. An agent can alter the route plan depending on the surrounding environment. The kernel of the agent contains its agent ID and local memory for operational and task related processes. The communication component manages all communication between the agent and external parties. This extends to: (i) access to host sites and (ii) interaction with other local agents once resident. The negotiation component is responsible for managing the bartering process between the agents. At this stage only a very rudimentary form of negotiation has been implemented, enabling agents to request resources for a connection with bandwidth, QoS and cost requirements, directly from the local object (i.e. a *NABManager*). Finally, the security component has not yet been implemented but exists as a key constituent of the agent. The intention is to introduce an authentication mechanism, whereby an agent can be assured that a potential host or neighbour agent is non-malicious and vice-versa.

5   Case Studies

A case study framework that demonstrates the usage of mobile agents is presented in this section.

5.1   Part A – Hunting for the Best Deal (e.g. Movie and Network Connection)

We start with a user who wishes to download a movie list from a remote video provider site, Video Service Provider (*VSPServer*) into his/her PC. A GUI is provided to specify preferences and is shown in figure 5.

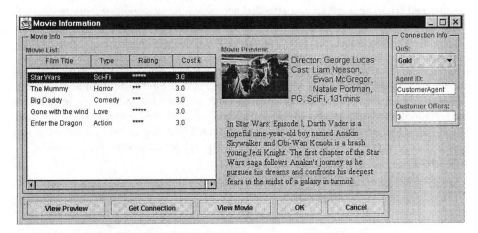

Fig. 5. Screenshot of Customer GUI

As illustrated in figure 6, the user's agent (*CAgent*) is launched into the network and arrives at the central service point, Access Service Provider (*ASPServer*) and contacts the local object (*ASPManager*) for movie list retrieval. After finishing the request, the *CAgent* returns to the *Customer* and waits for notification from *ASPAgent* launched by the *ASPManager*, the communication channel between the customer and *ASPServer* is closed during the waiting period. It processes the request and sends an agent (*ASPAgent*) to the *VSPServer* to contact the local object *(VSPManager)*. The request is passed to the *VSPManager* that stores/processes the request and hands over the movie list to the *ASPAgent*. Before notifying the customer that the movie list has been retrieved, it returns to the *ASPServer* to store the information and data.

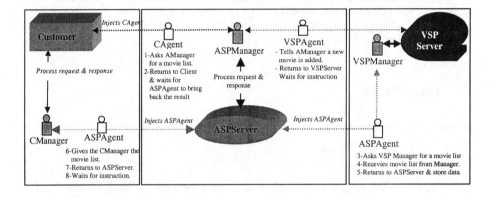

Fig. 6. Getting Movie List

To get a network connection, a user's mobile agent that contains a set of predefined criteria e.g. maximum afforded per view price and video quality is injected into the network. Upon arriving in the *ASPServer* as illustrated in Figure 7, the agent contacts the local object, the *ASPManager* for a service request. The request is then processed, stored and tasks are then delegated to an agent to query/negotiate the external entities (*NABManager*) which is resided in the remote host for that particular service. The *ASPAgent* visits all the *NABServer (A, B and C)* to get the respective quotations for the connection price according to its route plan. The inter-agent negotiation processes are shown in figure 8 and are recorded into the agent's virtual base. The agent will return to the customer when the best deal is found. Currently, the predefined criteria for service request, e.g. the network connection, are embedded in the agent.

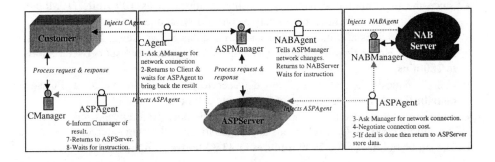

Fig. 7. Getting Network Connection

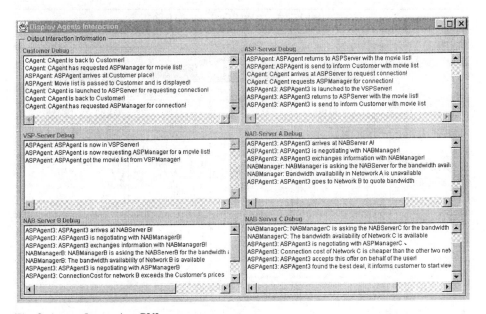

Fig. 8. Agents Interaction GUI

5.2 Part B - The Impact of Pricing Strategy on Network Loads

In part B, we present an associated scenario where the three network operators compete to sell their bandwidth by constantly changing their prices. This has an impact on their network loads. The specifications for this scenario are:

- Users request for video services such as MPEG2 video stream from the *VSP* through *ASP*.
- For preliminary studies, **10 units** of fixed bandwidth (*BW Unit*) were allocated for each connection. 1 *BW Unit* can be abstractly defined as 1kbps, 100kbps or 1Mbps. In this case, 1 *BW Unit* can be considered as 100kbps.

- Since the maximum BW capacity per link is allocated as **100 units** (10Mbps), therefore the maximum number of active connections at one time is **10**.
- The *VSP* provides all kind of video services such as MTV, News, Documentaries, Movies etc. Hence the duration for each connection is randomly set between **5mins to 180mins.**
- Users' requests arrival rate for video services are according to a **Poisson distribution** and has an average of **7 requests per hour**.
- The users will choose the cheapest network as the only preference at this stage. In the future scenarios, QoS selections and negotiations will be considered.
- The simulation is run for **2000 minutes** or **33.33 hours** (simulation time).

Figure 9 shows the changes in price and the corresponding number of connections (i.e. network loads) for each network operator, over the period of time.

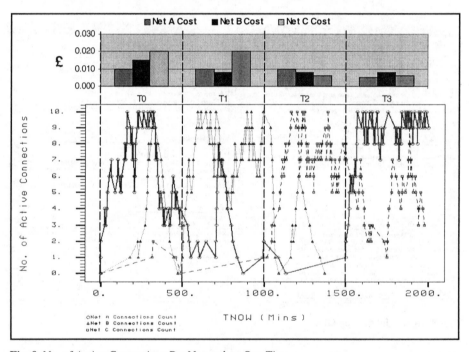

Fig. 9. No. of Active Connections Per Network at One Time.

The initial costs for all the networks are shown in time interval T_0. The changes in price take place at the beginning of time intervals T_1, T_2 and T_3 as displayed on the top of the graph. During time interval T_0, Network A dominates the connection counts as shown because it is offering the cheapest connections. At $T_0 \cong 200$mins, Network B begins to get connections due to network reaching link saturation. Towards $T_0 +500$, Network B begins to win custom through price reduction. This pattern continues with each network winning connections and revenue from one another through a combination of price reduction and competitor link saturation [Greenwood, 2]. From this rather straightforward pricing strategy, a question arises as to how many more price reductions can each network afford in order to constantly win/maintain customers. A consensus is therefore necessary regarding profit margin, pricing policy, etc. Furthermore, winning too many customers at one time will cause network congestion and consequently lose potential future customers due to link congestion and QoS degradation. Figure 10 shows the breakdown of revenue generated by each network.

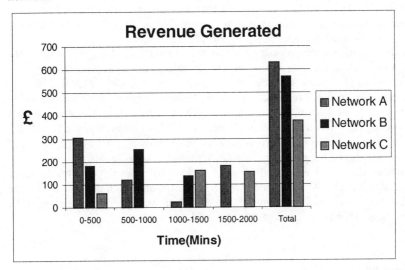

Fig. 10. Revenue Generated at Time Intervals

Figure 10 shows the importance of setting the right price at the right time. Being the most expensive will result in losing all the customers. By contrast, offering the cheapest connection will lead to network congestion and a drop off in profit. In this simulation, users only pay at the end of the session. The calculation for revenue per connection is based on:

Revenue = Cost/BW Unit * BW Used * No. of Links * Session Length (mins)

6   Conclusions and Future Work

In this paper we have discussed how mobile agent technology can facilitate brokering activities for network resources and services in the future open network marketplace. A prototyping framework, which consists of a multi-operator network model and a number of associated real-time agents, has been developed to enable analysis. The results show that the use of mobile agents introduces a much greater dynamic to the provision of network services. It is clearly seen that mobile agents have taken over most transactions and negotiation tasks, and therefore increase the speed of decision making according to user preferences. Furthermore this dynamic is accentuated by the increased competition (on a per service basis, rather than on per day or per week) which can be introduced into the market by agents representing new or aggressive network operators.

The interactions between a customer's agent and the static object in the remote hosts relies heavily on the usage of the *"future call back"* method which is documented on the Voyager ORB version 3.0's user guide [10]. Each agent has a set of tasks, and each task can be performed at the server, which provides a local object to interact with the mobile agent. The communication link between the Network Access Brokers and the network model is based on socket processing which is the only method to maintain the communication channel. In our current system, the *'rule-based approach'* is applied to enable the agents to negotiate.

Further developments are proposed for this framework. For the agent system, the communication between agents limits the syntax and semantic representation for a real negotiation to take place. In order to overcome this limitation, KQML [16] will be implemented to enable a realistic bargaining between agents. Security is another concern for the agent system; it prevents any hostile agent from violating another agent's privacy. This introduces the implementation of a common place for all the agents to perform trading (i.e. a "virtual network marketplace"). Also, some forms of federation policy can be adopted to enable fair-trading between agents.

The simulation will be enhanced to offer alternative paths through the network when congestion occurs. At the user level, service attributes such as QoS for movie viewing can be categorised in terms of Gold, Silver and Bronze. These may be provided as value added services in our future model.

Acknowledgement

Funding from Fujitsu Telecommunications Europe Ltd. is acknowledged. Input from Dr Colin South and Dr Dominic Greenwood is also appreciated.

References

1. Foreword, Ian Morphett, Managing Director, BTUK, Products and Services, *BT Technology Journal*, Vol. 17 No.3, July 1999.
2. C. Philips and P. Kirby, 'ATM and the future of telecommunication networking', *Electronics & Communication Engineering Journal*, June 1999, pp.108-124.
3. D. Greenwood, D. Chieng, A. Marshall, I. Ho, G. Parr, 'Brokering Marketable Network Resouces Using Autonomous Agents', Submitted to *World Telecommunications Congress/International Switching Symposium (WTC/ISS2000)* conference, Birmingham, U.K., May 2000.
4. M.A.Gibney, N.R.Jennings, N.J. Vriend and J.M.Griffiths, 'Market-Based Call Routing in Telecommunications Networks using Adaptive Pricing and Real Bidding', *IATA'99*, Stockholm, Sweden, 1999.
5. Mark S. Miller, David Krieger, Norman Hardy, Chris Hibbert, E. Dean Tribble, 'Chapter 5: An Automated Auction in ATM Network Bandwidth', *Market-Based Control: A Paradigm for Distributed Resource Allocation,* Edited by S.H. Clearwater, World Scientific Publishing Hardcover, 1994, pp. 96-125.
6. R.G Davision, J Hardwick and M.Coz, 'Applying the agent paradigm to network management', *BT technology Journal*, Vol 16 No 3, July 1998, pp. 86-93.

7. D.Gavalas, M Ghanbari, D.Greenwood, M.O'Mahony, 'A Hybrid Centralised Distributed Network Management Architecture'.
8. Andrzej Bieszczad, Bernard Pagurek, Tony White, 'Mobile Agents for Network Management', *IEEE Communications Survey magazine*, Sept 1998.
9. http://agents.www.media.mit/edu/groups/agents/projects/
10. http://www.objectspace.com/voyager
11. http://www.trl.ibm.com/aglets
12. http://www.generalmagic.com/technology/odyssey.html
13. http://www.meitca.com/HSL/Projects/Concordia/
14. http://www.infc.ulst.ac.uk/~phhwk/publication/IZS2000.htm
15. BONeS DESIGNER® Ver 4.01, Alta Group™ of Cadence Design Systems, Inc.
16. T.Finin, "Specification of the KQML as an Agent Communication Language", http://www.cs.umbc.edu/lait/papers/kqmlspec.ps

Legal and Administrative Services through Electronic Commerce

Silvia Llorente [1]

Media Technology Group, S.L.
Barcelona, Spain
Silvia.Llorente@mtg.es

Jaime Delgado

Universitat Pompeu Fabra (UPF)
Barcelona, Spain
jaime.delgado@tecn.upf.es

Abstract. This paper presents the issues related to the implementation of electronic commerce of services. To illustrate the main features of electronic commerce of services, we have selected legal and administrative services. We describe in the paper the actors who take part in these services, the system architecture and the workflow that defines the interactions and interchange of information among actors. With these elements we have implemented an electronic commerce application inside the TRADE project [1] that demonstrates how these services can be offered in an electronic manner.

1 Electronic Commerce of Legal and Administrative Services

This section attempts to describe in a general way the characteristics of electronic commerce of services. In particular, we have selected legal and administrative services to show the advantages of providing services through electronic means.

1.1 Electronic Commerce of Services Vs. Electronic Commerce of Products

When we talk about electronic commerce, we are usually thinking about electronic commerce of products, but there is also another kind of electronic commerce, the electronic commerce of services. Whilst electronic commerce of products consists on selecting the items we want to buy from a catalog and paying for them, electronic commerce of services consists on searching for a professional or company that offers the service and contracting them. During service development, there can be information request, payment or contacts with other users.

1.2 Characteristics of Legal and Administrative Services

We have studied in detail the legal and administrative services in order to describe them. These are their main features:

- There is a purchase of services. The result of these services (normally documents) is delivered after off-line development.
- Many official documents must be interchanged.
- There are business-to-consumer and business-to-business aspects.

---

[1] Working in her Ph.D. at the Universitat Politècnica de Catalunya (UPC).

[2] This work has been partly supported by the European Commission (TRADE, ACTS project 328) and the Spanish government (TEL98-0699-C02-01).

- Payments can be requested in different moments (in advance, after service, periodically, etc) and several mechanisms can be used (credit card, cheque, bank transfer, etc).
- Confidentiality and non-repudiation is needed.
- Many actors are involved in the services. These actors have a lot of interaction among them.

1.3   Advantages of Providing these Services Through Electronic Commerce Means

In legal and administrative services today, many documents are interchanged among the users involved. These documents are usually delivered by fax, courier, post or in person. If there is an error, the document must be resent, involving a waste of time and money. If legal and administrative services were offered in an electronic way, the professional could send documents from his own PC, after writing them. The document could be resent as many times as needed with just a mouse-click. Here we have the first advantage of providing these services in an electronic way.

Most of the steps of legal and administrative services have deadlines that lawyers or administrative consultants must control. If we have an electronic commerce application that controls service development, it could also control deadlines for each case of a lawyer or administrative consultant and notify the involved users that some actions must take effect.

Official documents normally have to be signed by several actors involved in the case. At the present moment, some laws and measures are being developed [8] in order to accept digital signatures as the handwritten ones. If this happens, electronic documents could be digitally signed and have the same value as the paper documents in front of administration.

Nowadays, we can send documents in an electronic way using e-mail or ftp. There are also centralized applications that control the workflow of legal and administrative services automatically. Our proposal is to define a distributed system that integrates all of these functionalities using electronic commerce, permitting all users involved in a case to have access to its information and status at any time, in a customized manner.

2   An Architecture for Electronic Commerce of Legal and Administrative Services

Based on the actors involved in legal and administrative services, we have defined the business and architectural models to facilitate the implementation of the services to be provided in an electronic commerce environment.

2.1   Business Model

In this section, we describe the actors involved in the legal and administrative services and their relationship.

2.1.1   Actors in Legal Services
In figure 1 we can see the actors who can take part in a legal service. The arrows represent the interactions among them.

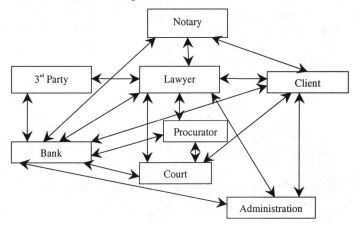

Fig. 1. Actors in legal scenario

The main actor in legal services is the lawyer. We will see in following sections that all the flow of information of a legal case passes through him.

2.1.2   Actors in the Administrative Services
In figure 2 we can see the actors who can take part in an administrative service. The arrows represent the interactions among them.

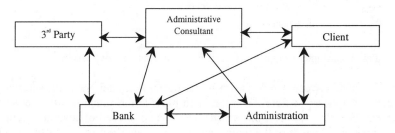

Fig. 2. Actors in administrative scenario

In these services, the administrative consultant is the main actor. He collects the information and is in charge of sending it to the corresponding actor.

2.2   Architectural Model

Our architecture is based on an agent that handles the interaction among the different actors. Each actor that appears in figure 3 corresponds to a network site. Only the agent is located in a fixed site, the rest of the actors can connect to the system at any site just using a browser.

In general, the users of an electronic commerce application are divided into customers and suppliers. We consider that an actor of the system is a customer if he has to pay for services received from other actors in the system. Taking this into account, lawyers and administrative consultants are customers and suppliers at the same time, because, apart from providing services to clients, they receive services from notaries, procurators or third parties. Figure 3 shows the customers-suppliers division for the legal and administrative services.

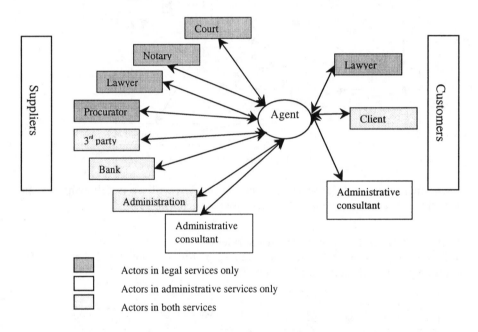

Fig. 3. Customer - supplier actors in legal and administrative services

The interactions between actors shown in the business model are made through a service agent, as appears in figure 3. In order to provide the functionality needed to connect customers and suppliers, this service agent is composed of several modules. For detailing the architecture of the three elements or levels (brokerage, customer and supplier) of our architectural model, we have adapted the architectural model defined in [1] to our particular case, as explained next.

The *Brokerage agent level* consists of the following building blocks:
Navigation access handler: Allows the users to navigate and browse in the service agent for information about services and suppliers.

General and management database: Manages the service agent database. In this database the information about services, customers, suppliers and all the information related to services contracted by customers is stored.

Supporting services: This block provides services for electronic brokerage that are required as supporting services, but are not part of the electronic brokerage. Some of these services are security services, electronic copyright management, etc.

Search module: Provides search functionality to help user access to services and suppliers. This module supports combined searches using several fields of information such as the speciality of the service or the city.

The *Customer level* is formed by:
Navigation access: Allows the user to browse and search for suppliers in order to contract a service. Once the service is contracted, it enables the user to query its associated information and to send or receive the documentation needed for the correct legal service provision.

The *Supplier level* is compound by:
Navigation access: Allows the supplier to request the information about the services that customers have contracted. It includes the sending and reception of documentation needed for the legal service provision. It also permits some kinds of users to create and start new services. The functionality it offers depends on the kind of supplier.

Figure 4 shows the different levels and the building blocks of each one.

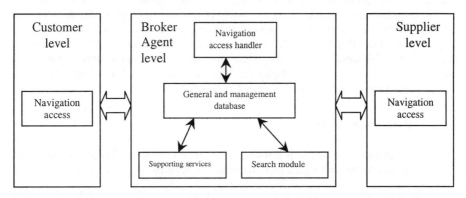

Fig 4. Customer, supplier and agent levels building blocks

3   Workflow and Information Management for Electronic Commerce in Legal and Administrative Services

In order to implement a system for electronic commerce of legal and administrative services, we need to complement the previous architecture with the definition of a workflow system to control, in the service agent, the flow of information and interactions between actors.

The workflow of legal and administrative services can be seen from two points of view:

- Business model point of view, that is, what the actors see during the legal service provision.
- Communications point of view. In other words, the underlying steps that are necessary to communicate between the service agent, customers and suppliers.

In the next section we describe the legal workflow from these two points of view. We do not describe the administrative workflow because it is a simplification of the legal one. More information about the workflow of these services is available in [2].

3.1   Legal Workflow

The workflow of a legal process can be divided in the following two phases:

- Preliminary workflow: In this phase only the client and the lawyer are involved.
- Case development workflow: In this phase, all the legal actors (notary, procurator, administration, court and 3rd party) depending on the kind of service contracted can be involved. This workflow can be divided in three parts: Document acquisition, procedures and result.

We describe the preliminary workflow and the case development workflow next.

For the preliminary workflow we show two schemas: one from the point of view of the business model and the other from the point of view of the communications.

For the case development workflow we show a complete schema from the point of view of the business model. As document acquisition, procedures and result workflows have a very similar schema from the point of view of the communications, we only show the schema of the document acquisition.

3.1.1   Preliminary Workflow

Figure 5 illustrates the steps followed by the client and the lawyer to make an agreement in order to start a case from the point of view of the business model.

Preliminary workflow: This workflow describes how a service is contracted. The client searches a lawyer and sends him an exposition of the problem. If they arrive to an agreement, the lawyer starts a case to provide the service requested by the client.

From the communications point of view, the schema of the *Preliminary workflow* will be as figures 6 and 7 show. We have separated the workflow in two figures for clarity. The numbers represent the order of the actions performed by the actors.

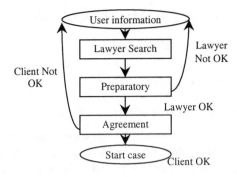

Fig. 5. Preliminary workflow

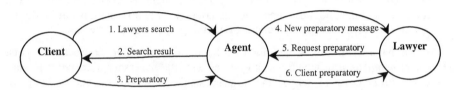

Fig. 6. Lawyer search and preparatory

1. The client requests the lawyer search from the service agent
2. The service agent returns the list of lawyers who accomplish the selected criteria
3. The client selects one of the lawyers and sends him an explanation of his problem
4. The lawyer receives a message from the service agent indicating that he has a new explanation to read
5. The lawyer requests the preparatory from the service agent
6. The lawyer receives the preparatory

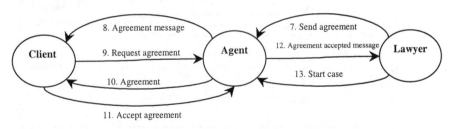

Fig. 7. Agreement and start case

7. The lawyer sends the agreement to the client's exposition
8. The service agent sends to the client a message to indicate that the lawyer's agreement has arrived
9. The client requests the agreement
10. The service agent sends the agreement to the client
11. The client reads and accepts the agreement

12. The lawyer receives a message from the service agent indicating that the client has accepted the agreement
13. The lawyer starts a new legal case

3.1.2 Case Development Workflow

The main actor in the case development workflow is the lawyer. He controls:

- the state of the case: He starts, finishes or adds steps to the case.
- the actors who are involved in the case, adding new ones when necessary (these new actors can be notaries, procurators, other lawyers, other clients, courts, administrations or third parties). The sender and the receiver of the information interchanged in the case define these actors.
- the information sending and retrieval. All the information of the case must be received or sent by the lawyer (for example, the client receives a document from the court and then he must send it to his lawyer). This information can be digital documents, forms, paper documents or anything necessary for the case.

As we have said, this workflow can be divided in three consecutive parts: *Document acquisition* in which the lawyer requests the necessary documents to start the case, *Procedures* in which the lawyer sends these documents to court or administration (depending on the kind of legal service) and finally the *Result*. We will represent the three consecutive workflows in only one schema for brevity.

Case development workflow

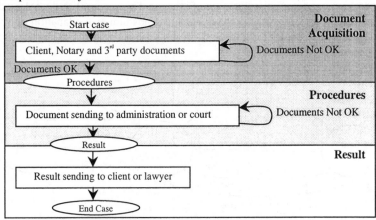

Fig. 8. Case development workflow

Document acquisition: The lawyer requests the documents he needs from all the actors involved in the case (at this point the actors can be: Client, notary and 3rd party).

Procedures: The lawyer sends the documents directly or through a procurator to court or administration (it depends on the specific legal service).

Court or administration carry on with the procedure depending on the legal service (the taken steps differ from service to service). They usually have to contact with different actors.

Result: The court or administration sends the result to the lawyer (directly or through a procurator) or the client (it depends on the specific legal service). The case is finished.

Fig. 9 Document acquisition workflow

1. The lawyer requests some documents from the client in order to start the legal case
2. The client receives a message from the service agent, indicating that he has to send some documents to the lawyer
3. The client requests the information of the documents he has to send
4. The service agent sends the information to the client
5. The client sends the requested documents to the service agent
6. The service agent sends a message to the lawyer telling him that the client has sent the requested documents
7. The lawyer requests the documents to the service agent
8. The service agent sends the documents to the lawyer

The procedures and result workflows are very similar to document acquisition workflow. One actor requests documents or information necessary to continue the case to another actor through the service agent.

3.2   Information Management in the Legal and Administrative Scenarios

Our service agent has to deal with different levels of information and store them in its database.

On the one hand, we have defined the general information needed to control the workflow of the legal and administrative services. This can be named services meta-information. Some samples of meta-information are the service *skeleton* (the basic steps that form the service), the more usual information for each step and the actors involved in general in the service.

On the other hand, we have the information of each contracted service. This information, taken steps, interchanged documents, people, may differ from the general case defined by services meta-information. It is necessary to store this information, as each user in a case always has to see the correct information of the status.

Finally, we have documents, forms and payment documents that users interchange. They are sent through the service agent but it only stores its status (sent, received, read, accepted, not accepted, etc) in its database.

4   An Implementation: the TRADE Project

The TRADE (TRiAls in the Domain of Electronic commerce) project implements the ideas explained in the previous sections. This project started in March 1998 and is planned to finish in February 2000. More information can be found in [3].

TRADE (a European Commission co-funded project in the ACTS programme [5]) implements a distributed system where lawyer and administrative consultants can offer their services by electronic means. The client can contract these electronic services.

TRADE not only implements legal and administrative scenarios applications but there are other applications and building blocks, as explained in [3]. Those building blocks and applications are not relevant for this paper.

4.1   Functionality

The functionality offered by the TRADE application is (in brackets the users who can perform each operation appear):
- User registration
- Service search
- Lawyer search   (Client only)
- Subscription search
- Negotiation and agreement (Client ⇔ Lawyer, Client ⇔ Administrative consultant)
- Status/Case information (All actors involved in a case)
- Periodical payments (Procurator, Lawyer, Client, Administrative consultant)
- Periodical administrative cases (Client, Administrative consultant and other actors involved in each particular case)
- Broadcast (Lawyer and administrative consultant)
- Send information to subscribers (Lawyer and administrative consultant)
- Payment (All actors)

4.2   Goals

The main goals of the implemented application are the following:
- Simple software needed by suppliers and customers. A user can access to the application simply using a browser (like Netscape Communicator or Microsoft Internet Explorer). It is possible to install a security module to cipher the communication, but the application can run without it.

- All the workflow of the cases is controlled by a program installed in the service agent.
- Use of standard languages, like XML [6], to express information with a defined format in real life (such as administration forms, receipts or bank orders). Doing so will facilitate its processing and integration with other existing and future deployed applications.

4.3   TRADE Application for Legal and Administrative Scenarios

Using the specifications made for legal and administrative services in TRADE project [2] we have developed a prototype application with the functionality listed in section 5.1.

We started the prototype development in mid-98 and the first version was running at the beginning of 1999. More functionalities have been added in the consecutive versions.

We have validated the application with real cases. They are stored in the application database, but the lawyer can create new services as needed.

The legal cases we have used for this test are:
- Divorce
- Dismissal
- Ordinary labour proceeding
- Proceeding involving claims less than a standard amount
- Undefended separation
- Contentious separation

The administrative cases used to test the application are the following:
- VAT declaration
- Personal income tax
- Vehicle number plate registration
- Sick leave certificate
- Driving license renewal

They are stored in the application database, but administrative consultant can create new ones.

The application and some demonstrations are available in [4]. Demonstrations allow users to familiarize with the program before using it.

4.4   Technical Aspects of the TRADE Project

In figure 10 we can see the software elements needed by users of the TRADE system (these users are customers and suppliers defined in the architectural model and in this figure must be located in the client side) and by the server (that is, the service agent).

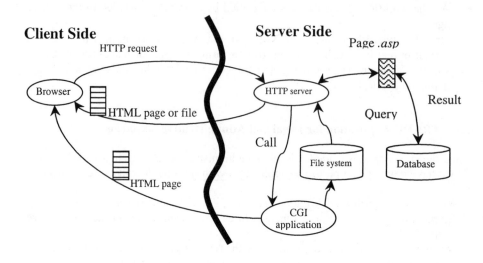

Fig. 10. TRADE system architecture

A normal sequence of actions between the TRADE server (the service agent located in the Server Side) and an actor (Client Side) can be summarized as follows:

1. The actor connects to the server using a standard web browser (no extra software is needed in his site).
2. He requests a web page with some information about services he is involved in.
3. The TRADE server receives the request and generates an HTML page using the ASP pages installed in the machine.
4. This HTML generated page is sent to the actor's browser.
Steps 2 to 4 are repeated until the actor disconnects from the server.

Apart from ASP pages, also XML files can be requested by customers. These files represent structured information, like administrative forms.

5   Conclusions

This paper has focused on the description of electronic commerce of services vs. the *traditional* electronic commerce of products. To illustrate electronic commerce of services we have selected an example, the legal and administrative services.

For legal and administrative services we have presented an architecture and workflow, that became the guide for the real implementation developed for the TRADE project.

Some of the described services in the paper can be sent to the administration using electronic ways. If we could express the information contained in the forms in a standard document representation language, such as XML, the integration with existing systems and the automation of the information interchange with the administration would be easier. Some work is being done in this area but for different scenarios, like medical or insurance scenarios [7].

References

1. GALLEGO I., DELGADO J. and ACEBRÓN J.J. "Distributed models for brokerage on electronic commerce", International IFIP Working Conference Trends in Electronic Commerce (TREC'98), 3-5 June 98, Hamburg (Germany). Springer, ISBN 3-540-64564-0.
2. TRADE Document AC328/COS/R&D/DS/R/015/a1 "Specification of TRADE applications", June 1998.
3. TRADE web page, http://trade.cosi.it/.
4. TRADE Legal and administrative scenarios web page, http://trade.ac.upc.es/.
5. ACTS Web Page, http://www.infowin.org/acts/.
6. W3C XML Web Page, http://www.w3.org/XML/.
7. EBES Web Page http://www.cenorm.be/ebes/.
8. Spanish Law, REAL DECRETO-LEY 14/1999 de 17 de septiembre, sobre firma electrónica, Boletín Oficial del Estado número 224, 18th September, pages 33593-33601.

Developing Communications Efficiency Through Management and Control

Tony Richardson

Systems Technology Solutions Ltd.
sts@anglianet.co.uk

Communication Service Providers are under continual pressure to improve the Quality of Service (QoS) they offer to their customers. Added to this are a number of additional business drivers, these include the need to provide services at ever reducing costs and a number of factors of convergence which are appearing within the communications industry.

These issues of convergence include increased use of general purpose IT technologies (such as CORBA) in the development of communication systems; moves towards the integration of switched voice and IP-based data services over a common communication media – such as an ATM infrastructure; and the convergence of fixed and mobile communications services. In addition to these items there are ever increasing demands by customers for communication services which are supported by Service Level Agreement (SLAs).

Central to these trends are the many Management and Control capabilities which will need to be deployed to support these QoS, convergence and SLA needs. Of particular importance will be capabilities which provide for increased communication efficiency.

This session looks at some of the management and control capabilities which provide for more efficient communications services in three diverse areas related to the areas of convergence outlined above.

The first, "Load Balancing for a Distributed CORBA-Based SCP" looks at issues surrounding the development of a distributed systems implementation of a Service Control Point (SCP). Here a particular concern is improvement of overall systems performance through appropriate choice of process load-balancing algorithms within the distributed CORBA-based environment. Simulation results show the benefits of an ant-based algorithm – which is presented in the paper.

The next "Measurement Based Connection Admission Control Algorithm for ATM Networks that use Low Level Compression" looks at the communications efficiencies which can be obtained through the use of data compression in ATM cells. Using the compression techniques described in the paper, coupled with an appropriate choice of ATM Call Admission Control (CAC) algorithm, greatly improved communication link efficiencies were demonstrated through simulation test results.

Finally, "Integrating Position Reporting, Routing and Mobility Management in Multihop Packet Radio Networks" considered the issues of developing improved routing and mobility management in dynamic ah-hoc mobile networks. A simulation model produced results which demonstrated the combined benefits of applying geographic position reporting and multiple zones to provide improved efficiency of communications routing. The resultant flat routing architecture will be of particular value in networks which are of large diameter and have diverse mobility patterns (e.g. battlefield or disaster / emergency situations).

Load Balancing for a Distributed CORBA-Based SCP

Conor McArdle[1], Niklas Widell[2], Christian Nyberg[2],
Erik Lilja[2], Jenny Nyström[2], Thomas Curran[1]

[1] Dublin City University, Glasnevin, Dublin 9, Ireland
mcardlec@teltec.dcu.ie

[2] Department of Communication Systems, Lund Institute of Technology,
P.O. Box 118, SE-221 00, Lund, Sweden
niklasw@tts.lth.se cn@tts.lth.se

Abstract. This paper examines load balancing issues relating to a distributed CORBA-based Service Control Point. Two types of load balancing strategies are explored through simulation studies: (i) A novel ant-based load balancing algorithm, which has been devised specifically for this type of system. This algorithm is compared to more traditional algorithms. (ii) A method for optimal distribution of the computational objects composing the service programs. This is based on mathematically minimising the expected communication flows between network nodes and message-level processing costs. The simulation model has been based on the recently adopted OMG IN/CORBA Interworking specification and the TINA Service Session computational object model.

1   Introduction

There is increasing interest in the use of object-oriented Distributed Processing Environments (DPE) as the infrastructure for new telecommunications service platforms as they promise the benefits of more flexible service design and service deployment, increased software reuse and increased interconnection capabilities with external resources such as the Internet and private databases. One such technology, the Object Management Group's (OMG) Common Object Request Broker Architecture (CORBA), has already gained acceptance in the industry for use in network management applications and there have been recent standardisation initiatives for its introduction into the Intelligent Network (IN) [3]. One of the first evolutionary steps towards the introduction of CORBA to the IN has been seen as the replacement of the Service Control Point (SCP) with a CORBA-based distributed system [6]. This approach allows investment in most of the legacy IN infrastructure to be preserved while bringing to bear the advantages of CORBA.

Although CORBA promises many technological and business advantages for this type of application, there are important performance concerns that need to be addressed so that distributed CORBA-based systems can provide the real-time performance and reliability characteristics that are required of telecommunications systems. Sharing of load between the nodes of the distributed system is one important issue that can greatly impact on overall performance and reliability. This paper examines this issue, taking into account both processing load due to service related tasks and processing load due to inter-node communications, which can be quite significant in CORBA-based systems. Two solutions to load sharing in this type of environment are examined. (i) The choice of location of service objects on nodes in the network can greatly effect performance. An optimal method is used, which minimises total processing costs. Although location of service objects provides a basis for combating loading problems, it is static i.e. determined at design time based on expected service demands and it does not account for queuing and stochastic effects in the network. (ii) More dynamic methods are required to cope with variations in the arrival rates of service requests. This paper presents an ant-based load-sharing algorithm along with several simpler algorithms that are used for benchmarking purposes. Both solutions have been incorporated into a CORBA-based SCP simulation model and results are presented in this paper.

Section 2 of this paper introduces the recent developments in the area of CORBA in the Intelligent Network. Section 3 describes the simulation model, which is based on the IN/CORBA Gateway and the TINA Service Session components and has been implemented using MIL3's Opnet Modeler. Three test services, Virtual Private Network, Ringback and Call Forwarding have been simulated. In Section 4, the performance issues relating to these types of networks are discussed. Section 5 presents an optimal service object placement method. Section 6 describes the load sharing algorithms that have been simulated and Section 7 presents simulation results.

2   CORBA-Based IN

Much of the investigation into the application of CORBA to IN systems has been initiated by the Eurescom P508 project [4], the goal of which was to determine the options for evolving from legacy systems towards TINA. A major result of the project was that the gradual introduction of a TINA DPE (i.e. CORBA technology enhanced with real-time capabilities) into the existing IN environment represents a fundamental prerequisite for such an evolution. During the course of the P508 project, White Papers [1] and [2] were produced and submitted to the OMG in order to support the then emerging activities on IN/CORBA interworking. These White Papers were targeted at providers of information technology solutions and had the purpose of stimulating their interest towards telecommunication operator specific needs. They analyse a specific element of the problem area: the introduction of CORBA into the Intelligent Network. The central idea put forward is to adopt the OMG CORBA standard, enhancing it to make it suitable for telecommunications systems, particularly IN. Subsequently, the work was continued within the Telecommunications Domain Task Force of the OMG, which has recently produced a standard [3]. This standard focuses on the interworking of CORBA-based systems with TC-User applications, such as traditional Intelligent Network and mobile systems.

The primary goal of the IN/CORBA Interworking standard is to provide interworking mappings and supporting CORBA services that enable traditional IN systems (such as a Service Switching Point (SSP)) to interwork with CORBA-based implementations of IN systems (such as a CORBA-based Service Control Point (SCP). In order to do this, the standardised interworking mappings produce IDL for a CORBA object model that provides interfaces to legacy IN systems from the CORBA domain and also provides interfaces to CORBA-based IN applications from legacy IN systems. In effect, this object model may be used to build a gateway that provides protocol conversion and alignment of execution semantics between the IN and CORBA domains, allowing CORBA-based services to be introduced to the IN.

Figure 1 below shows the main IDL interfaces defined by the standard and how they interact to provide an interworking function (gateway) between the IN and CORBA domains. A more complete description of the standard may be found in [5].

A legacy SSP interacts with a CORBA-based SCP using the IN/CORBA object model defined in the Interworking specification. The objects shown in grey are CORBA objects whose interfaces are defined in OMG IDL in accordance with standard. The Gateway Administration object (GWAdmin) performs the functions of name translation and object location between the two domains. Messages arriving from a legacy SSP are addressed to a particular SS.7 Application Entity (AE), identified by a particular AE title. The GWAdmin provides an interface for translating the AE title to the CORBA object reference of a Service Interface Factory object, which may create instances of the appropriate Service Interface Object. This Service Interface Object acts as a proxy for the CORBA-based SCF. In order to represent the SSP in the CORBA domain, a SSF Proxy object is required. This object provides an IDL interface for invocation of INAP operations on the SSF from the CORBA domain and performs the protocol translation and communication with the SS.7 stack. The SSF Proxy Factory provides a standardised means of instantiating a SSF Proxy. The Service Interface Object provides a complementary IDL interface for invocation of INAP operations from the SSF to the CORBA-based SCP. Protocol translation for these invocations is provided in the gateway.

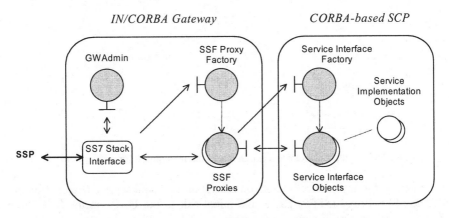

Fig. 1. Main elements of the IN/CORBA Interworking Gateway

The Service Implementation Objects are not defined by the specification and may be implemented by some arbitrary set of fine-grained CORBA objects, which provide the functionality required for service execution. One such model based on the TINA Service Session, which forms the basis for the simulation studies, is detailed in the next section.

2.1   TINA-Based Computational Objects

With this approach, the IN service logic and data, residing on the CORBA-based SCP, are modeled as a subset of the computational objects composing the TINA Service Architecture. An approach given in [6] is adopted, which defines methods for modeling IN services executing in a TINA environment. Here it is assumed that all calls originate and terminate on the IN side so that neither the calling nor called party uses a TINA end-system and thus, is not modeled as a TINA user. This is appropriate for the CORBA-based SCP scenario as all SSTs resides in the IN domain and these are the only originators of calls. As a result, the IN service capabilities may be encapsulated entirely within the TINA Service Session COs. That is, the TINA Access Session is ignored and the COs that provide this functionality are not required. All calls are established through the IN Service Switching Function (SSF) under the supervision of the TINA Service Session Manager (SSM).

With this approach, the service capabilities are modeled within a User Application (UAP), interacting with an Service Session Manager (SSM), which makes use of a service specific IN Service Support Object (SSO), e.g. a database containing number translation tables. As there is no call-party specific access session, the User Agent (UA) is anonymous and acts on behalf of all IN users. The Provider Agent (PA) is also generic in this case. Figure 2 below shows the COs required for an implementation of a Virtual Private Network (VPN) service.

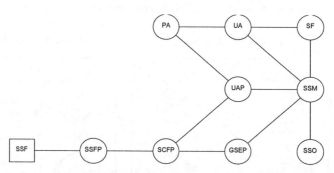

Fig. 2. Computational Objects and the interactions required for a VPN Service

Generally, for any service session, on receipt of the initial service request from the SSF, the SSF Proxy (SSFP) passes the initial call to the UAP via the SCF Proxy (SCFP), which in turn initiates a corresponding TINA service session via the PA. The PA interacts with a UA in order to perform a generic access session for service session establishment. Once the SSM has been created and initialised by the Service

Factory (SF), a control relationship is established between the IN SSF and the TINA SSM. The interactions between components are thence dependent on the specific service in execution.

Note that the IN/CORBA Interworking is modeled by considering only the core objects necessary for communication between the IN and CORBA domains during a service session i.e. the Proxy objects. The SSF Proxy object accepts INAP operations from the SSF over SS7 and translates them to CORBA invocations on the SCF Proxy. The SCF Proxy accepts INAP IDL invocations from the UAP and GSEP, transfers them to the SSF Proxy object which translates them to the corresponding INAP operations on the SSF. Proxies for the Intelligent Peripheral (IP) will also exist if required by the service. IP Proxies act in an identical manner to SSF Proxies.

3   CORBA-Based SCP Simulation Model

This section provides an overview of the distributed CORBA-based SCP model which has been developed to provide the basis for simulation studies of likely performance bottlenecks and for study of suitable strategies for load balancing in this type of environment.

In this scenario, the Intelligent Network Service Control Function (SCF) and Service Data Function (SDF) are no longer encapsulated within single functional entities but are decomposed into fine-grained Computational Objects (COs) which use the CORBA Object Request Broker (ORB) for communication. These objects communicate with entities in the legacy Intelligent Network via the IN/CORBA Gateway. Thus, the service logic programs and data that normally reside at the SCP and SDP are distributed across a multi-node network. Figure 3 shows the general network configuration of the CORBA-based SCP scenario and how it interconnects to a legacy Intelligent Network.

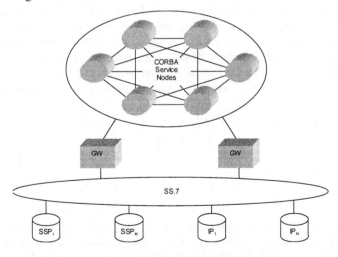

Fig. 3. Network Scenario for CORBA-based SCP

The scenario under study consists of a network of ten CORBA Service Nodes and two IN/CORBA Gateway Nodes. Gateway Nodes may communicate with all CORBA Service Nodes and all Service Nodes may communicate with each other i.e. the Gateway Nodes and Service Nodes form a fully connected network which is connected to the SS.7 Network at the Gateway Nodes. The network connection scenario in the CORBA domain is intended to represent machines interconnected over a local area network. The number of Service Nodes has been chosen so that adequate processing power is available to replace the processing power provided by a legacy SCP. It is assumed that individual Service Nodes have considerably less processing capacity than a legacy SCP and that service execution requires considerably more processing due to distribution. It is assumed that two Gateway nodes are required so that there is an element of fault tolerance within the system. The Gateway Nodes execute the functionality required for interworking between SSPs, IPs and the CORBA-based SCF, which is a distributed application executing on the CORBA Service Nodes. It is assumed that the Gateway function consists of the standard IN/CORBA interworking components described in Section 2. Thus, each Gateway Node executes CORBA Proxy objects, which provide an interface for invocation of IN operations on the SSP and IP from objects in the CORBA domain. The Gateway Nodes also execute the functionality that translates incoming messages from the legacy IN entities to CORBA invocations on SCF Proxies, which reside in the CORBA Service Node network. All other COs required to complete service execution reside on the CORBA Service Nodes. The legacy IN entities (SCP and IP) and the SS7 network are not modeled explicitly but are viewed as the source and sink of messages arriving to and departing from the Gateway Nodes.

3.1   Execution Model

The processing time for a service is decomposed into processing times for the set of messages passed between COs that are required to complete service execution. Each message passed between two COs has associated with it a CORBA marshaling (protocol encoding) time on the client-side node, a CORBA demarshaling (protocol decoding) time on the server-side node plus a processing time for completion of some service specific task on both the client and server side nodes.

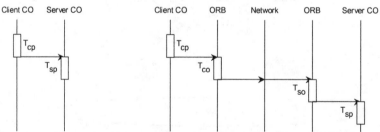

Fig. 4. Execution times for messages passed between COs. In the right hand figure, two COs are executing on different processors (i.e. COs are distributed). The processing times T_{CP} (service processing time) and T_{CO} (marshalling time) give the total processing time at the client node associated with this message. Similarly, T_{SO} (demarshalling time) and T_{SP} give the total

processing time at the server node. In the left hand figure both COs are executing on the same processor so the total processing time is given by T_{CP} and T_{SP}.

Marshaling, demarshaling and processing times remain constant for a particular message over all sessions of a service. If the communicating COs are located on the same node, the marshalling and demarshalling times are not included in the overall processing time for the message as CORBA is not required for communication.

The marshalling and demarshalling times used for simulation are based on times measured on a commercial ORB (Visibroker 3.3 running on a Sparc Ultra 5). The IDL used for determining timing measurements is based on the IN/CORBA specification and the TINA Ret Reference Point specification so that each message has associated with it the appropriate marshaling and demarshalling times. Processing times for actual service related tasks are based on the processing times for the service executing on a legacy SCP.

An asynchronous invocation mechanism is assumed, such as the CORBA Messaging Service. Thus, a CO making a CORBA method call does not block the process while waiting for a response from the server side. As a result, it is also assumed that all CORBA objects on a node execute in a single thread of execution and that the servers are modeled as a single FIFO job queue.

It is assumed that delays in network transmitter queues and transmission times on the network are negligible compared to delays due to marshalling and demarshalling of CORBA method calls between nodes. Experiments have shown that marshalling and demarshalling times for the IDL used for this model are typically an order of magnitude greater than transmission times over IP on a fast network, such as 100Mb Ethernet. It is also assumed that the order of messages is preserved in the network and that there is no message loss.

3.2   Test Services

Three different test services have been chosen to execute on the CORBA-based SCP in order to study the performance issues:

- Service A   Virtual Private Network
- Service B   Ringback
- Service C   Restricted Access Call Forwarding

The COs required for Service A and their intercommunications are shown in Figure 5 (above) Service B and Service C have been similarly defined. The duration of User Interaction A1 (phone ringing) is drawn separately for each service session from a negative exponential distribution with a mean of 5 seconds. The duration of User Interaction A2 (Conversation period) is drawn separately for each service session from a negative exponential distribution with a mean of 100 seconds. It is assumed that service users never abandon ongoing service sessions and thus the messaging for handling these cases does not need to be defined.

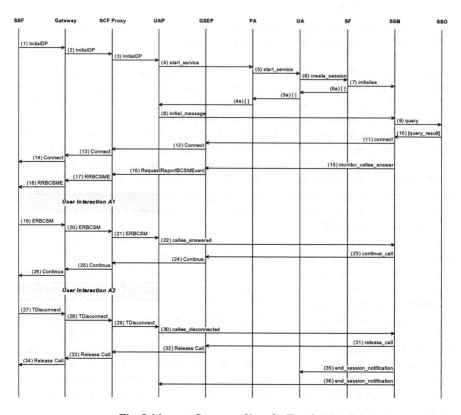

Fig. 5. Message Sequence Chart for Test Service A

4   Performance Issues

One of the most important benefits of distributed computing is the ability to split computational tasks among multiple processors. However, the distribution of software objects across different physical nodes can cause severe performance problems such as synchronisation of memory or databases, as well as much inter-object communication between nodes, which creates a large computational overhead. Thus, it is important to be aware of how the distribution of software objects affects performance. This section first describes some areas where performance issues are likely to appear, then discusses three related areas that can be taken into account to reduce these problems.

4.1   Problems

The performance of a telecommunications system is important for several reasons. Firstly, the users of the system have several expectations, such as fast system response times and reliability. Secondly, the network operators require their systems to operate as efficiently and cost effectively as possible.

One major factor in the real time performance of a CORBA-based system is the processing overhead caused by inter-node communication. Typically the transmission times of a message are very low when compared to the time required by protocol wrapping and unwrapping (marshalling and unmarshalling in CORBA) at the sending and receiving node. Thus, too much, or simply inefficient distribution of objects on physical nodes can easily cause overload due to protocol overhead alone.

The characteristics of traffic within the CORBA-based SCP are different from the traffic in normal non-telecom CORBA-applications. The real-time characteristic of teletraffic is more "now or never" than normal traffic which has more strict rules for finishing a task than finishing it within a given time frame. The "now or never" property of teletraffic allows us to block incoming calls if they would result in degraded performance for the already accepted ones.

4.2 Problem Solutions

Solving the performance problem requires several different techniques and strategies. Below is a list of three different areas where large performance gains are likely to be found.

1. Object distribution: This concerns the placement of objects on different nodes. Different configurations can cause very different communication patterns and, as communication is very computationally expensive, we want to make sure that we have no more communication than is necessary. While object distribution can be static, meaning that objects stay where they were placed at design time, it is possible to move objects around during run time and thus dynamic distribution schemes are possible. One must note here, however, that since moving objects around is a computationally intensive operation it is not likely to be feasible solution to solve short-term traffic transients.

2. Load Balancing: When a object distribution has been decided upon, there must be some kind of mechanism that directs the object remote procedure calls if there are multiple nodes that offer the same object type. It is the purpose of the load sharing mechanism to direct these procedure calls so that they, if possible, keep some nodes from being overloaded while others are almost idle. Load sharing is difficult since the object and node which we chose might have favorable conditions at the decision time, but these conditions can change quickly and we must therefore take some account of future load as well.

3. Load control: When load sharing is not enough to handle the offered amount of traffic, then some kind of load control mechanism is necessary. Load control is not usually used in distributed systems, since the traffic must get through if the application is to work. As it was said earlier, this is not necessarily true in a telecommunications environment, where it is more important to finish some work in time rather than to finish all work long overdue.

This paper is concerned with solutions 1 and 2. Future work will explore solution 3.

4.3   Measuring Performance

Measuring performance in this type of environment can be difficult due to the very flexible nature of the application. Below are a number of possible considerations which have been used in the past for evaluating the effectiveness of performance controls:

1. Throughput: A traditional comparison, throughput measures the number of finished service requests per time unit.
2. Scalability: Scalability is very important for an algorithm, since we want our algorithms to work for any network size.
3. Transient survival: Due to the high reliability requirements on telecommunication networks, the algorithms must be able to quickly adapt to changing conditions such as traffic peaks or node failures, which might cause rapidly changing traffic patterns.
4. Algorithm Complexity: Since the system is very complex to start with, we want the algorithms to be simple and easily implemented. Also, simple algorithms tend to be fast and have little overhead.

In this paper we have chosen to use two simple measurements for evaluation of our performance controls:

1. Mean Service Completion Time: This is an important factor for a realtime system. It also gives an indication of the level of queueing delays in the network.
2. Maximum Load : This is the mean load on the highest loaded node in the network and gives an indication of the effectiveness of the load balance amongst network nodes.

5   Optimised Computational Object Distribution

The choice of location for service objects on nodes in the network can greatly effect performance. If objects requiring large amounts of processing are clustered on a small number of nodes then queues lengths increase on these nodes or worst, an overload can occur. Conversely, if objects are distributed too much then large amounts of unnecessary protocol processing is incurred which lengthens the service time. This section describes an optimal method that allows processing capacity in the network to used efficiently under normal loading conditions. The problem is stated below.

With the exception of the Proxies, there is no restriction on assignment of COs to CORBA Service Nodes. The SCF and IP Proxies reside only on the two Gateway Nodes, as these objects need to communicate directly with the SS.7 stack. It is assumed that an instance of a service, initiated through an SCF Proxy on a particular node, may use only SCF and IP proxies on that same node for the duration of the service execution. All other COs may be duplicated across many nodes. However, COs are assumed to be atomic i.e. may not be decomposed and distributed between nodes. The assignment of these COs to network nodes is determined by minimising the total processing cost on all nodes given the expected user demands from the two Gateway nodes for each service and given the maximum load allowed on each of the Gateway and CORBA Service nodes. This approach is similar to that found in [7] where the communications costs between COs are to be minimised.

The problem may be formulated as a Linear Programming problem in the form given in Equation (1). The variable for minimisation (vector x) denotes the processing costs associated with the total traffic flow during a service session between each communicating component pair, relative to the processing cost due to the traffic offered to the SSF Proxy CO by the SSF for that service. Note that it is assumed in the formulation of problem that copies of all COs reside on all nodes. If in the solution, the costs associated with a particular CO copy on a particular node are all zero, then that CO need not exist on the node.

In the objective function, C is simply the unit matrix as it is required that all processing costs in the network are minimised and that all costs are of equal importance. This form of the objective function determines that the problem is linear.

The A matrix and b vector in the constraint inequality are determined by: (i) the number of units of input traffic load offered by each SSF, for each service. These are equality constraints; (ii) the relative processing costs between associated component pairs. These are also equality constraints with each constraint expressing the processing cost associated with a component pair relative to the processing cost associated with one other component pair. An adequate number of constraints are required to associate all components in the service graphs. The relative processing costs are derived by summing the processing times for all messages passed between each pair of COs during a service session. When both COs reside on the same node the costs express the sum of all client and server service processing times for interactions between these COs. When the COs reside on different nodes then the costs also include ORB processing costs for both client and server; (iii) the limit on processing capacity for each node. These are inequality constraints that limit the sum of all costs associated with a node. These constraints may be set to give a component distribution that is optimal at a particular operating point, for example, to give a maximum of 40% loading on all the nodes.

$$
\begin{aligned}
\min \quad & C^T x \\
s.t. \quad & Ax \le b \quad . \\
& l \le x \le u
\end{aligned}
\tag{1}
$$

The bounding inequality is defined to constrain x to be positive. There is no upper bound on x as the limiting factor is the total processing cost associated with each node, which is expressed as part of the constraints ((iii) above).

The solution to this minimisation determines the optimal placement of COs in the network. That is, if the costs associated with a particular copy of a CO on a node are all zero, then the CO is removed from that node. Having removed all such null COs, the remaining COs are optimally placed in the network. The solution also determines the relative flows between each CO and copies of the COs with which it communicates. That is, the routing probabilities for requests between COs are determined by the solution. These routing probabilities may be used as the basis for a load balancing algorithm which aims to minimise overall network load. Such an algorithm is presented in Section 6.

6   Load Balancing Algorithms

In more tightly coupled systems, generally, two approaches to load balancing have been taken: an idle processor may request more work from other processors or a busy processor may send excess work to idle processors. These approaches do not work very well in CORBA-based distributed systems since it costs too much in terms of protocol processing to move jobs. In this type of systems, load balancing can be done when an instance of a service component exists at more than one node. The load balancing algorithm is required to choose the most suitable node on which the component shall be executed, given certain data relating to the current state of the network nodes.

The main purpose of our work is to investigate so called *ant* algorithms, where mobile agents are used to find the most suitable node, in terms of low load, on which to execute the required component. To allow evaluation of this strategy, two simple benchmark algorithms, described below, have also been implemented. In order to describe the operation of these algorithms, the following notation is introduced. Let **R** be the set of all nodes which contains service component **r**. If the next service component needed is **r** we have to choose one of the nodes in **R** where **r** will be executed.

6.1   Benchmark Algorithms

The benchmark algorithms have been chosen to allow the lower bounds (or close to the lower bounds) of the performance measures to be established. These algorithms are not intended to be viable as a practical solution to the load balancing problem but allow the ant based algorithms to be evaluated against theoretically near-optimal solutions. The benchmark algorithms are as follows:

1. **Shortest queue**: the node in **R** with the shortest processor queue is chosen. If nodes with the same queue lengths are found, the lowest numbered node is chosen. We assume that all nodes have instantaneous knowledge of the queue lengths in the nodes in **R**. This assumption obviously renders the algorithm impractical. However, the results are expected to give close to the lower bound of the Service Completion Time as queuing delays at nodes are maintained at a low level.

2. **Random**: one node is chosen randomly in **R**. The probability for choosing a particular node is derived from the static routing scheme determined by optimisation (see Section 5). This assumption renders the algorithm impractical as it does not respond to transients and drifts in service mix. However, assuming constant service mix, the algorithm it expected to maintain loading at a near-optimal level.

6.2 Ant Algorithms

Ants are simple agents that are sent out by the nodes to probe the load status of all nodes in the system. An ant is sent away from a node (called the sending node) to another node (called the receiving node) and then returns to the sending node. The sending and receiving node may be the same node. In our investigations we have compared three load status parameters: queue length, load of the receiving node and the roundtrip time. The load of a node is measured as the proportion of an interval the server is busy. The length of these intervals is 0.1 seconds. In our model the ants are handled like this:

1. First the ant is created in the sender node. It is put back in the high priority processor queue of the sender and after queuing it is wrapped into a CORBA protocol and sent to the receiving node.
2. At the receiving node the ant is queued and its CORBA protocols are unwrapped. After this it is queued once more and the queue length or load is put into the ant. Then the ant is queued once more after which it is wrapped into the CORBA protocols again and sent back to the sending node.
3. When the ant has returned to the sender it is queued in a high priority queue, its CORBA protocols are unwrapped, it is queued once more after which its load status information is stored in a load status table as described below.

We assume that each node in the network keeps a load status table with information about all nodes in the network. When an ant returns to the sender, the load status for the receiving node is updated in the table.

The values used can be either the receiving node's queue length, its load or the roundtrip time of the ant. For all these measures we have that the higher the load status value, the fever calls should be sent to the node. The load status table is used in this way to chose a node in **R**: Assume that the load status of node i is $l(i)$. Observe that $l(i)$ may be queue length, load or roundtrip delays. Then we calculate probabilities for all nodes in **R** as follows:

$$p(i) = \frac{l(i)^{-k}}{\sum_{i \in R} l(i)^{-k}} \ . \tag{2}$$

With probability $p(i)$ node i is chosen. In this way there is a larger probability of sending a call to a node with a low load status value. The k in the calculation is the factor that describes the randomness of the weights. $k=0$ is the case where every node is chosen with equal probabilities, larger k's give larger weights to values on the limits.

Ants are generated in a node according to a Poisson process with rate λ. The generation rate is an important parameter. If it is too low, the values in the load table will not be updated fast enough which could lead to oscillations. If it is high, the ants themselves will increase the load of the system. When an ant has been generated, it must be decided where to send it. We have chosen the following algorithm: with probability α the receiving node is chosen randomly with equal probability for all nodes, with probability $1-\alpha$ the node with the lowest value in the load status table is chosen, i.e. the node with lowest load. Thus we have two parameters λ and α that must be tuned.

7   Simulation Results and Analysis

Several simulations were run using the model defined in Section 3. Both the benchmark algorithms and the three ant-based algorithms were simulated. The simulation parameters are listed in Table 1 and the simulation results are presented in Table 2.

As indicated in Table 1, all algorithms were simulated under a low loading and high loading scenario. The service mix is balanced in each case, i.e. the mean arrival rates for all three services are equal in each case. These arrival rates were chosen to give an average load of about 30% in the low load case and about 80% in the high load case. The services executing in the simulation are Virtual Private Network (Service A), Ringback (Service B) and Restricted Access Call Forwarding (Service C), described in Section 3.2. The distribution of service objects was obtained using the method described in Section 5. This distribution is optimised for the given arrival rates. The ant spawning intervals for the ant-based algorithm were tuned for each variation and are given in Table 1.

Table 1. Simulation Parameters

Simulation Parameters	
Arrival Rate Low Load	$225 \ s^{-1}$
Arrival Rate High Load	$675 \ s^{-1}$
Object distribution	Optimised for balanced service mix
Service Types	Services A, B and C
Service Mix	Balanced
Ant Spawning Interval (Load Query)	$0.05 \ s^{-1}$
Ant Spawning Interval (Round Trip)	$0.01 \ s^{-1}$
Ant Spawning Interval (Shortest Queue)	$0.015 \ s^{-1}$

Table 2. Simulation Parameters

		Benchmark		Ant-based		
		Random	Shortest Queue	Load Query	Round Trip	Shortest Queue
Low Load	Service Time	12.4 ms	11.4 ms	12.4 ms	13.0 ms	12.9 ms
	Max Load	25.8 %	55.3 %	25.6 %	30.4 %	28.9 %
High Load	Service Time	35.2 ms	25.1 ms	34.6 ms	40.9 ms	40.5 ms
	Max Load	78.7 %	87.0 %	74.5 %	81.7 %	81.5 %

As indicated in Table 2, The *Mean Service Completion Time* was measured for each algorithm and averaged over all three services. Note that the User Interaction times, indicated in Section 3.2, are excluded from this measurement as they are independent of loading in the network. The *Max Load* value indicated is the load on the most heavily loaded node in the network.

Considering the performance of the benchmark algorithms, as expected the *Shortest Queue* algorithm performs best overall in terms of minimising the service time. This is due to the fact that queue sizes are kept as short as possible. This method will not perform as well when there is a large difference in the processing required from message to message, as queue size will not accurately indicate how long an arriving message will need to wait for service. However, for the service simulated, there is not a large difference in message processing and thus the Service Time is close to the minimum and gives a good benchmark. The uneven loading, indicated by the high *Max Load* value for the *Shortest Queue* benchmark algorithm is due to the fact that the node with the lowest number is chosen when several queues have the same length. This condition will occur frequently at low loading levels when there is a significant probability that the queue length is small.

The *Random* benchmark algorithm gives a low *Max Load* value as expected, however, this is not the lowest overall value. This is because of the static nature of the algorithm. The algorithm will give minimum loading on all nodes only at exactly the intended operation point i.e. perfectly balance service mix with deterministic arrivals. To avoid this problem, future work will consider combining this algorithm with a more dynamic algorithm. The service times observed for the *Random* benchmark algorithm are increased compared to the *Shortest Queue* benchmark due the constraint on node loading. Because the optimisation is constrained to maintaining load below a particular level on all nodes, service execution is more distributed and thus service times increase due to increased protocol processing costs. Future work will consider "softening" this constraint to allow load to be less balanced but so that shorter overall service execution times can be obtained.

The simulation results show that the *Ant-Based* algorithms compare quite favorably with the benchmark results. In the low load condition, the service times are comparable to the service time for the *Shortest Queue* benchmark. The *Max Load* for the *Ant-Based* algorithms is also close to the value for the *Random* benchmark for low loading. For high loading, the service times increase considerably compared to the *Shortest Queue* benchmark. Further work is required to refine the Ant-based algorithms. In the results presented here, the weighting factor (k) has been set to 1. Results are required to investigate the performance with higher weighting factors and to investigate tuning of the ant spawning interval.

Overall, further work is required in a number of areas. The results presented here were generated under stable network conditions. In order to fully assess the algorithms, the behavior under transient traffic conditions needs to be studied. Overload protect has not been considered here and needs to be investigated and incorporated into the algorithms. The algorithm computational complex and robustness also needs some consideration.

8   Conclusions

This paper has presented a number of approaches for improving the performance of a distributed CORBA-based Service Control Point. Although distributed systems technologies can contribute greatly to this area by allowing processing requirements to be divided among a large number of less expensive processors, it is unwise to assume that increasing processing power or memory sizes of network processors ad infinitum will alone guarantee high performance. The solutions offered in this paper aim to increase the efficiency and cost effectiveness of resources with a view to making CORBA-based solutions more suitable for high performance, reliable systems required by telecommunications environments.

Acknowledgements

This work has been partially sponsored by the Commission of the European Union under the project MARINER, project number AC333 of the ACTS program. This work has also been partially sponsored by the Swedish Research Council for Engineering Sciences (TFR) under Contract 271-97-203.

References

1. Object Management Group, "Intelligent Networking with CORBA," OMG DTC Document:telecom/96-12-02, December, 1996
2. Object Management Group, "White Paper on CORBA as an Enabling Factor for Migration from IN to TINA: A P508 Perspective," OMG DTC Document: telecom/97-01-01, January, 1997
3. Object Management Group, "Interworking between CORBA and TC Systems," OMG document telecom/98-10-03, August, 1998
4. EURESCOM Project P508, "Introduction of Distributed Computing Middleware in Intelligent Networks White Paper," OMG DTC Document: 97-09-01, September, 1997
5. Nilo Mitra, Rob Brennan, "Design of the CORBA/TC Inter-working Gateway," Proceedings 6th International Conference on Intelligence in Services and Networks, Han Zuidweg et. al. (eds.), ISBN: 3-540-65895-5 Springer-Verlag, April, 1999
6. U. Herzog, T. Magedanz: "From IN toward TINA – Potential Migration Steps", in: "Technology for Cooperative Competition", Fourth International Conference on Intelligence in Services and Networks", Springer Publishers, Cernobbio, Italy, May, 1997
7. Anagnostou, M.E., "Optimal Distribution of Service Components,", Proceedings, IS&N'98, 5th International Conference on Intelligence in Services and Networks, Antwerp, Belgium, May 1998
8. Widell, N., Kihl, M., Nyberg, C., "Measuring Real-time Performance in Distributed Object Oriented Systems", Proceedings, Performance and Control of Network Systems III, SPIE Photonic East '99, Boston, USA, September 1999
9. Kihl, M., Widell, N., Nyberg, C., "Load balancing strategies for TINA networks", Proceedings, 16th International Teletraffic Congress, Edinburgh, Scotland, June 1999

Measurement Based Connection Admission Control Algorithm for ATM Networks that Use Low Level Compression

Ioannis Papaefstathiou[1]

University of Cambridge, Computer Laboratory
Ioannis.Papaefstathiou@cl.cam.ac.uk

Abstract. Since ATM switches are intented to be simple and inexpensive, a significant part of the network cost is in the cost of the links. A way of increasing the traffic we can send over these expensive links is to transmit compressed ATM cells. This idea, although it seems very simple, is a new one for ATM and as we show it can signicantly increase the useful bandwidth of a typical ATM network. In this paper the CAC algorithm that can be effectively used in a network which uses this compression scheme is also described and the results that can be achieved when this algorithm is used in a real network are presented. This CAC algorithm is based on Large Deviation Theory: the large deviation rate-function (entropy) of compressed (and thus more bursty) ATM traffic can be estimated from measurements of trffic activity. The entropy can be used to determine the bandwidth requirement of the traffic.

1 Introduction

An important application of ATM is in the interconnections of private LANs, through WAN links. A typical such a network is shown in Figure 1. ATM is used as the backbone infrastructure in such networks, due to its scalability, speed and potential fault tolerance. One of the main characteristics of these networks is that the cost of the WAN links is comparatively high. By increasing the network traffic we can send over these links we can increase the effectiveness of the whole network. This can be done by compressing the data transmitted. In particular, ATM cells can be compressed and then encapsulated over newly formed standard ATM cells. In this paper we show the effectiveness (especially in this kind of interconnection networks) of this simple idea.

The compression scheme we use is dynamic and understands ATM cell structures, it compresses only the payload of the cells, and adapts the payload technique depending on the type of traffic the different flows carry (i.e. whether they carry real-time, non-compressible traffic or not). In this way, each VC will rapidly acquire a compression context that suits its data. For example, traffic with a timing requirement and low compressibility (such as compressed video) will become transmitted

---

[1] Supported by a Marie Curie Research Training Grant under TMR activity 3.

uncompressed whereas compressible files, such as web pages and wordprocessor documents will experience good compression.

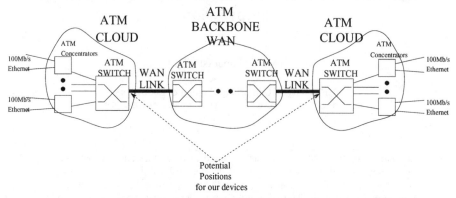

Fig. 1. Typical application of our compression scheme

In particular our algorithm works as follows: In the gateway of the LAN (where the WAN link is attached to) we try to collect all the cells that have the same header, we remove all the headers and form a long packet. Then before, we actually transmit this long packet we apply our compression algorithm on it and then encapsulate the compressed stream into ATM cells. So for each packet we fragment the compressed stream in 48 byte quantities and attach to them the original header. By doing that these newly formed cells are routed over the WAN network just like the original ones and they are decompressed in the gateway of the destination LAN. The major advantage of this approach, in comparison with our older approach described in [2],[3], is that the WAN network involved can be a typical one (no compression units needed). A more detailed description of the compression scheme can be found in [4].

Since these newly formed cells are ordinary cells in terms of their format a standard ATM Call Admission Control (CAC) algorithm can be used for determining whether or not to accept a new connection to the network.

In the literature a lot of CAC algorithms are proposed. For all of them perhaps the most difficult aspect is determining whether accepting a call will affect the existing traffic contracts for existing calls or in other words the *impact* of a new call on the already admitted ones. In recent years, the notion of the *effective bandwidth* of a traffic source has been developed to describe the effective resource requirement of a source. Although different denitions of the term exist, it has been widely used to express the trade-off between instantaneous bandwidth requirements and buffer space for a source (or collection of sources), given its traffic characteristics and QoS constraints [7],[8],[9].

The effective bandwidth of a source depends strongly on the statistical properties of its traffic as well as its QoS requirements. It is thus essential to have an accurate characterization of each source which contains precisely the information required to compute its effective bandwidth. Two approaches to obtaining this information are possible. The first, which has been adopted to date by the majority of publications in the area, insists that the source should provide the network with a full statistical characterization of its behavior. Typically this is achieved by constructing and analyzing a parametric model of the source, based on precise or statistical descriptions of its behavior. Parameters of the model may be fitted from actual measurements of the traffic, made in isolation [10],[11].

The second approach argues that it is unreasonable to expect traffic sources to provide these parameters. Instead, its supporters argue that the network is capable of deriving the trafficc characteristics in a concise and efficient way. This approach, based on on-line measurements of traffic activity, was pioneered by Courcoubetis et al.[12].

Because the behavior of the compressed traffic is much less predictable than the one of the original data (since it depends heavily on the very difficult to determine compression factor and the cross-talk between virtual circuits) we argue that the second approach should be used in a network that uses our compression scheme. In particular we argue that an algorithm presented by Crosby et al.[13], and briefly described in the following sections, based on recent theoretical developments in the field of large deviations[14],[15] produces near optimal results.

To briefly outline the CAC algorithm consider a network carrying a multiplex of compressed and uncompressed traffic connections, and receiving requests to set up new connections. These will have some simple user declared parameters associated with them (even if this is only the peak rate). This CAC algorithm combines estimation of the bandwidth requirement of the existing load, using on-line measurement, with a prediction of the bandwidth requirement of the offered traffic (after this traffic has been compressed) based on its declared parameters. If the sum of these bandwidths exceeds the line rate, the offered traffic is rejected; otherwise, it is accepted. Both the estimators and the predictors used make essential use of mathematical properties in the field of large deviations.

2 Compression Algorithm

The algorithm we use for compression is a variation of the LZ(Lempel Ziv) algorithm [16], which is probably the most commonly used compression algorithm.

In an encoder/decoder pair, at the two LAN gateways, there are two data structures which are initialized to the same known state, and they are updated in an identical fashion. The encoder does this using the cells to be transmitted while the decoder, generates an identical stream of cells at the destination.

The actual compression process consists of examining the data that are ready for transmission, to identify any sequences of strings of data bytes which already exist in the encoder history. If an identical such history is available to a decoder, this matching string can be encoded and output as a 2 element string, containing a byte count and a history location. It is then possible for a decoder to reproduce this string exactly by copying it from the given location in its own history. If an incoming byte of data does not form part of a matching string, a one element string, containing this embedded value, is encoded and then transmitted to explicitly represent this byte. In other words it is a byte-oriented compression scheme.

A decoder performs the inverse operation by first parsing a compressed data stream in the two element strings (encoded string of more than one byte) and one element strings(encoding for just one character).

In our particular implementation different dictionaries for the different ATM flows are used. Each flow consists of all the cells with the same VPI/VCI bits. So according to the VPI/VCI bits, the corresponding dictionary for the compression of the payload is accessed. At the other end by looking at the VPI/VCI bits a decision of what dictionary to use for the payload is made. The reason for using different dictionaries for different flows is that, as it is known, the more related the data stored in a dictionary are, the better the compression ratio. With very high probability all the cells belonging to the same VP/VC, are parts of the same large unit of data. This is not the case for the current IP over ATM protocols like LANE and CLIP but we chose the multiple VC/VP approach (like AREQUIPA [17]) since the QoS issues on the ATM network can be supported more efficiently when such protocols are used. In the case of the LANE our compression scheme causes a slightly smaller -than in the AREQUIPA case- but still signicant improvement on the efficiency of an ATM network as Tables 1 and 2 show. These tables demonstrate the compression ratio achieved when the two different IP over ATM schemes are used.

Table 1. LAN scenario, "AREQ" shows the results when the multiple VP/VC approach is used, "LANE" the results when LANE is used: *Compression ratio = compressed size / raw size.*

IP			TCP			UDP		
DATA SIZE (MB)	COMP RATIO AREQ	COMP RATIO LANE	DATA SIZE (MB)	COMP RATIO AREQ	COMP RATIO LANE	DATA SIZE (MB)	COMP RATIO AREQ	COMP RATIO LANE
1362	0.374	0.404	1332	0.362	0.390	1249	0.424	0.462
1418	0.384	0.418	1250	0.379	0.4.5	1301	0.394	0.436
1364	0.399	0.420	1296	0.412	0.449	1205	0.435	0.468
1610	0.379	0.399	1396	0.395	0.428	1421	0.411	0.446

Table 2. WAN scenario, "AREQ" column shows the results when the multiple VP/VC approach is used, "LANE" the results when LANE is used: *Compression ratio = compressed size / raw size.* These traces consist of flows with either source or destination address outside the University of Cambridge.

HTTP (non-local)			FTP (non-local)		
DATA SIZE (MB)	COMP RATIO AREQ	COMP RATIO LANE	DATA SIZE (MB)	COMP RATIO AREQ	COMP RATIO LANE
1470	0.6023	0.6368	1226	0.8239	0.8370
1278	0.6286	0.6574	1148	0.8277	0.8419
1273	0.6224	0.6227	1178	0.8692	0.8826
1151	0.6623	0.6973	1144	0.8479	0.8734
1186	0.6529	0.6852	1108	0.8602	0.8913

As it can easily be derived from these tables the multiple VC/VP solution gives better results than the single VC/VP one. The main reason behind it is that, as it was described earlied, each VC/VP flow is associated with a single dictionary in the compressor. Is is also known that the more relevant the data compressed by a dictionary the better the compression ratio achieved. Thus in the case of the multiple VC/VPs, the data compressed in every dictionary are probably parts of the same junk of data and thus the compression ratio achieved is higher than in the case that one dictionary is used for all the data departing from a certain TCP source. Due to hardware limitations the main hardware prototype, described in the next section supports up to 4095 4KB dictionaries (for 4095 flows). The choice of the size of the dictionaries is supported in [18]. Whenever the first cell of a new ow enters the switch, its flow is associated with the next available dictionary. If there is none available, it uses the dictionary of one of the least recently used ones, according to some heuristics which are also used by the unit at the other end of the link for choosing the new dictionaries.

The reasons for using this LZ algorithm are mainly the following:

- Since both the decoder and the encoder at any given time maintain the same dictionary, there is no need for transmitting this large piece of data from the encoder to the decoder.

- It has an acceptable behavior for real-time data.

- It can be combined with a simple error recovery scheme for delivering reliable communication.

In comparison with the more sophisticated LZ-78:

- It produces a much higher compression ratio when applied to real network traces.

- The decoding process is faster.

- It is easier to implement in hardware.

3 Hardware Architecture

In Figure 2 the block diagram of the compression/decompression chip is shown.
As in every dictionary based compression device the core consists of the dictionary and the comparison circuits around it. Thus the speed of the device depends heavily on the memory throughput and the comparisons' latency. In the architecture proposed the speedup implementation techniques of pipelining, parallelism and repetition of information have all been used in order to accelerate this core. The block diagram of the main unit is shown in Figure 3. In general the compressibility table determines if a flows should be compressed/decompressed or should bypass the main unit and the header memory, and the concat and bypass circuits ensure the cells will be formed and sent over the transmission link correctly.
The core circuit of the device is organized in a 256-stage pipeline. Each pipeline stage consists of a memory bank of 31 bytes for each dictionary supported and 60 comparators that can operate in parallel.

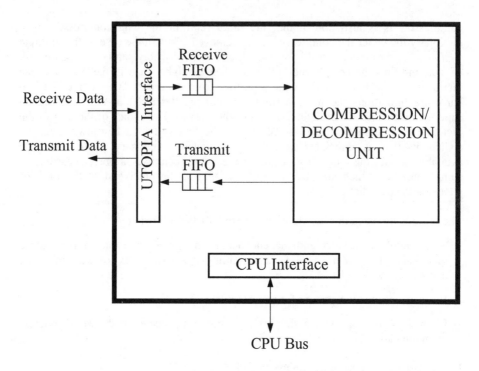

Fig. 2. Block Diagram of the compression/decompression chip

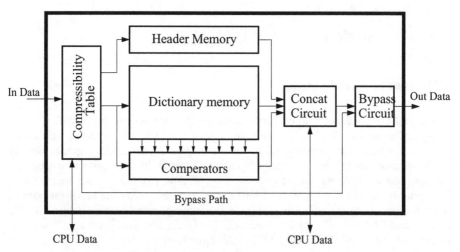

Fig. 3. Compression/Decompression Unit Block Diagram

So the main characteristics of the architecture are:
– 256-stage pipeline.

- 60 comparisons in parallel at each pipeline stage and thus 15360 comparisons at the same time throughout the chip.
- Memory repetition (100% more memory used) for higher memory throughput.

Using the above speedup techniques the network data can be compressed at speeds up to 1Gb/sec and the latency introduced is within the acceptable limits for network data (3 cell times). The latency can be further reduced if more hardware resources are to be used. A more detailed description of the hardware prototype can be found in [1].

4 CAC Algorithm

The CAC algorithm used, named Measure, makes use of on-line measurements to develop an estimate of the thermodynamic *entropy* of the traffic. Our approach is based upon the theory of Large Deviations: the large deviations rate function, or entropy, of bursty traffic can be estimated from measurements of traffic activity. The entropy can be used to determine the bandwidth requirement of the traffic, subject to user or network imposed Quality of Service (QoS) constraints. The key benefit of our approach is that a connection or flow need only be parameterised in a straightforward way (for example by its *peak rate*).

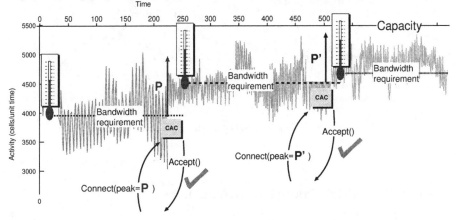

Fig. 4. Operation of the Measure CAC algorithm

The behavior of Measure is shown graphically in Figure 4. Given a current multiplex of calls, with a particular (bursty) traffic activity (comprised of both compressed and uncompressed traffic), the system makes an estimate of the bandwidth requirement of the multiplex. This is obtained from the entropy estimator (depicted by a thermometer). A new call attempt declares its peak rate P when requesting admission. The CAC algorithm essentially sums the peak rate P, and the current estimate of the effective bandwidth; if the total is less than the link capacity, then the connection can be accepted without violating the QoS of any of the currently multiplexed calls. As soon as the new call commences, the estimator will begin to revise its idea of the current resource requirement, producing in turn a new estimate of the bandwidth requirement of the sources. When the next call attempt arrives the procedure is repeated, as shown. If a new call attempt arrives before the estimator has developed an accurate estimate of the new effective bandwidth, as shown in Figure 5, the algorithm acts *conservatively*. It uses the most recent stable estimate of the effective bandwidth of the multiplex, plus the sum of the peak rates of all subsequent calls. Thus, in Figure 5, the second call will be rejected, because the sum of the two peaks P + P∪ and the first effective bandwidth estimate exceeds the link capacity. The technical specications of the implementation of the Measure algorithm are described in [6].

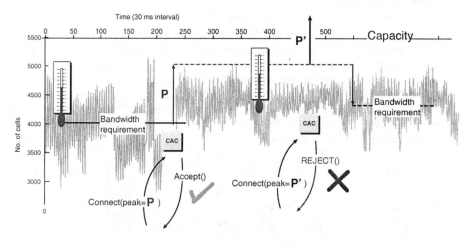

Fig. 5. Conservative Call Acceptance using the Measure CAC algorithm

5 Simulation of the Compression Scheme

Our compression scheme was simulated using an adapted version of the NIST ATM Network Simulator [19]. The most important advantage of this simulator is that it has some specic components for dealing with IP traffic transmitted over ATM. These components were used for loading the simulator with real IP traffic.

The data traces used in the simulator were collected using UNIX's tcpdump from an SDDI ring which is the backbone which interconnects all the different sites at the University of Cambridge network. Since this network is the backbone which interconnects different LANs, and it is used by 10000 people with a variety of backgrounds and ages, its data can probably be described as representative for a LAN over WAN interconnection network.

These data were suitably parsed and loaded to the ATM network simulator. A particular characteristic of this parsing is that each TCP flow is associated with a different VP/VC flow[2]. In order to have as general results as possible, large amounts of a wide range of data types were collected. These results can be found in Table 3 and Table 4.

Table 3. LAN scenario: Simulation Results, *Compression ratio = compressed size / raw size.*

IP		TCP	
TRACE SIZE	COMP RATIO	TRACE SIZE	COMP RATIO
1362543480	0.3012	1332884548	0.2974
1418068578	0.3122	1250395541	0.3022
1364950357	0.3218	1296691959	0.3539
1610230148	0.2998	1396224640	0.3588

Table 4. WAN scenario: Simulation Results, *Compression ratio = compressed size / raw size.* These traces consist of flows with either source or destination address outside the University of Cambridge.

HTTP (non-local)		FTP (non-local)	
TRACE SIZE	COMP RATIO	TRACE SIZE	COMP RATIO
1470465052	0.6374	1226539198	0.8422
1278311624	0.5630	114.509094	0.8332
1273319843	0.5687	1178827100	0.8963
1151159980	0.7015	1144921233	0.8725
1186316184	0.7001	1108130352	0.8920

As it can be derived from these tables the kind of networks that would be perfect for applying this compression scheme to, will be, as we have already mentioned, the LAN interconnection networks over WANs (See Figure 1). In these networks the high compression ratio of the LAN traffic[3], and the high cost of the WAN bandwidth are combined. So our compression scheme will be much more effective there since someone will be able to send 155Mbps of data over a 50Mbps leased WAN link, at the additional cost of a couple of our inexpensive devices. Since the hardware device, described in the last section, introduces a delay of at most 3 cell times, the latency of our compression scheme does not cause any problems either. A remark for the data of Tables 3 and 4 is that there is a variation in the compression ratios of even the same kind of traffic.

The reasons for that are mainly the following:

- In some cases we have more pre-compressed traffic than in some others.
- The size of the TCP flows in some cases is smaller than in some others. Since the smaller the size the lower the compression ratio, the size of the flows can probably result in this variation.

---

[2] As it is known currently LANE doesn't do this but there are plenty of other schemes that do.
[3] Consider that what we define as LAN traffic here is mainly this interconnection LAN traffic since the network from which we collected our traces is such a network.

6 CAC Simulation

This section presents the results of simulation experiments using the CAC algorithm of Sections 3. The aim of the experiments was to evaluate the performance of our approach mainly with respect to the number of calls our network can service without violating the QoS requirement. A mcuh more detailed description of the test environment can be found in [5].

Simulation Model. In each of our simulations we model a single output buffer and transmission link of an ATM switch. The link speed used is 100 Mb/s, which corresponds to the TAXI transmission rate for the Fairisle ATM network [20] at Cambridge University.

We present results for simulations using the compressed and the uncompressed http traffic[4]. The compressed one was produced by the simulation of Section 4. Consider again that these traces were derived from real networks. The buffer size was 100 cells and the CLR constraint was 10^{-5} for all theresults. The Peak rate for both types of traffic is 10Mb/sec whereas the mean rate was 1.32Mb/sec for the uncompressed one and 0.8Mb/sec (uncompressed rate * compression ratio = 1.32/sec * 0.6) for the compressed one.

Call Model. We study a scenario in which calls of a particular traffic type arrive according to an exponential inter-arrival time distribution, an assumption which appears to be well founded [21]. In the absence of realworld data we have used call lengths which are exponentially distributed. We present results for *long calls*. In both cases calls arrive at a high (Poisson) rate with mean 5 calls/s. Blocked calls are lost, but the high arrival rate means that the system is continually faced with new call attempts. We thus expect the system to remain close to maximum utilization.

Calls have an exponentially distributed length with a mean of 60 seconds. Each accepted call transmits an independent *trace*. This trace is derived by randomly selecting a start point in the traffic trace. Calls are not correlated in any way.

For the results which compare the performance of the different algorithms, the same *random seed* was used with each different algorithm, resulting in precisely the same call arrivals process in each case. This allows us to compare not only the statistical properties of the algorithms, but also their dynamic behavior.

Results. Figure 6 shows histograms of the average number of connections in progress during a simulation run, for both the Measure algorithm and the Peak-Rate one. The left-most histogram in the plot was made using peak-rate admission control; this allows for an average of approximately just 10 connections in the system and it is the same for the compressed and the uncompressed data. The central histogram was made using the Measure algorithm applied to the uncompressed traffic. The advantage gained by exploiting statistical multiplexing is clearly apparent as the average number of calls in progress is now approximately 32. The right-most histogram shows the effectiveness of our compression scheme when used together

---

[4] We chose to demonstrate the results of the http traces throughout this paper since these trigger the worst case scenario for our scheme (e.g. the other traces gave us better results with respect to all the issues discussed).

with the Measure Algorithm. As the graph shows the average number of calls in progress is approximately 54 or in other words 70 calls can be routed over the same network. Since this increase in the amount of data transmitted comes without any violation of the QoS the calls get, we argue that our scheme is extremely useful in real ATM networks[5].

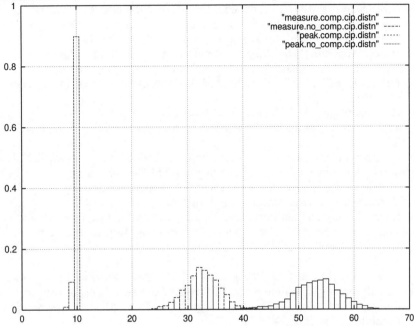

Fig. 6. Histograms of the average number of calls in progress over a simulation run.

References

1. I. Papaefstathiou, Hardware implementation of a Gb/sec network compressor, White Paper, http://www.cl.cam.ac.uk/ ip207/hard.ps.
2. I. Papaefstathiou, Accelerating ATM: On-line compression of ATM streams, 18th IEEE IPCCC'99, Phoenix, Arizona, 10-12 February 1999.
3. I. Papaefstathiou, Compressing ATM streams, IEEE Data Compression Conference 1999 (DCC'99), Utah, 29-31 March 1999 .
4. I. Papaefstathiou, Complete Framework for low level, on-line ATM Compression, Submitted for publication, http://www.cl.cam.ac.uk/ ip207/comp.ps.
5. Andrew Moore and Simon Crosby, An experimental conguration for the evaluation of CAC algorithms, Performance Evaluation Review, Vol 27:3, December 1999.

---

[5] Consider also that our results for the other types of traffic (IP,TCP etc.) are even better.

6. Andrew Moore and Raymond Russell, Evaluation of a Practical Connection Admission Control for ATM Networks Based on On-line Measurements, in preparation, 1999.
7. Joseph Y. Hui, Resource Allocation for Broadband Networks,IEEE JSAC, 6(9):1598{1608, December 1988.
8. N.M. Mitrou and D.E. Pendarakis, Cell Level Statistical Multiplexing in ATM Networks: Analysis Dimensioning and Call Acceptance Control w.r.t QoS Criteria, In Broadband Technologies, New Jersey, October 1990.
9. James Appleton, Modelling a Connection Acceptance Strategy for ATM Networks, In Broadband Technologies, New Jersey, October 1990.
10. F.P. Kelly, Effective Bandwidths at Multi-Class Queues, Queueing Systems, 9:5-16, 1991.
11. R. Guerin and H. Ahmadi and Naghshineh, Equivalent Capacity and its application to bandwidth allocation in high speed networks, IEEE JSAC,9:968-981, 1991.
12. C. Courcoubetis and G. Kesidis and A. Ridder and J. Walrand and R. Weber,Admission Control and Routing in ATM Networks using Inferences from Measured Bu er Occupancy, Technical Report, University of California, Berkeley, EECS Department, UCB, California, CA94720, 1991.
13. Simon Crosby and Ian Leslie and John Lewis and Raymond Russell and Fergal Toomey and Brian McGurk, Practical Connection Admission Control for ATM Networks Based on On-line Measurements, Computer Communications, January 1998 .
14. N. G. Duffield and J. T. Lewis and Neil O'Connell and Raymond Russell and Fergal Toomey, The Entropy of an Arrivals Process: a Tool for Estimating QoS Parameters of ATM Traffic, Proceedings of the 11th UK Teletraffic Symposium, Cambridge, March 1994.
15. N.G. Duffield and J.T. Lewis and N. O'Connell and R. Russell and F. Toomey Entropy of ATM Traffic Streams, IEEE Journal on Selected Areas in Communications, Special issue on advances in the fundamentals of networking - part 1, 13(6), August 1995.
16. J. Ziv and A. Lempel, A Universal Algorithm for Sequential Data Compression, IEEE Trans. Information Theory, Vol IT-23, No 3, May 1978, pp 337-343.
17. Ecole Polytechnique Federale de Lausanne, Application REQUested IP over ATM, Technical Report, http://lrcwww.ep.ch/arequipa/, July 1997.
18. E. Fiala and D. Greeve, Data Compression with Finite Windows, Communications of the ACM, Vol 32, No 4, April 1989, pp 490-505.
19. National Institute of Standards and Technology, NIST ATM/HFC Network Simulator: Operation and Programming Guide , March 1995.
20. I. M. Leslie and D. R. McAuley, Fairisle: An ATM Network,Computer Communications Review, 21(4):327-336, September 1991.
21. V. Paxson and S. Floyd, Wide-Area Traffic: The Failure of Poisson Modeling, Proceedings ACM SIGCOMM 94, London, UK, August 1994.

Integrating Position Reporting, Routing and Mobility Management in Multihop Packet Radio Networks

Konstantinos Amouris[1], Symeon Papavassiliou[2*], and Sheng Xu[2]

[1]The MITRE Corporation, Battlefield Systems Division,
145 Wyckoff Road, Eatontown, NJ 07724, USA

[2]New Jersey Institute of Technology
New Jersey Center for Multimedia Research
Electrical and Computer Engineering Department
University Heights, Newark, NJ 07701, USA
papavassiliou@adm.njit.edu

Abstract. In this paper we propose an integrated proactive routing and mobility management strategy that makes feasible the realization of a flat single-tier routing architecture in multihop packet radio networks. The integration of the routing and mobility management functions is achieved via the use of the geographic position reporting mechanism and the generalization of the zone routing concept. The underlying principle behind the proposed routing strategy is to reduce the network routing overhead by making the accuracy of the routing information in each node "inversely proportional" to its distance from any other node in the network. The proposed integrated scheme provides for a flat routing architecture with no hierarchical entry/exit points, where every node can act as a router, therefore increasing the network's routing flexibility and robustness. Finally we show through modeling and simulation that the proposed routing protocol is a bandwidth-efficient routing mechanism that can be applied across large-scale networks.

1 Introduction

Mobile wireless networking has enjoyed dramatic increase in popularity over the last few years. The advances in hardware design, the rapid growth in the communications infrastructure, and the increased user requirement for mobility and geographic dispersion, continue to generate a tremendous need for dynamic ad hoc networking. Multihop packet radio networks (or mobile ad-hoc networks) are an ideal technology to establish "instant" communication infrastructure for military and civilian [1,2] applications in which both hosts and routers are mobile. There are many existing military networking requirements for robust communications in a variety of potentially hostile environments that may require the rapid deployment of mobile radio networks (commonly referred to as packet radio networks) [3,4], as well as

* This work is supported, in part, by New Jersey Institute of Technology under Grant No 421050.

future military applications and requirements for IP-compliant data services within mobile wireless communication networks [5]. Moreover mobile ad hoc networking technology can provide extremely flexible method for establishing communications for operations in disaster areas resulting from flood, earthquake, fire, or other scenarios requiring rapidly deployable communications with survivable efficient dynamic networking. Some other applications of mobile ad hoc networking technology could include industrial and commercial applications involving cooperative mobile data exchange. In addition mesh-based mobile networks can be operated as robust, inexpensive alternatives or enhancements to cell-based mobile network infrastructures. In such mobile ad hoc networks there are no dedicated base stations as in conventional commercial cellular networks, and all nodes interact as peers for packet forwarding. This distributed nature eliminates single points of failure and makes those packet radio networks more robust and survivable that the commercial cellular networks. The vision of mobile ad hoc networking is to support robust and efficient operation in mobile wireless networks by incorporating routing functionality into mobile nodes. Such networks have dynamic, sometimes rapidly changing, random, multihop topologies. The goal of mobile ad hoc networking is to extend mobility into the realm of a set of wireless mobile nodes, where themselves form the network routing infrastructure in an ad hoc fashion.

A multihop packet radio network results from the fact that not every pair of nodes are within the transmission range of each other. In this case a packet must relayed over several hops before reaching its final destination, and therefore routing problems and issues emerge [6]. Choosing the proper routing strategies is very important for efficient network operation. Additional issues associated with the routing strategies in multihop packet radio networks stem from the fact that those networks are plagued with problems such as, variable quality of the links, the hidden terminal problem, and particularly the inaccuracies in the routing information due to the mobility of the nodes. In such mobile ad hoc networks a routing strategy (or system) can be defined as a set of several component functions including the following: monitoring network topology; locating end-points and performing mobility management; distributing this information for use in route construction; constructing and selecting routes. An efficient peer-to-peer mobile routing mechanism (protocol) in a purely mobile, wireless domain must: provide for effective operation over a wide range of mobile networking environment with its corresponding set of characteristics; provide algorithms and methods for allowing newly arrived nodes to be incorporated into the network and become an integral part of the network automatically, without manual intervention (self-organizing); react efficiently to topological changes and traffic demands while maintaining effective routing in a mobile networking environment; and address effectively the issue of scalability..

Implementing hierarchic routing in a highly dynamic network is complicated by the following issues: a) The hierarchy must be defined dynamically (because nodes that are now close together may later be far apart), b) Routing algorithms must adapt to changes in hierarchic connectivity, as well as to changes in radio connectivity, and c) Nodes must be able to determine the "hierarchical address" of a destination node. On the other hand some ad hoc mobile networks may be able to route messages using position (e.g. latitude and longitude) rather than topologically-derived information about nodes. Position based routing strategies, since they do not require the exchanges of routing tables, are especially attractive in highly mobile environments where topological changes are frequent and routing tables become obsolete very quickly [7].

In order to implement these protocols we must assume that a node is able to determine its location in some way. There are several systems for position location and/or position updates [8,9,10]. For example the NAVSTAR Global Positioning System (GPS) [10] is a position location system which is capable of global coverage. This paper deals with the issues related to the routing and mobility management in multihop packet radio networks. Specifically we propose an integrated proactive routing and mobility management strategy that makes feasible the realization of a flat single-tier routing architecture in such networks. The integration of the routing and mobility management functions is achieved via the use of the geographic position reporting mechanism and the generalization of the zone routing concept [11].

This paper is organized as follows. In section 2 we describe the network model that is used throughout our paper. Section 3 provides a detailed description of the integrated routing and mobility management mechanism, while in section 4 we provide some performance evaluation results based on modeling and simulation. Finally section 5 concludes our paper.

2   Network Model

In ad-hoc wireless networks nodes are equipped with wireless transmitters and receivers using antennas which may be omnidirectional (broadcast), highly directional (point-to-point) or some combination thereof. At a given point in time, depending on the nodes' positions and their transmitter and receiver coverage patterns, transmission power levels and co-channel interference levels, a wireless connectivity in the form of random, multihop graph exists between the nodes. This ad hoc topology may change with time as the nodes move or adjust their transmission and reception parameters. Throughout this paper, we assume that every node is aware of its position with the aid of a reliable position locating system (i.e. GPS). Therefore the topology of such networks is modeled as a directed graph $G=(V,E)$, where V is the set of nodes and E is the set of edges connecting the nodes. In a wireless network, a node can have connectivity with multiple nodes in a single physical radio link. For routing purposes a node A can consider another node B to be adjacent ("neighbor") if there is link-level connectivity between A and B and A receives update messages from B reliably. Accordingly, we map a physical broadcast link connecting multiple nodes into multiple point-to-point bidirectional links defined for these nodes. A functional bidirectional link between two nodes is represented by a pair of edges, one in each direction and with a cost associated that can vary in time. Finally we assume that a router is capable of detecting within a finite time the existence of a new neighbor, the loss of connectivity with a neighbor, and the reliable transmission of packets between neighbors.

We also assume that the terrain is divided into several overlapping and dynamically changing virtual zones. For each node X a set of y such zones are defined (initially concentric) where zone-i represents a circle around node X with radius $R(i)$, such that: $R(i)<R(i+1)$. Specifically: let $Z(X,i,t)$ denote the node's X zone-i at time t and $C(X,i,t)$ the corresponding center. For sake of simplicity in the description of the routing strategy let $R(0)$ (that is the radius of zone-0) be measured in number of hops while $R(i)$ $(i>0)$ is measured in meters. Furthermore with each node X there is a set of timers associated, where $ZRT(X,i)$ denotes the Zone Refresh

Timer of node X for zone-i, and ZRT_Residual(X,i,t) denotes at time t the residual time until ZRT(X,i) expires. The purpose of the Zone Refresh Timer is to periodically 'refresh' node X's position in the position databases of all nodes included by Z(X,i+1,t). In the following we refer to zone-0 of a node X as the core zone of node X. We also denote by L(X,t) the position of node X at time t.

3   Routing Strategy

In this section we provide a detailed description of the proposed routing strategy. The underlying principle behind the proposed routing strategy is to reduce the network routing overhead by making the accuracy of the routing information in each node proportional to its distance from any other node in the network. The use of multiple zones provides a method of implementing that principle in a discrete way. The proposed routing strategy consists of two main elements: 1) Routing Policy (protocol) that describes the generation and propagation process of the appropriate routing information, and 2) Routing Mechanism that describes how the routing messages created by the routing policy is used in order to forward the data packets in the network. The routing policy itself may be divided in to two subprocesses: a) Link State Update Propagation used within the core zone of each node, and b) Position Update Generation and Propagation used for the generation and propagation of routing updates within any zone-i, such that i>0.

3.1 Routing Policy

3.1.1 Link State Update Propagation

This element refers to the routing policy used within the core routing zone. The functionality of the core zone in our protocol is similar to the functionality of the routing zone in the Zone Routing Protocol [11]. The core zone of node X includes the nodes whose minimum distance in hops from X is at most some predefined number, which is referred to as the Core Zone Radius. Our protocol supports the proactive maintenance of the core routing zone through its link-state [12] Core-zone Routing Protocol (CRP). Through the CRP, each node learns the identity of and the least-cost route to all the nodes in its core routing zone. The choice of the link-state CRP is arbitrary; however, the chosen link-state protocol needs to be modified to: a) ensure that the scope of its operation is restricted to the radius of a node's core routing zone, and b) a field containing the node's geographic position is added to the link-state CRP protocol packets. Moreover our core zone link-state routing protocol is augmented to support Multipoint Relaying (MR), a mechanism designed to minimize the overhead of packet "flooding" throughout the network, by optimizing/reducing the number of relays/retransmissions. Our core zone link-state routing protocol supports multipoint relaying in the following manner: each link-state packet transmitted by a node in response to a link status change contains not only the new link-state information, but it also contains a flag indicating whether the node at the other end of the link has been selected as a MultiPoint Relay (MPR) by the node issuing the link-state packet. An

MPR is a router that has been selected by a one-hop neighbor to forward or retransmit that neighbor's packets. Each router has one or more neighbors that have selected it as an MPR. In addition, a set of Core-zone Border Routers (CBR) is defined for each node; CBRs are nodes whose minimum distance to the node in question is equal exactly to the core zone radius. The nodes in the CBR set are sorted according to their angular distance from the node's geographical y-axis.

3.1.2 Position Update Generation/Propagation

This element describes when the position update messages are generated, how they are propagated within the network and how the multiple zones are interrelated and change dynamically as the nodes continuously move. Specifically when node X's position relative to $C(X,i,t)$ changes by more than $R(i)$ meters, or $ZRT(X,i)$ expires then: a) a routing/position update $GPU(X,i)$ is generated and propagated throughout $Z(X, i+1, t+Dt)$, and 2) a new geographic zone $Z(X, i, t+Dt)$ is defined for node X with center $C(X,i,t+Dt)$, where $(t+Dt)$ denotes the time that the above event occurred. As we can see each time a $GPU(X,i)$ is generated the corresponding zone-i for node X is redefined therefore making the smaller zones to move within a larger zone. In consistency with the notation defined in the beginning of this section, this operation is described by he following algorithm:

IF {Distance[$L(X, t)$, $C(X, i, t)$] > $R(i)$ OR ZRT_Residual$(X,i,t) = 0$ }
a) $C(X, i, t+Dt) := L(X, t)$; A new zone $Z(i)$ has just been defined (or refreshed) for node X
b) Construct $GPU(X,i)$ packet with dissemination area equal to $Z(X, i+1, t+Dt)$

As can be seen, the geographic zone radii and the values of the corresponding Zone Refresh Timers are set so that the accuracy of the knowledge of a node's position is proportional to the geographic distance from that node. Figure 1 provides a graphical illustration of the dynamic evolution of the multiple zone interrelation.

In order to implement an efficient GPU propagation mechanism the $GPU(X,i)$ packet should contain the following fields: 1) current position of node X, 2) source ID of node X, 3) GPU type, 4) sequence number, 5) $C(X, i+1, t+Dt)$. At any time t, a node F that received (from a neighbor G) a $GPU(X,i)$ packet (generated by node X), will propagate that packet if all of the following conditions are true:
1)   Node F has been selected as an MPR by neighbor G
2)   Distance[$L(F, t)$, $C(X, i+1, t)$] < $R(i+1)$, and
3)   Received GPU sequence number > latest GPU sequence number associated with node X.

3.2 Routing Mechanism

In the following we present a complete view of the routing mechanism operation by outlining the main steps implemented by the router module of any node (current_node) in the network that needs to route a packet. The first node (current_node) that implements the following procedure is the source node of the packet (node that originally generated the packet).

Step 1: The current_node checks whether the destination is within its core routing zone.

If so, the path to the destination is completely specified via the link-state CRP, and no further route processing is required; go to step 4.

Otherwise go to step 2.

Step 2: The current_node
a) looks up the destination's geographic position,
b) picks the CBR with the minimum angular distance from the imaginary line joining the current node and the destination, and
c) forwards the packet towards the selected CBR via the least-cost path specified by the current node's link-state CRP. This node is the next hop recipient (next_hop_node).

Step 3: The next hop recipient (next_hop_node) of the forwarded packet, executes the same procedure. That is: current_node:=next_hop_node; go to step 1.

Step 4: Stop.

Steps 1, 2 and 3 may be repeated until the packet reaches the boundaries of the destination's core routing zone; at this point, the packet is picked-up by the link-state CRP instance running in the vicinity of the destination. Figure 1 graphically illustrates the operation of the proposed overall routing strategy. Assume that node X is the destination, and node S is the source, which is physically located inside $Z(X,4,t)$ (not shown) and outside $Z(X,3,t)$. Given the above, node X's perceived position at node S is equal to $C(X,3,t)$. Thus, the packet is being forwarded towards $C(X,3,t)$, according to the routing mechanism described above. The dashed, overlapping circles that are superimposed on top of the routing trajectory represent overlapping core zones, where local, link-state knowledge is available. As soon as the packet reaches a router within $Z(X,3,t)$, it is re-directed toward $C(X,2,t)$, since nodes within $Z(X,3,t)$ are closer to X, and therefore have more accurate knowledge of X's geographic position. Similarly, as the packet enters $Z(X,2,t)$, another course-correction happens: the packet is re-directed towards $C(X,1,t)$. As soon as the packet reaches the borders of $Z(X,1,t)$, it is picked up by the core zone link-state protocol running in the vicinity of node X.

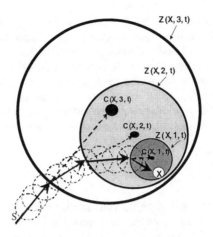

Fig. 1. Graphical Illustration of the Routing Mechanism

4   Performance Results

In this section we evaluate the protocol's performance via modeling and simulation. Specifically we study the protocol's routing effectiveness and scalability measured by routing overhead as a function of the network size (number of nodes) for varying user mobility. The specific metric used throughout this study is defined as follows:

- Routing Overhead: Average number of routing packets per node

It should be noted that the performance results presented in this section constitute a preliminary subset of results of our performance evaluation process. A more in-depth and systematic performance analysis of our routing protocol is still in progress in order to evaluate the protocol's operation with respect to a complete set of performance metrics (i.e. include delay measurements, study the effect of different design parameters related to the protocol operation etc.). In the following subsection 4.1 we describe the simulation framework, models and assumptions used throughout the simulation, while in subsection 4.2 we present some preliminary numerical results of our experiments.

4.1 Simulation Model

The performance evaluation of our protocol is accomplished via modeling and simulation using the Optimized Network Engineering Tool (OPNET). A mobile ad-hoc wireless network is modeled in the simulation. Each node is modeled by a store-and-forward queuing station. Specifically each node consists of: a data source module, a data sink module, physical layer transmitter and receiver, and a routing module. A link can be modeled as a FCFS queue with service time as the transmission time. All nodes are assumed to have adequate buffer capacity for buffering packets waiting to be forwarded. Since the main objective of the paper is to develop and evaluate an efficient routing strategy our simulation study is mainly limited to network layer details, while simplified models, that do not affect the validity of the qualitative analysis of our results, are assumed for the data link layer. The details about the node mobility model, network topological changes and link layer assumptions that are used for the simulation purposes are discussed in the following sub-sections.

4.1.1 Mobility Model

Mobile nodes are assumed to be moving around throughout a closed rectangular region of size Y m x Y m according to a mobility model. The mobile nodes are assumed to have constant radio range of Z m. The mobility model used throughout the simulation is as follows : at any point in time mobile movement is characterized by two parameters, its velocity vector v (value v and direction ϕ) and its current position (x,y). The velocity value v and the movement direction ϕ are updated every time interval Δt, according to the following model:

$$v(t+\Delta t) = \min\{\max[v(t)+\Delta v, 0], v\_max\} \,. \tag{1}$$
$$\phi(t+\Delta t) = \phi(t)+\Delta\phi \,. \tag{2}$$

where:

Δv is a uniformly distributed random variable in $[-\alpha*\Delta t, \alpha*\Delta t]$,

v_max is the maximal mobile velocity, α is the maximal mobile node acceleration (deceleration),

$\Delta\phi$ is a uniformly distributed random variable in $[-\Delta\phi\_max*\Delta t, \Delta\phi\_max*\Delta t]$ and $\Delta\phi\_max$ represents the maximal angle change per time unit.

Each mobile recalculates a new position (new move) every Δt time units. After each such move the mobile node X computes its neighborhood and possibly generates the appropriate GPU(X,i) packet for routing update purposes.

4.1.2 Wireless Link Status

An important issue associated with the routing strategies in mobile ad-hoc network stems from the variable quality of the wireless links that may result to link failure or appearances. In general the link status in such an environment can be modeled in many different ways such as: two state (on/off) continuous time Markov process, periodic link status sensing by using hello messages etc. Since no link layer details are modeled here as explained in the following sub-section, throughout our simulation we assume that a link fails or reappears as a node goes out or in transmission range of another node, due to the mobility of the nodes.

4.1.3 Link Layer/Access Layer Model

Since our study is limited to network layer details no link layer details (such as MAC protocol, link errors, frame retransmissions etc.) are modeled here. The operation and functionality of an effective and reliable link access layer protocol is simulated as follows. Each time a node has to transmit a packet to any of its neighbors we assume error free transmission in any direction (within the node's transmission range). To ensure that no blocking occurs due to spectrum limitations, we assume that the system has large enough number of channels. Since a detailed link access layer protocol has not been directly implemented in our simulation, it is possible for a node to receive multiple packets from different neighbors simultaneously. Those packets are placed in the node's queue to be served by the node (in an order determined by OPNET).

4.2 Numerical Results

In this section we present numerical results of our simulation based on the models and assumptions described in the previous section. Multiple topologies are used in order to vary the network size. Since our primary goal in this study is to evaluate the net effect of the network size on the protocol's routing overhead (associated with the scalability issue), we obtain the following numerical results by keeping fixed most of the protocol and network parameters throughout our simulation and varying only the network size. The default values used in our test are: v_max = 72 (km/hr), Δt=20 (secs), $\Delta\phi$_max = 0.125π (rad/sec), a = 2 (m/sec^2), Node Transmission Range = 500(meters), radius of zone[i] = $500*2^i$ (meters) for i=0,1,2,3,4. In the current implementation, the radius of the core zone is set equal to the transmission range of a node (core zone radius =1 hop). Therefore, for destination nodes (either final destinations or intermediate destinations) within the core zone of a node A, we assume ideal direct routing and communication. In addition, the Zone Refresh Timers are not utilized in the current implementation (i.e., timers are set to infinite values). The number of mobiles ranges from 100 to 1000 while we maintain the network density (number of mobiles per surface unit) constant and equal to 10 mobiles/Km^2.

In figure 2 we present (vertical axis) the Average Number of Routing Packets (GPUs) per Node per time unit (sec) for each zone as well as the Average Number of Total_GPU packets versus (horizontal) the network size (number of nodes in the network) for the mobility scenario described above. As we see in this figure the average number of routing packets transmitted by a node is practically insensitive to the network size. This clearly indicates that our protocol possesses the scalability property which is very critical in multihop packet radio networking environments, without having to resort to complicated and vulnerable hierarchical approaches.

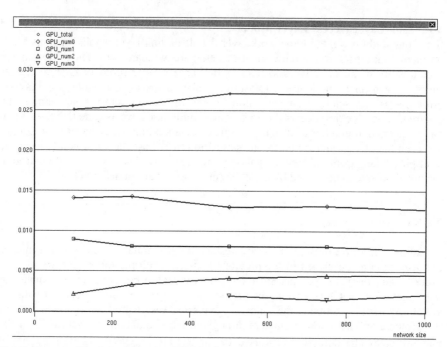

Fig. 2. Average Number of Routing GPU Packets (for each zone and total) per Node per time unit (sec) versus the network size (number of nodes)

Note, that the actual number of routing packets, generated and relayed throughout the operation of our protocol, depends heavily on a) the mobility model implemented, and b) the "flooding" technique used to propagate the GPU packets to the appropriate network area. It should be noted here that, throughout our simulation, we have not implemented the MR mechanism described in section II that minimizes the number of relays. Therefore, although a GPU packet can be forwarded only once per each link, every node will forward a GPU packet that has not been received before.

5   Conclusions and Further Research

In this paper, we proposed an integrated proactive routing and mobility management strategy that makes feasible the realization of a flat single-tier routing architecture in multihop packet radio networks. The proposed strategy is mainly intended for networks with large network diameters and diverse mobility patterns. The proposed approach provides an alternative, simplified way of localizing routing information overhead, without having to resort to complex, pyramid-like, hierarchical routing organization schemes. This is achieved by merging the functions of routing and mobility management via a) the use of geographic position, and b) the generalization of the routing zone concept. The proposed strategy provides for a flat routing architecture with no hierarchical entry/exit points, where every node can act as a router, therefore increasing the network's routing flexibility and robustness. Moreover it controls routing overhead generation and propagation by making the overhead generation rate and propagation distance directly proportional to the amount of change in a node's geographic position.

We have also shown via modeling and simulation that our routing mechanism is a bandwidth-efficient one that can be applied across large-scale networks. We are currently performing a more in-depth analysis of our routing protocol in order to evaluate the protocol's operation with respect to a complete set of performance metrics (i.e. include delay measurements, study the effect of different design parameters related to the protocol operation etc.) and we plan to compare our routing strategy and its effectiveness (i.e. number of hops to final destination, delays, etc.) against the corresponding performance of other traditional solutions.

References

1. C.E Perkins and P. Bhagwat, "Highly dynamic destination-sequenced distance-vector routing (DSDV) for mobile computers", Proc. ACM SIGCOMM, London, UK, 1994, p.p. 234-244.
2. W. Diepstraten, G. Ennis and P. Berlanger, "DFWMAC: distributed foundation wireless medium access control", IEEE Document P802.11-93/190, November 1993.
3. J. Jubin and J.D. Tornow, "The DARPA packet radio network protocols", Proc. IEEE 75(1), 1987, p.p. 21-32.
4. N. Shacham and J. Westcott, "Future directions in packet radio network architectures and protocols", Proc. IEEE 75(1), 1987, p.p. 83-99.
5. B. Adamson, "Tactical radio frequency Communication Requirements for Ipng," RFC 1677, Aug.1994.
6. Z.J. Haas, "On the Relaying Capability of the Reconfigurable Wireless Network", VTC'97, Phoenix, AZ, May 1997.
7. Ting-Chao Hou and Victor O.K. Li, "Position updates and sensitivity analysis for routing protocols in multihop mobile packet radio networks", Proc. IEEE GLOBECOM, 1985, p.p. 243-249.
8. J.A. Kivett, "PLRS-A New Spread Spectrum Position Location Reporting System", Proc. IEEE PLANS, 1976, p.p. 223-230.
9. J.D. Olsen and R.G. Sea, "Network Management for the PLRS-JTIDS Hybrid (PJH) System", Proc. Nat. Telecomm. Conference, November 1980, p.p. 41.5.1-41.5.11.
10. B.W. Parkinson and S.W. Gilbert, "NAVSTAR: Global Positioning System – Ten Years Later", Proceedings of IEEE 71, 10, October 1983, p.p. 1177-1186.
11. Haas, Z. J. and Pearlman, M. R., "The Zone Routing Protocol (ZRP) for ad-hoc networks (Internet-Draft), http://www.ee.cornell.edu/~haas/Publications/draft-zone-routing-protocol-01.txt" August 1998
12. Garcia-Luna-Aceves, JJ and Spohn, M., "Scalable Link-State Internet Routing", Proc. IEEE ICNP 98, Texas, October 14-16, 1998

Architectures for Multimedia Communications

Han Zuidweg

Alcatel, Belgium
johan.zuidweg@alcatel.be

The days that telecommunications were equivalent to making a phone call between two people are behind us. The personal computer, mobile telephony and the Internet have changed telecommunications for good. It is now technically feasible to offer multimedia services in public networks. But whereas the bandwidth and basic technology are there, multimedia services are not yet enjoying common acceptance. The truth is that multimedia is not yet well understood in terms of market, in terms of actual services, even in terms of technology.

Today there is still a sharp distinction in circuit switched voice networks and packet switched data networks. Apart from the pure transmission aspect, traditional voice and data networks differ in where the service features are controlled and managed. Circuit switched telephony networks have the intelligence built into their core, managed by the network operator. Voice telephony terminals are very simple devices with no or only local intelligence. In contrast, data networks usually only provide basic routing features while services are provided entirely outside the network by sophisticated terminals – usually clients and servers.

But the boundaries between circuit switched networks and packet based data networks are blurring. As a result telecommunications services are becoming increasingly complex, covering multi-media, multi-party services and involving personal and terminal mobility.

It is often said that multimedia applications and services will contribute to improvements in quality of life, employment, and regional development. But what are these applications and services? Who will build them, who will deploy and provide them, and who will make money from them? Is there a market for multimedia services? What services will exist beyond video conferencing and video-on-demand?

'Multimedia' is a vast area covering applications, services and networks. The technical issues in creating, deploying, operating, and delivering multimedia services are myriad. This section contains four technical articles that address distinct aspects of multimedia services.

The first two papers present two alternative approaches to controlling multimedia services. The first paper, entitled "Towards Mobile Agent Based Provisioning of Voice Over IP Services", proposes Mobile Agents as a technology for implementing advanced IP telephony services. Although the paper focuses on value added services for IP telephony, the proposed solution is also applicable to multimedia services because it uses SIP as its protocol base. The authors from IKV++ and GMD Fokus in Germany implemented a prototype of their architecture on the Grasshopper environment for Mobile Agents.

The second paper is entitled "Supporting Advanced Multimedia Telecommunications Services Using the Distributed Component Object Model". The authors of this paper, from Surrey University in the UK and from Hellenic Telecommunications Organisation (OTE) in Greece, propose a framework for the control of multimedia communications based on Microsoft's Distributed Component Object Model (DCOM). The DCOM interfaces allow applications to set up and control continuous media streams (or just 'streams' in short) between defined points.

The first and the second paper rely on two quite different computing models. The Mobile Agents approach of the first paper is based on the mobile code paradigm. Autonomous pieces of code, called 'Agents', move between execution environments. They execute and communicate with other agents in the 'visited' locations. The DCOM approach of the second paper is based on the object oriented version of Remote Procedure Calls (RPC). In this case the code does not move, but is remotely invoked by sending messages. The first two papers are highly representative for the two streams of research that are being conducted world-wide and within the ACTS and IST programs.

The third paper addresses a very specific problem in the delivery of Video-On-Demand (VOD). One of the important problems in VOD systems is the scalability. Providing VOD services to tens of thousands of users requires diligent management of resources. The author from the Technical University of Hamburg proposes a group of 'near-client' caching algorithms that can increase the efficiency of a VOD system by an order of magnitude. He presents performance results of several simulation scenarios (home, local area) that demonstrate the use of these algorithms.

The last paper in this section addresses the management side of multimedia services. By offering a Service Level Agreement (SLA) to its customers, a service provider commits itself to providing services with a certain quality at a certain cost. Whenever an SLA is being compromised by network problems, it is very important that the cause of the trouble is spotted immediately so that action can be undertaken. This is especially challenging in an 'open' network environment where several network providers may be involved in the delivery of a service. The authors of the fourth paper in this section from GMD Fokus in Germany, discuss what ITU, TMF, Eurescom and TINA-C have contributed to solving this problem. Based on this, they describe a Trouble Report System for a TINA based open service platform.

Multimedia services will probably drive the telecommunications market in the near future, and are therefore a key topic for the IS&N 2000 conference. It seems that we are only beginning to uncover the potential of multimedia services. Our research must lead to a further uncovering of this new ground. So that the end, *multimedia* will be more than just the sum of the individual media: audio, video and data.

Towards Mobile Agent Based Provision of Voice over IP Services

W. Chang, M. Fischer, T. Magedanz

GMD FOKUS, Kaiserin-Augusta-Allee 31, D-10589 Berlin, Germany
Email: chang@fokus.gmd.de
IKV++ GmbH, Bernburger Str. 24-25, D-10963 Berlin, Germany
Email: {fischer|magedanz}@ikv.de

Abstract. Mobile Agent technology forms today the basis for the realisation of many innovative and flexible telecommunication architectures. After its emerging application in telecommunications management, intelligent networking and mobile communications, the increasing adoption of internet technologies in the telco environment suggests a new promising application domain: Voice over IP environments. This paper presents a mobile agent based VoIP service provision architecture. This architecture is based on the integration of the Grasshopper agent platform, a state of the art mobile agent platform which has been developed for the implementation of advanced telecommunication service environments, and an IETF based VoIP protocol suite, i.e., the Session Initiation Protocol (SIP) and related IETF protocols. The outlined architecture, in which most of the SIP entities are realised by mobile agents enables a more flexible and distributed realisation of VoIP services providing better load balancing, and particularly user mobility.

1 Introduction

With the exponential growth of data services, the future telecommunications environment will be increasingly based on internet technologies. Therefore new architectures for Voice over IP (VoIP) transport are currently emerging, such as H.323, SIP, and Megaco [5] [6] [4]. Initially these new architectures concentrated on the provision of basic telephony capabilities, where the provision of supplementary and value added services, such as Intelligent Network (IN) [1] services (including mobility) has not been addressed in a comprehensive way. Therefore, substantial work is currently underway to extend the application of IN capabilities also to the emerging VoIP domain, where so-called VoIP "soft switches" (such as Gatekeepers, SIP proxies, or Media Gateway Controllers) enable access to IN as a backend (service) system, in order to provide value added services.

However, in order to overcome the limitations of existing IN systems in regard to extensibility, scalability and openness, the evolution of IN [33] [30] [22] is driven by the increasing incorporation of new software technologies, such as distributed object technologies and most recently mobile agent technologies [34] [35], including Java, towards an open, extensible and distributed IN architecture. In this regard, the IN architecture evolves from centralised service control nodes towards distributed application servers, service nodes and enhanced switching systems where service provisioning moves increasingly towards the edges of the network. Therefore the traditional telecommunication environment is getting closer to the internet, which is based on the paradigm of having service intelligence at the network edges, i.e. end systems.

Looking at the distributed architectures of VoIP systems, they enable also end users to develop and run their VoIP supplementary services on dedicated servers or locally at their own end systems. On the other hand emerging mobile agent based distributed IN architectures, as developed in the ACTS MARINE (Mobile Agent based IN Environments) project [2] [30], enable the flexible development and deployment of IN services in form of Mobile Agents. These service agents can move after service creation and subscription to the most appropriate centralised servers and distributed service nodes/switches within this enhanced IN architecture. By means of optional replication/cloning of service agents, a fully distributed service provisioning environment is implemented, decreasing signalling network and server load, enhancing reliability and performance.

Based on this, we propose in this paper a mobile agent based VoIP environment, and describes in detail a flexible VoIP service provision architecture. This architecture is based on the integration of the Grasshopper agent platform, a state of the art mobile agent platform which has been developed for the implementation of advanced telecommunication service environments, and an IETF based VoIP protocol suite, i.e., the Session Initiation Protocol (SIP) and related IETF protocols [7].

We will demonstrate in this paper, that the outlined architecture enables a flexible and distributed realisation of VoIP services, where through the realisation of all major SIP entities in the form of mobile agents the network servers and end systems can be dynamically configured and upgraded. This enables dynamic service deployment, load balancing, and particularly user mobility, which is not yet possible in traditional VoIP developments. As this paper extends the idea of intelligence on demand investigated by the ACTS MARINE project [2] [30] in the context of MA-based IN services for switched circuit networks, unified provision of advanced telephony services becomes possible across traditional circuit switched and packet data networks. In principle, all services are deployed dynamically as mobile agents in enhanced (soft)switches, and if needed can move between these switches and even into the user end systems.

Our proposed MA-based VoIP architecture is based on the Session Initiation Protocol (SIP), so the following section will first introduce SIP and related protocols for VoIP. Section 3 describes the mobile agent platform Grasshopper which forms the basis for our proposed mobile agent based VoIP architecture, which is presented in section 4. Afterwards service examples are given in section 5, whereas section 6 displays a prototype implementation. Related work is outlined in section 7 and section 8 concludes the paper.

2   Voice Over IP based on the Session Initiation Protocol

Signalling and voice transfer are separated in VoIP and have their different paths. Firstly signalling messages (such as set-up, tear down) are exchanged among Internet telephony end systems. When the signalling succeeds, voice media are then packetized into IP packets and transferred over IP network. Because IP network is a best effort network, packetized voice messages may take different paths in IP network.

There are three signalling protocols in VoIP: MGCP [4], H.323 [5], and SIP [6]. MGCP is a protocol for control of a Media Gateway (MG) from a control unit called Media Gateway Controller (MGC). In this paper we will not introduce it in detail. H.323 which was developed originally for multimedia conference by ITU-T is a set of standard protocols: H.225, H.235, H.245, H.246, T.120, H.26x, G.7xx, and Q.931 etc. It is then used for Internet telephony. H.323 is based on TCP/IP and defines four components: Gateways, gatekeepers, multipoint control units (MCUs), and terminals. Gateways connects PSTN and IP networks and are responsible for signalling between IP networks and PSTNs and conversion and transfer of media. Gatekeepers are responsible for mapping of H.323 aliases to IP addresses, entrance control to make calls, etc. MCUs are made of two parts: Multipoint Controllers (MCs) and Multipoint Processors (MPs). MCs are responsible for negotiation of capabilities and combination of different terminal types. MPs are for mixture and distribution of media. Terminals can be computers or PSTN- or IP-telephones. Another project based on H.323 is TIPHON [32].

The Session Initiation Protocol (SIP) is another Internet telephony standard. SIP is developed by the IETF Working Group "Multiparty Multimedia Session Control" (mmusic) [7]. SIP is based on HTTP and a client-server and text-based application control and signalling protocol. Several servers and clients are involved in a call. It can be used with other protocols for example RTP (Real-time transport protocol) [8], RTCP (Real-time control protocol) [8], RSVP (Resource reservation protocol) [9], SAP (session announcement protocol) [10], and SDP (session description protocol) [11] to complete a call. It supports name-mapping, personal mobility, and invitation of new services into a call. SIP is independent of the underlying transport protocol and can run on UDP, TCP, X.25, ATM, and PPP.

Callers and callees are identified by SIP-addresses in SIP. An SIP-address is a URI. SIP define its own URI, but its header fields can carry other URI such as http, mailto, phone.

There are four types of SIP servers: Proxy servers, redirect servers, user agent servers, and register servers. A proxy server is an intermediary program that can take actions as a client and a server. A participant in a call can generate requests or receive responses. An SIP-enabled end system includes a user agent client and a user agent server. The user agent client generates call requests and sends the requests to a proxy server or a redirect server. A request to a server is generated by a client and the server processes the request and sends back a response to the client. A request and response form a transaction. A SIP request can traverse many proxy servers which receive requests and forward them to a next hop server which can be a proxy server or a final user agent server. A server may also be a redirect server. A redirect server informs a client of the address of the next hop server (for example a next hop proxy server or final user agent server), so that the client can contact the next hop server directly. The user agent server responds to a request based on human interaction or some kinds of inputs. A proxy server and a redirect server can not accept and reject a request. Only a user agent server can do so. A client can register himself at a register server. A register server is used to locate users.

A client request invokes methods on a server. Requests and responses have header fields to convey call properties and service information. SIP defines several methods:

INVITE, BYE, OPTIONS, ACK, CANCEL, REGISTER. INVITE method invites a user to a call. BYE terminates a connection in a call. OPTIONS ask for information about capabilities. ACK is a reply to invitations. CANCEL terminates a search for a user. REGISTER conveys information about a user's location to a SIP server. There are two types of responses: temporary responses and final responses. A temporary response is used by proxies or redirect servers to describe a processing state when the process is still on. A final response is a final answer from a user agent server. SIP uses many HTTP header fields in requests and responses, for example a request uses header fields to convey information about a call, such as To, From, Subject, Call-ID, Contact and etc. These headers in SIP can be used to provide advanced services.

H.323 and SIP are both standards used for Internet Telephony (for a comparison of H.323 and SIP please refer to [12]). SIP is simple and flexible in comparison with H.323 and can be used on many networks such as IP, ATM etc. Another reason for choosing SIP in our implementation is the possibility to program all SIP components with Java (or as Grasshopper Java Agents).

Additionally SIP Common Gateway Interface (SIP CGI) [13] and Call Processing Language (CPL) [14] are developed for SIP service creation. SIP CGI is specifically for SIP and CPL is independent of the underlying signalling protocol. SIP servlet [15] is being studied, it is like Java servlet and used for SIP, so that more services can be introduced into SIP. With SIP CGI, CPL, and SIP Servlet services can be easily created. With SIP headers services can be easily realised. So our architecture is based on SIP, CPL SIP, and CGI or SIP servlet.

SIP CGI is similar to HTTP-CGI [16]. In the web CGI is a flexible mechanisms to create dynamic content. Like traditional HTTP-CGI, a SIP CGI script is invoked when a SIP request arrives at a server. The server passes the body of the message to the script through its standard input, and sets environment variables containing the information on the message headers, user information and server configuration. The script performs some processing, and generates some data which is written to the standard outputs of the script. The data is read by the server, and the script terminates. The intelligence locates at Internet telephony server or advanced clients. Since a CGI script is a normal process, it has access to network and system resources, such email, web etc. Because SIP CGI can access network and system resources, it is used by trusted providers.

There is another method which provides flexible services in Internet telephony too: Call Processing Language (CPL). CPL is a new language designed for users to write scripts to define his own services, e.g. whose telephone call he does not want to accept. These scripts can be uploaded to the service provider and automatically placed in networks. Because of security concern, the CPL has limited functions, e.g. there is no loops in this language. CPL is based on XML [17] and the structure of this language includes: switch, location, signalling actions, non-signalling actions. Switch is used to test the condition. Location refers to a location where signalling actions take place. Signalling actions concern signalling.

In SIP there is no uniform service architecture which can deploys, provides and manages VoIP services. We should provide an architecture which will enable flexible service creation, deployment, provision, and management of VoIP services. In this architecture SIP shall be able to work well with SIP CGI or SIP servlet and CPL, so that services can be flexibly introduced in SIP, while this issue is not well addressed in SIP. Secondly in SIP a proxy server becomes easily a bottleneck when there are too many requests of users for it to process. In this case it needs a mechanisms which distributes services into several other proxies and the proxy server itself becomes a redirect server and directs the requests of users to these other proxy servers. Additionally corresponding services shall be able to be distributed to where they are needed on the fly and loads in network shall be enable to be balanced. Finally user references can stay at end systems and servers. They shall be able to migrate to proxy or redirect servers when the user's machine is off.

Considering these problems in SIP we want to present a mobile agent based service provision architecture for VoIP. A mobile agent is a software component which can migrate to places where it is needed and performs a specific task autonomously on behalf of a person or an organisation. Mobile agent technology provides flexibility, distribution, scalability, and robustness, so it is widely used in many areas. Because of the benefits of mobile agent technology in many fields we investigate the issue of mobile agent in VoIP. We will illustrate our MA-based architecture for VoIP in section 4 and describe the used Mobile Agent platform in section 3.

3   Grasshopper – Basics and Application for Advanced Telecommunications

Grasshopper [18], which has been developed by GMD FOKUS and IKV++ GmbH, is a mobile agent development and runtime platform that is built on top of a distributed processing environment. Thereby an integration of the traditional client/server paradigm and mobile agent technology is achieved. Grasshopper is implemented in Java, based on the Java 2 specification. Most importantly Grasshopper has been designed conformant to the first industry mobile agent standard, namely the Object Management Group's Mobile Agent System Interoperability Facility (MASIF) [19]. In addition, the newest Grasshopper version is also compliant to the specifications of the Foundation for Intelligent Physical Agents (FIPA) [20].[1]

Figure 1 shows the structure of a Grasshopper-based system environment which consists of several Grasshopper agent systems (agencies), grouped within a domain (region), such as an intranet.

---

[1] The first Grasshopper version has been released in summer 1998. Since February 1999, Grasshopper Release 1.2 is available. Grasshopper 2 is available since spring 2000.

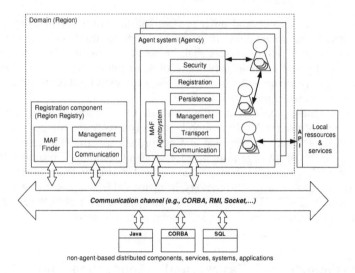

Fig. 1. Structure of a Grasshopper-based Environment

Agent System (Agency)

An agency is a Java process that enables and controls the execution, management, transport, communication, etc., of Grasshopper agents. Each agency covers the following services/components:

- *Security*: Since a mobile agent must be considered as "alien element" on the visited hosts, a fundamental task of an agency is to protect the hosts from unauthorised agent access. On the one hand, Grasshopper provides *external security mechanisms* that allow the encryption of agents (code and state) during their migration, using the Secure Socket Layer (SSL) protocol. If required, also remote interactions between Grasshopper entities (agents, agencies, non-agent-based components) can be protected via SSL. By means of certificates, an agency is able to identify and authenticate agents. On the other hand, *internal security mechanisms* enable access control inside an agency.

- *Registration*: Each agency registers all locally running agents in order to monitor and control the entire agency-internal processing and in order to enable agents to find each other for information exchanges.

- *Persistence*: In order to save agent-related data in case of a system crash, the agency's persistence service can be used to periodically store the internal states of all locally running agents. In this way, the agents can continue their tasks after restarting the agency.

- *Transport*: The transport service enables the serialisation of an agent's state, the transfer of the agent to its destination agency (by accessing the agency's communication service), and the restart of the agent at its destination.

- *Communication*: On the one hand, the communication service enables the transfer of agents between different agencies, triggered by the transport service. On the other hand, the communication service is responsible for managing interactions between remote agents and non-agent-based entities. In this way, as explained above, Grasshopper achieves an integration of the traditional client/server approach (using remote communication) and mobile agent technology (using agent migration).

- *MAFAgentSystem*: This component realises an interface that is part of the OMG MASIF standard and that increases the interoperability between Grasshopper and other standard-compliant agent platforms.

Registration Component (Region Registry)
Optionally, an agency can register itself at a registration component, called region registry. The region registry is a central information service that maintains information about all registered agencies as well as all agents running inside these agencies. The set of all agencies registered at a single region registry builds a region.

The registration of agents is automatically performed by the hosting agency and is completely transparent for the agents. If an agent migrates, its location information is automatically updated inside the registry. In this way, an administrator or agent is always able to get up-to-date information about the region or to locate specific agents. The region registry provides the following services/components:

- *Management*: The management service is responsible for locating agents inside a region.
- *Communication*: Equivalent to an agency's communication facility, this service enables interactions between the region registry and remote entities (agencies, agents, ...). Via a textual user interface, administrators can monitor and control the registry's processing.
- *MAFFinder*: This component realises an interface that is part of the OMG MASIF standard and that increases the interoperability between Grasshopper and other standard-compliant agent platforms.

Grasshopper Integration
An important characteristic of Grasshopper is the possibility to "fit" the platform into existing computer systems in order to take advantage of agent technology without the need for system or hardware modifications (Figure 2). Traditional computer systems provide an operating system (e.g., Windows, Unix, etc.). This operating system offers different interfaces (APIs) to the system processes. A limitation of traditional systems is their strongly restricted ability to react to the case of missing information or software. In this context, a Grasshopper-enhanced system can put things right: Grasshopper is installed on top of the operating system, and parts of the system applications are realised by means of mobile agents. Whenever information or processing logic is not available, it can be dynamically installed in terms of further agents, in this way enhancing the existing (traditional) system. Grasshopper can be used in many different application contexts, such as telecommunications, electronic commerce, telematics, etc.. Historically, the tele-communications domain is one of the most prominent application areas. Here Grasshopper has been used in several European projects, such as ANIMA, EURESCOM P815, CAMELEON, FACTS, MARINE, MARINER, and MIAMI. Most of these projects belong to CLIMATE [21], the Cluster for Intelligent Mobile Agents in Telecommunication Environments, which is part of the European Research Programme on Advanced Communication Technologies and Services (ACTS).

As we will see in the following section, it can be also used in our mobile agent based service provision architecture for VoIP.

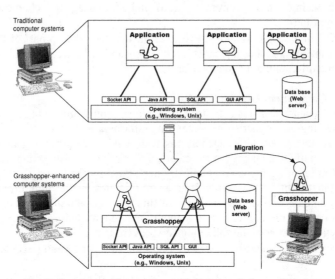

Fig. 2. Grasshopper Integration with Existing Software Solutions

4   The Mobile Agent-Based VoIP Architecture

Based on the Mobile Agent Technology introduced in section 3 we have designed an architecture which will be described now. Our architecture presents a flexible mobile agent based service provision for VoIP in SIP. Some possible ways to support IN Service Integration are shown also.

4.1 Architectural Overview

In Figure 3 an overall overview of our architecture is given. All mobile agent components and the interactions of mobile agents are shown. The following Mobile Agents are supported:

SIPSA: A SIP Server Agent (SIPSA) is a full SIP server and can act as a SIP Proxy Server or as Redirect Server. During his lifetime a SIP Server Agent can change his role if needed.

USA: A User Service Agent (USA) stores the user generated CPL script (or scripts) and service features to which a user has subscribed.

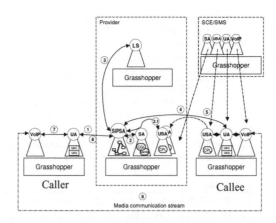

Fig. 3. SIP Implementation as Mobile Agents

User Agent (UA): A User Agent is a full SIP client and can act as a User Agent Client (UAC) which initiates call requests (caller) or as a User Agent Server (UAS) which receives call requests (callee). Only the SIP signalling functions are implemented in the UA. For the voice media stream communication a VoIP Agent is used.

VoIP: A VoIP mobile agent is used for the media communication. With the Java™ Sound API [23] and the upcoming Java™ Media Framework 2.0 API [24] it should be possible to implement such a tool in pure Java™ (with full GSM en- and decoder on a computer) and communication in real-time over TCP/IP.

LS: A Location Server mobile agent holds the information of the current locations of the SIP UA. If only one SIPSA exists these information can be stored directly in the SIPSA. If there exists more than one SIPSA (e.g. in high traffic/high load operations) a Location Server is required.

SA: Service Agents (SA) are units of services for example IN service features. A Service Agent can store SIP CGI scripts or SIP Servlets. Each supported IN service is implemented as one SA. On request the needed SA is transferred to the necessary location (e.g. where the SIPSA is located).

4.2 Mobile Agent Interactions

The following table shows the interactions of mobile agents which are supported by our architecture. First an SIPSA must be started and configured to find and use his Location Server (LS). The SIPSA starts in Proxy Server mode.

Table 1. Basic operations

A User Agent (UA) is invoked	• The User Agent sends his REGISTRATION message to the SIPSA. • The SIPSA stores the location information of the user in his Location Server (LS). • The VoIP Mobile Agent is queried for his media capabilities.
The User wants to change (or create) his CPL settings	• The SIPSA is connected and asked if a USA exists for this user. If it exists it is cloned and a copy migrate to the users agency. The user can change his settings and send the USA back to the SIPSA. • Additionally he can subscribe to one or more offered services (for example IN services). These settings are also stored in the USA • If the USA arrives at the SIPSA agency, he connects the needed Service Agents (SAs) and these agents are send to the SIPSA agency.
A call arrives at the SIPSA (INVITE message)	• The SIPSA searches for a USA of the wanted user. • If a USA is found the corresponding actions (based on the found CPL scripts and subscribed IN-Service Features) are executed. • If no USA is found, the LS is contacted for a more precise location of the user and a new INVITE message is send to the UA.. • ... (corresponding to the normal SIP actions) • After a successful invitation, the necessary parameters for a SIP session are determined and the callers VoIP MA can connect to the callees VoIP MA to have a conversation.

The VoIP Agent is used to easily support different signalling protocols (e.g. H.323). The call signalling is done by the UA and the media communication is done by the VoIP agent. So it is possible to support VoIP on systems, where only Grasshopper is installed and no VoIP-Tool is available.

4.3 Performance and Mobile Agent Benefits

If the traffic on the SIPSA rises up, he clones himself (and all the registered USAs and SAs) and send several copies to new agencies. The old SIPSA acts furthermore only as a SIP redirect server and the new SIPSAs act as SIP proxy servers. All incoming INVITATIONs at the first SIPSA (which acts now as a SIP redirect server) are responded with the alternating addresses of the new SIPSA Proxy Servers. If the traffic goes down, the new SIPSA Proxy Servers can stop and the first SIPSA redirect server acts as a SIP Proxy Server again.

5 Service Examples

In this section we will present several examples to illustrate our architecture namely: Call distribution, Follow-me diversion, and virtual private network. In IN many advanced services can be provided such as call forwarding, call distribution, conferencing calling etc. These IN services can also be realised with SIP [25]. The purpose of our architecture is to show the benefits of using mobile agent technology in SIP and with this architecture new services such as IN services can be more flexibly provided. It is an open issue and further studies must be taken.

5.1 Call Distribution

Call distribution is a kind of IN services. It allows a subscriber to have calls routed to different destinations, according to an allocation law which may be managed in real-time by the subscriber. This service can be also provided in our architecture. After a subscriber has written his call preferences in CPL and subscribed to the wanted services, his USA migrates to a suitable SIPSA. When an incoming call comes in, the UA of the caller contacts the SIPSA and the SIPSA connects the LS or communicates with USA to execute appropriate service features which in this case is where the call should be distributed. After that the call is routed to the right end user. Our architecture can also provide more advantages in call distribution:

1. Balance of the overload of an SIPSA when there are too many calls to the SIPSA. The SIPSA can clone himself to several SIPSA Proxy Servers and changes himself to a redirect server, so that overload on one SIPSA can be reduced. The efficiency of call distribution can be increased.
2. With our architecture users can communicate with each other with different media communication methods. When a UA of a caller contacts an SIPSA, he gives the capability and characters of the caller VoIP mobile agent. After the SIPSA executes wanted services, SIPSA contacts the UA of the wanted end user. The SIPSA conveys the capability and characters of the UA of the caller to the UA of the end user and the UA of the end user can select an appropriate VoIP mobile agent to communicate with the VoIP mobile agent of the caller. In this way portability and scalability is achieved.
3. To overcome the complexity of the possible configurations, a Personal Call Logic Editor can be used to configure the settings without the need to program it directly in CPL. The PCL Editor offers a simple GUI and outputs a CPL script of the corresponding settings.

5.2 Follow-me diversion

Follow-me allows subscriber to remotely control his call forwarding capabilities from any point in the network. Call forwarding refers to a call that is forwarded to another number when this service is activated.

This service can be easily realised in our architecture. A user can remotely change his CPL agent (USA) from any place and this agent migrates to the SIPSAs of the user's domain which forwards calls for this user. As the benefits illustrated in section 5.1 the UA and SIPSA can exchange the capabilities of end systems and UAs can select suitable media transfer tools for media transfer. When the end system is an Internet telephone (not only a computer), audio is transmitted between the two users although one of them is computer and can transmit videos.

5.3 Virtual Private Network

A virtual private network is a network which uses public network switches to form a network for a corporation. In a virtual network, private numbers and security etc. can be provided. Virtual private Internet telephony network can also be established. We illustrate the flexibility of our architecture in providing virtual Internet telephony network. Virtual private network includes many aspects. We exemplify in this section only one small part namely: closed user group. Closed user group allows a user to be a member of a set of VPN users who are normally authorised to make and receive calls only within the group. This service can be easily realised in our architecture. A user wants to join a set of VPN users. He first configures his USA, then this USA migrates to an SIPSA in the domain. Outgoing calls must go first to SIPSAs. On an SIPSA there are services which may be SIP Servlets or SIP CGIs and identifies the incoming calls and routes the authorised calls to the numbers of the set of VPN users.

In this section three services are used to exemplify the benefits of our architecture in provision of services for VoIP. Because of intelligence of end systems in IP networks and many advanced IP software tools, more services can be provided in VoIP. We will, in the future, investigate how to provide more advanced services with our architecture.

6   Prototype Implementation

Our prototype is realised on the Grasshopper platform which is introduced in section 3. VoIP services are mobile agents in the Grasshopper platform in our architecture. SIPSA and UAs are also realised as mobile agents. Because the Grasshopper platform and agents on it are realised in Java, so we will use Java in our overall prototype. In addition because we use SIP Servlet or SIP CGI and CPL, we need to realise a CPL Java Parser and provide SIP CGI or servlet engine functions also in Java. The overall Implementation work is still going on. The upcoming Sun JAIN™ APIs [31] should make easier a similar implementation in the future.

7   Related Work

Besides our work there are also research people at Ericsson corporation and people at Carleton University Ottawa [26] who will study the problem of the use of mobile agent technology in VoIP. At Carleton University an advanced service architecture for H.323 protocol telephony is being studied [26]. This architecture is based on H.323. Mobile agent technology, Java Bean [27], and Jini [28] technology.

8   Conclusion

This paper has described a mobile agent based VoIP environment, where in particular a SIP implementation has been integrated with the Grasshopper agent platform. The presented architecture and the related prototype implementation have revealed promising results in the context of adding value to the current SIP-based internet telephony services. These added values relate to more flexibility, mobility, load balance and uniform service creation, deployment, and management. Users can describe their preference for the process of calls and calls can be processed according to the user's will. Advanced services can be dynamically inserted into VoIP network. The load of Proxy or redirect servers can be balanced among several servers, so that the call process efficiency is increased.

It has to be noted that the presented approach is not limited to SIP-based internet telephony. Rather, the outlined architecture is general enough to be adapted to the H.323 and Megaco [4] [29] architectures. However, the authors believe, that SIP and the related IETF protocols have a big potential in the merging VoIP market, as they originate from the real internet community and therefore embody the main spirit of this open communications environment.

In summary, the presented architecture represents a projection of an MA-based distributed IN architecture for switched circuit networks, as developed within the ACTS MARINE project, onto packet data networks. In this regard the presented approach provides an interesting starting point for the development of future integrated service architectures for converged networks. The development of such an integrated service architecture based on the Grasshopper agent platform, will be undertaken in the IST project STARLITE [22] [30].

References

1. T. Magedanz , R. Popescu-Zeletin: "Intelligent Networks - Basic Technology, Standards and Evolution", International Thomson Computer Press, ISBN: 1-85032-293-7, 1996
2. MARINE project homepage: http://www.italtel.it/drsc/marine/marine.htm
3. M. Breugst, T. Magedanz: "Mobile Agents - Enabling Technology for Active Intelligent Networks", IEEE Network Magazine, Vol. 12, No. 3, Special Issue on Active and Programmable Networks (1998) 53-60
4. F. Cuervo, C. Huitema, K. Kelly, B. Rosen, P. Sijben, E. Zimmerer: "MEGACO Protocol", IETF Internet draft, July 1999, Work in progress
5. ITU-T Rec. H.323, "Packet-based multimedia communications systems", 1998
6. M. Handley, H. Schulzrinne, E. Schooler, J. Rosenberg: "SIP: Session Initiation Protocol", IETF RFC 2543, March 1999
7. R. Lang, E. Schooler, M. Handley, J. Ott: "Multiparty Multimedia Session Control (mmusic)", homepage: http://www.ietf.org/html.charters/mmusic-charter.html, June 1999
8. H. Schulzrinne, S. Casner, R. Frederick, V. Jacobson: " RTP: A Transport Protocol for Real-Time Applications", IETF RFC 1889, January 1996
9. R. Braden, L. Zhang, S. Berson, S. Herzog, S. Jamin: "Resource ReSerVation Protocol (RSVP) - Version 1 Functional Specification", IETF RFC 2205, September 1997
10. M. Handley, C. Perkins, E. Whelan: "Session Announcement Protocol", IETF Internet draft, August 1999, Work in progress
11. M. Handley, V. Jacobson: "SDP: Session Description Protocol",IETF RFC 2327,April 1998
12. H. Schulzrinne, J. Rosenberg: "A Comparison of SIP and H.323 for Internet Telephony", http://www.cs.columbia.edu/~hgs/papers/Schu9807_Comparison.pdf, July 1998

13. J. Lennox, J. Rosenberg, H. Schulzrinne: "Common Gateway Interface for SIP", IETF Internet draft, May 1999, Work in progress
14. J. Lennox, H. Schulzrinne: "CPL: A Language for User Control of Internet Telephony Services", IETF Internet draft, Feb. 1999, Work in progress
15. A. Kristensen, A. Byttner: "The SIP Servlet API", IETF Internet draft, September 1999, Work in progress
16. K. Coar, D.R.T. Robinson: "The WWW Common Gateway Interface - Version 1.1", IETF Internet draft, April 1999, Work in progress
17. XML, ORG homepage: http://www.xml.org
18. IKV++ GmbH – Grasshopper homepage: http://www.ikv.de/products/grasshopper
19. OMG (1995), "Common Facilities RFP3", Request for Proposal OMG TC Document 95-11-3, Nov. 1995, http://www.omg.org; MASIF specification is available through http://ftp.omg.org/pub/docs/orbos/97-10-05.pdf
20. FIPA homepage: http://www.fipa.org
21. CLIMATE homepage: http://www.fokus.gmd.de/research/cc/ima/climate/climate.html
22. F. Zizza, et.al.: "Towards a Distributed Intelligent Network", International Conference on Intelligence in Networks (ICIN 2000), Bordeaux, France, January 18-20, 2000
23. Sun's Java Sound API homepage: http://java.sun.com/products/java-media/sound
24. Sun's Java Media Framework 2.0 API: http://java.sum.com
25. J. Lennox, H. Schulzrinne, T. La Porta: "Implementing Intelligent Network Services with the Session Initiation Protocol", Tech-Report Number CUCS-002-99
26. J. Tang, T. White, B. Pagurek, R. Glitho: "Advanced Service Architecture for H.323 Internet Protocol Telephony", ftp://ftp.sce.carleton.ca/pub/netmanage/infocomm.ps.gz, Submitted to INFOCOM '99
27. Sun's Java Beans homepage: http://java.sun.com/beans
28. Sun's JINI homepage: http://www.sun.com/jini
29. N. Greene, M. Ramalho, B. Rosen: "Media Gateway control protocol architecture and requirements", IETF Internet draft, September 1999, Work in progress
30. T. Magdanz, et.al.: "Towards an Integrated Architecture for the Harmonisation of PSTN and Internet Services", International Conference on Intelligence in Networks (ICIN 2000), Bordeaux, France, January 18-20, 2000
31. JAIN™, APIs for Integrated Networks-Homepage: http://java.sun.com/products/jain
32. ETSI TIPHON Homepage: http://www.etsi.org/tiphon
33. Magedanz, T., Popescu-Zeletin, R. (1996) "Towards Intelligence on Demand - On the Impacts of Intelligent Agents on IN", Proceedings of 4th International Conference on Intelligent Networks (ICIN), Bordeaux, France, December 2-5, pp. 30-35
34. Magedanz, T., Karmouch, A., (Eds.), (2000) Special Issue on Mobile Agents for Telecommunication Applications, Computer Communications Journal, Elsevier Publishers, Vol. 27, No.1
35. Chess, D., Harrison, C.G., Kershenbaum, A. (1998) "Mobile agents: Are they a good idea?", in G. Vigna (ED.), Mobile Agents and Security, LNCS 1419, pages 25-47. Springer Verlag, 1998.

Supporting Advanced Multimedia Telecommunications Services Using the Distributed Component Object Model

Dionisis X. Adamopoulos[1], Prof. George Pavlou[1], Dr. Constantine A. Papandreou[2]

[1] Centre for Communication Systems Research, University of Surrey, UK
{D.Adamopoulos, G.Pavlou}@ee.surrey.ac.uk
[2] Hellenic Telecommunications Organisation (OTE), Greece
kospap@org.ote.gr

Abstract. The demand for a great variety of sophisticated telecommunications services with multimedia characteristics is increasing. This trend highlights the need for the efficient creation of distributed programs with multimedia data exchanges over distributed processing environments. Therefore, it is necessary to support the object-oriented development of distributed multimedia applications in a flexible manner. This paper recognises Microsoft's Distributed Component Object Model (DCOM) as a key potential technology in the area of service engineering and examines a structured approach to enhance it for the handling of continuous media streams through the design and implementation of a collection of suitable multimedia support services. The proposed approach focuses on the modelling of continuous media communications in DCOM and is validated through the design and implementation of a multimedia conferencing service.

1   Introduction

Driven by technological advances, market growth and deregulation, the global telecommunications industry is rapidly adopting a highly dynamic and open character, which, in combination with the evolving synergy between information and telecommunication technologies, provides a wide range of opportunities for the delivery of advanced multimedia telecommunications services (also referred to as *telematic* services). Based on recent developments in object orientation and distributed computing, these telecommunications services are designed, realised and deployed as multimedia applications operating on distributed computing platforms [1][12]; the latter are also known as Distributed Processing Environments (DPEs).

Despite the fact that multimedia support has been considered in general terms in the ISO/ITU-T Reference Model for Open Distributed Processing (RM-ODP) [6][9], it has not yet been considered in Microsoft's Distributed Component Object Model (DCOM) [3], and is not yet mature in the Object Management Group's (OMG) Common Object Request Broker Architecture (CORBA) [8][13]. Recently, a wide range of new telecommunications services are becoming increasingly popular by employing video to convey information and to enhance communication among human users (e.g. videoconferencing, etc.). Therefore, in the emerging multi-vendor, multi-stakeholder telecommunications environment, it is necessary to facilitate the rapid and flexible deployment of a great diversity of multimedia, multi-party services by providing support for continuous media in DPEs. The role of DCOM is expected to be important as it is one of the promising distributed object platforms for service engineering [2]. Key advantages are its ubiquity and the fact that it supports key DPE features such as multiple interfaces per object and object groups [17].

This paper presents an approach which extends DCOM to an environment suitable for the development of advanced multimedia telecommunications services. The proposed approach is validated through the design and implementation of a multimedia conferencing service. Finally, experiments are conducted in order to assess the flexibility and efficiency of the proposed approach and conclusions are drawn.

2   Modelling Multimedia Telecommunications Services

Multimedia communication involves the interaction of devices which can deal with networked suppliers and consumers of various types of digitally represented information. The tasks broadly involved in this process can be divided into the coding and transport of the different media, and into related control aspects, such as how to locate services, request transfer, establish and maintain connections, ensure integrity and timeliness, and handle presentation issues during the delivery of multimedia information. These control aspects are the concern of this paper, since they are particularly important for the realisation of the full potential of distributed object platforms [11]. Another important requirement is the ability to hide the heterogeneous low-level aspects of dealing with streams through high-level Application Programming Interfaces (APIs) and to provide abstractions which could be easily dealt with by non-network programmers. For the rest of this section, we examine other research and standardisation work related to the flexible handling of multimedia streams.

The model of object interaction conventionally adopted in distributed object platforms (i.e. remote method invocation) is inappropriate for continuous or dynamic media, i.e. media which contain a temporal element, such as real-time audio or video. For these media types, a streaming, i.e. continuous mode of interaction is required rather than a request / response method invocation model [4][10].

This is in full agreement with the RM-ODP's multimedia computational model, which uses stream interfaces to build streaming interaction over the primitive notion of a signal [6][9]. Streams are also present in several other distributed computing architectures. Initially, they appeared in the Multimedia Systems Services (MSS) architecture, which was proposed by the Interactive Multimedia Association (IMA). The IMA MSS is currently being adopted and extended by ISO in its PREMO standard. Furthermore, much of the initial work on streams, which influenced greatly the RM-ODP, took place trying to address the requirements of multimedia support in the Advanced Network Systems Architecture (ANSA) [4]. The current ANSA Phase III Distributed Interactive MultiMedia Architecture (DIMMA) project is based on the ANSAware distributed systems platform, which has been enhanced with a modular protocol stack and a flexible multiplexing structure.

OMG has recently addressed the need for streams and real-time services in CORBA by issuing a request for proposals (RFP) for the control and management of audio / video streams, and by summarising submissions in [13]. However, this RFP does not examine the implementation of streams in CORBA. Such implementation issues were addressed by specific Object Request Broker (ORB) vendors [8], and by the ACTS ReTINA project, which designed a distributed object platform based on CORBA, enhanced with streams and Quality of Service (QoS) extensions. Another CORBA version 2.0 compliant implementation considering multimedia support is the TAO ORB, which runs on real-time operating system platforms, and is primarily designed for strict real-time applications.

In all the above architectures, the modelling of continuous media communications through a flexible, high-level but efficient infrastructure are crucial. More specifically, modelling mainly involves the choice of suitable and sufficient abstractions, their efficient implementation upon a target DPE, and the adoption of appropriate interaction patterns and semantics. Flexibility is a general property referring to the way that modelling concepts and artefacts are used for the design, development and deployment of open telecommunications services. The great variety, inherent complexity, and the increasing demand for customisation of such services raise the importance of flexibility.

Based on these assumptions, we propose a generic platform for the handling of continuous media in DCOM. This consists of primitive COM objects or services (multimedia support services), which can facilitate the construction of telematic services with multimedia characteristics. More specifically, the multimedia support services, and the associated COM objects, are compatible with RM-ODP in the sense that they adopt related concepts and functionality, and thus enable a wide degree of information sharing and application interoperability. Furthermore, these services can be reused and customised, and their interfaces have been designed to allow flexibility and efficiency in achieving their implementation. This is important for telecommunications services which manipulate multimedia objects, where performance is critical.

While this paper deals with the flexible modelling of multimedia streams in a DCOM-based DPE, it should be noted that besides supporting modelling aspects, the control and management software of new telecommunications infrastructures needs also to support a range of QoS characteristics, the synchronisation of continuous media, and the management of underlying resources [4][12][18]. These aspects are *not* within the scope of this paper.

3   Supporting Continuous Media in DCOM

DCOM is the distributed extension to COM (Component Object Model) that builds an Object Remote Procedure Call (ORPC) layer on top of DCE RPC to support remote objects. In general, DCOM provides all the necessary facilities for the integration of heterogeneous components in a distributed environment [3][7]. However, DCOM does not satisfy the more complicated and stringent requirements of handling multimedia streams. To enable DCOM to be the basis for new telecommunications services which require the handling and control of continuous media, extra features are necessary. The most obvious requirement is that the concept of streams should be added to the DCOM object model through the introduction of stream interfaces, since at present only operational interfaces are defined.

Before focusing on DCOM, it has to be noted that COM handles multimedia information through the Microsoft DirectShow architecture (previously Microsoft ActiveMovie architecture), which incorporates the notion of streams. Apparently, the use of this notion is restricted to the environment of stand-alone multimedia capable computers with Microsoft Windows operating systems (9x, NT 4.0, 2000), i.e. DirectShow is not a distributed architecture.

In subsection 3.1 below we present the key modelling abstraction, in 3.2 we present the stream communication algorithm and in 3.3 we discuss some important implementation details.

3.1   The Proposed Approach

We propose here a multimedia support platform with the introduction of a number of support services in the DCOM architecture. These services (which are used in conjunction with existing DCOM services) provide new functionality without requiring any changes to the basic underlying DCOM architectural model. The new services consist principally of two types of COM objects: devices and stream binders. These are both seen by the higher layers of (software) abstraction as normal services with standard abstract data type interfaces, but they encapsulate the control and transmission of continuous media.

COM object devices are an abstraction of physical devices, stored continuous media or software processes. They may be either sources, sinks or transformers of continuous media data ("modules"). A source is a media producer and is normally an abstraction of a media-generating hardware device, such as a camera or a microphone. A sink is a media consumer and is normally an abstraction of a media-rendering hardware device, such as a framebuffer / VDU or a loudspeaker. Finally, a module is both a media producer and consumer, as it accepts incoming data, processes it in some way, and produces output.

Most devices present a device dependent interface, a generic control or chain interface (IChain), and an endpoint interface (IEndpoint). The device dependent interface contains operations specific to the device modelled and is used for the management of the device. For example, a camera might have operations such as focus, pan or tilt. Furthermore, it has to be noted that all devices have to inherit from the IUnknown interface, which provides functionality required by all COM objects. This is common in most distributed object platforms e.g. in CORBA and Java Remote Method Invocation (Java-RMI).

A piece of continuous media can be visualised as a chain comprising a sequence of segments or links, each of which represents an atomic unit specific to the media type in question (e.g. a frame of video) [4]. Thus, a chain is an abstraction over a continuous media source or sink that focuses on the control of the production and consumption of continuous media data. Based on this abstraction, the IChain interface provides generic operations for controlling continuous media devices and managing continuous media transmissions. It is a device- independent interface which is common to all continuous media devices. The IChain interface is summarised in Fig. 1 using (a simplified variation of) Microsoft's Interface Definition Language (M-IDL), which is an extension of the Distributed Computing Environment's (DCE) IDL.

```
interface IChain : IUnknown
{    typedef enum {in, out, inout} DeviceType;
     HRESULT GetDeviceType([out] DeviceType* DType);
     HRESULT Start();
     HRESULT StartEx([in] int NumberOfSegments);
     HRESULT Stop();
     HRESULT Suspend();
     HRESULT SuspendEx([in] int Time);
     HRESULT SuspendEx1([in] int NumberOfSegments);
     HRESULT Resume();
     HRESULT Skip([in] int NumberOfSegments);
     HRESULT GetPosition([out] int* SegmentNumber); };
```

Fig. 1. The IChain Interface

More specifically, the GetDeviceType operation returns the type of the device under examination (producer, consumer or module), while the Start and Stop operations switch the device's information flow on and off accordingly. The functionality of the two last operations, which are the most important of the IChain interface, is based on the use of a virtual pointer (CurrentSegment) that moves through the media chain as it is played or recorded.

There are also operations for suspending and resuming the activity of a device (Suspend and Resume respectively). After a Suspend operation the production / consumption of segments in a device stops until the Resume operation is called or in the case of SuspendEx / SuspendEx1 for the time period specified (explicitly or implicitly) by the parameters of the operation. In both cases the value of the CurrentSegment pointer is preserved. Finally, this pointer may be located and moved using the GetPosition and Skip operations.

Another interface which is common to all continuous media devices (device independent interface) is the IEndpoint interface. An endpoint is a connection point (a port) for a stream, and the IEndpoint interface is thus the "stream interface" of a device [11]. The IEndpoint interface abstracts over all aspects of a device which are concerned with the transport of continuous media. Essentially, as it can be seen in Fig. 2, it presents a pair of operations, GetSegment and PutSegment through which segments can be read (from the OutputSegment buffer of a producer device) or written (to the InputSegment buffer of a consumer device) respectively. With this approach, the content of a stream is not considered and it is viewed purely as a byte transport mechanism. The two other operations of the interface (SetCharacteristics and GetCharacteristics) refer to a number of transmission related characteristics used for QoS issues.

```
interface IEndpoint : IUnknown
{    HRESULT GetSegment([out] BSTR* Segment);
     HRESULT PutSegment([in] BSTR* Segment);
     HRESULT SetCharacteristics ([in] long ChrSize,
               [in, size_is(ChrSize)] long* ChrArray);
     HRESULT GetCharacteristics([in, out] long* ChrSize,
               [out, size_is(ChrSize)] long* ChrArray);};
```

Fig. 2. The IEndpoint interface

The operations inside the IChain and the IEndpoint interfaces of a specific device must take place in an acceptable and semantically correct order (e.g. for the same device a Stop operation can not be followed by a Suspend operation) to avoid unexpected results / errors. To ensure such an order, each device has a state (DeviceState), which is checked before an operation is executed. The proposed and anticipated transitions between the states of a device can be seen in the state transition diagram of Fig. 3, which also depicts the possible states of a device (idle, ready, active, and suspended).

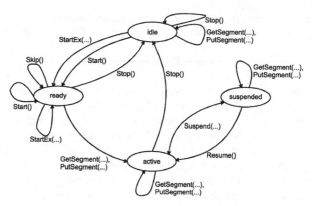

Fig. 3. The state transition diagram of a device

In order to be able to control streams the binding process must be made explicit. This is done through the introduction of a binding COM object (StreamBinder). StreamBinder represents the connection between bound COM objects and, as can be seen in Fig. 4, provides an operational interface (IStreamBinder) that hides continuous media transmissions, which can be optimised by using dedicated transport protocols, entirely distinct from those used to convey control messages.

```
interface IStreamBinder : IUnknown
{ HRESULT StartSource([in] IUnknown* SourceGroup);
  HRESULT StartSink([in] IUnknown* SinkGroup);
  HRESULT Connect&Transfer([in] IUnknown* SourceGroup,
                           [in] IUnknown* SinkGroup);
  HRESULT StopSource([in] IUnknown* SourceGroup);
  HRESULT StopSink([in] IUnknown* SinkGroup);
  HRESULT SuspendSource([in] IUnknown* SourceGroup);
  HRESULT SuspendSink([in] IUnknown* SinkGroup);
  HRESULT ResumeSource([in] IUnknown* SourceGroup);
  HRESULT ResumeSink([in] IUnknown* SinkGroup);
  HRESULT DestroyConnection([in] IUnknown* SourceGroup,
                            [in] IUnknown* SinkGroup);}};
```

Fig. 4. The IStreamBinder interface

The binding action can be initiated by a COM object involved in the binding or by a completely separate object. In general, client COM objects wishing to initiate continuous media transfer, initiate a request from the StreamBinder to start the appropriate source and sink devices (StartSource, StartSink). Then, the StreamBinder establishes a stream connection between these devices and activates the transmit function (Connect&Transfer). The resulting stream can be managed by suspending (SuspendSource, SuspendSink), resuming (ResumeSource, ResumeSink), and stopping (StopSource, StopSink) the participating devices and it can be destroyed when desired (DestroyConnection)

The StreamBinder, in the most general case, supports multiple stream connections, as it allows M sources to be connected to N sinks (without necessarily M=N), by establishing the appropriate streams between them. When it is desirable to start, stop, establish, and generally perform control operations on a number of streams simultaneously, the notion of object groups simplifies greatly the necessary code (calls to the StreamBinder operations). Additionally, it eases considerably the process of ensuring that the code reflects the correct / intended semantics, as it decreases the possibility of missing, wrong, or out of order operations on devices. This is due to the fact that errors can now appear only during the formation of object groups; an activity which corresponds to a relatively small and well structured piece of code that can easily be examined. Two typical errors that can be avoided in this way are; the execution of an operation on a device that belongs to a different stream than the one intended; and the execution of Connect&Transfer and / or Destroy Connection on two devices that (are intended to) participate in different streams.

```
interface IObjectGroup : IUnknown
{   HRESULT Join([in] IUnknown* refiid);
    HRESULT Leave();
    HRESULT Use([out] IUnknown* refiid);
    HRESULT Reset(); };
```

Fig. 5. The IObjectGroup interface

Conceptually, object groups are modelled using the COM class ObjectGroup, which collects in a group a set of related COM objects. Actually, it maintains a list of the interface references (REFIIDs) of the COM objects that belong to a specific group. The IObjectGroup interface can be seen in Fig. 5. Join and Leave operations allow new members to join the group and existing members to leave the group respectively, while Use and Reset provide access to the group current membership list.

In a typical scenario, two instances of the ObjectGroup COM class are used: a SourceGroup and a SinkGroup (which are actually the interface references of the two instances). The two lists that are maintained by these two groups contain, at corresponding positions, the interface references of the sources and sinks that are going to be engaged in stream communication. Thus, the use of (the interface references of) these two groups as parameters in the operations of the IStreamBinder interface allows the invocation of (corresponding) operations on a number of COM objects (sources / sinks) at the same time. However, it must be noted that in order to increase the flexibility and support application semantics, where the simultaneously establishment and control of multiple streams is not desirable, the use of object groups in the operations of the IStreamBinder interface is not mandatory. Interface references to simple COM objects (sources / sinks) can also be used as parameters.

The COM objects examined so far constitute the proposed multimedia support services for DCOM, and should be reused during the development of specific multimedia services. They may therefore have to be customised according to the specific service requirements. This activity, which is very important as it determines the practical value of the proposed approach, is supported through the use of either *containment* or *aggregation*, as DCOM allows only interface and not implementation inheritance [7]. It has to be noted that aggregation can only be used for in-process COM object servers (i.e. DLL server modules). Therefore, because some of the COM objects of the proposed API may be, either local servers (implemented as EXEs) or remote servers (executed on a remote server machine), when reusing or customising these COM objects, the containment method is the preferred way since it enables the resulting component to operate under all possible COM server types.

3.2   The Stream Communication Algorithm

The proposed multimedia support services described in the previous section can be used for the establishment and control of stream communication in DCOM in a structured fashion. To illustrate this approach, a possible scenario is examined. According to Fig. 6, which depicts the configuration of the COM objects involved in the example scenario, two source devices (e.g. video cameras) are connected via a StreamBinder to two sink devices (e.g. VDUs), and two different streams are established between the source and sink devices.

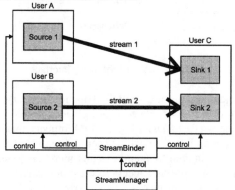

Fig. 6. An example scenario for the multimedia support services

The necessary steps that have to be followed in order to realise the two video connections (stream 1 and stream 2) between the sources and sinks of Fig. 6 using the proposed multimedia support services are the following:

Step 1 *Obtain the necessary interface references*: The interface references (REFIIDs) of the two sources (Source1UserA and Source2UserB) and the two sinks (Sink1UserC and Sink2UserC) involved in stream communication are obtained. Device dependent operations are also performed if necessary.

Step 2 *Create new instances of required services (COM objects)*: A StreamBinder instance is created and the related interface reference is obtained. Additionally (if required), two ObjectGroup instances are created and the related interface references are also obtained (SourceGroup and SinkGroup).

Step 3 *Form the appropriate object groups (if required)*: Taking into account the streams that are desirable to be established (or actually considering the source and sink devices that need to be connected by streams), the REFIIDs of the sources become members of the SourceGroup [Join(Source1UserA), Join(Source2UserB)], and the REFIIDs of the sinks become members of the SinkGroup [Join(Sink1UserC), Join(Sink2UserC)].

Step 4 *Start the devices*: The sink and source devices are started [StartSink (SinkGroup), StartSource(SourceGroup)].

Step 5 *Establish connections between source and sink devices*: Associate the appropriate sources and sinks and initiate continuous media transfer between them [Connect&Transfer(SourceGroup, SinkGroup)]. Steps 4 and 5 can also take place in the opposite order.

Step 6 *Stop the devices*: When the interaction is finished the sink and source devices are stopped [StopSink(SinkGroup), StopSource(SourceGroup)].

Step 7 *Destroy connections and services*: The connections established between the appropriate sources and sinks are destroyed [DestroyConnection(SourceGroup, SinkGroup)]. Then, the StreamBinder and the ObjectGroup instances created in step 2 are also destroyed.

The above described steps that constitute a kind of algorithm, i.e. a stream communication algorithm, for establishing and controlling stream connections in DCOM. Two more steps can be added to this algorithm depending on the functionality required by some applications. More specifically, between steps 5 and 6 (i.e. while all the devices are active), the sink and source devices can be suspended [SuspendSink (SinkGroup), SuspendSource(SourceGroup)] and then, on a consecutive step, they can be resumed [ResumeSink(SinkGroup), ResumeSource(SourceGroup)].

The stream communication algorithm utilises the operations of the IStreamBinder interface to create and manage bindings between the appropriate sources and sinks. These bindings do not have to be controlled directly by the COM objects involved in the binding (i.e. the sources and the sinks), but may instead be created by third party COM objects which obtain references to interfaces owned by those COM objects. This facility eases considerably the configuration and structuring of potentially complex multimedia telecommunications services containing many per-media COM objects. A similar situation is described in Fig. 6, where the StreamManager COM object interacts with the StreamBinder and performs all the steps of the stream communication algorithm.

From these remarks is evident that the structure and the behaviour of the StreamManager depends on the requirements of a specific application, and on the way that this application handles streams. On the contrary, the interfaces and the functionality of the (COM objects used to model) devices and the StreamBinder are application independent and thus suitable for reuse. Actually, these interfaces (and the corresponding multimedia support services) can be considered as a high level API for the handling of continuous media in DCOM.

The advantages of this high level API are highlighted when taking into account that the main alternative approach for stream handling in DCOM requires the use of low level native Windows APIs (such as Win32), which is characterised by excessive low level details that divert the attention of the developers from the more crucial (broader) application-related semantics and the program structure. This can raise the potential for errors, increase the learning effort required, and hinder the development of complex applications, together with the continuous re-discovery and re-invention, in an ad hoc manner, of incompatible higher-level programming abstractions that seriously hampers programming productivity and code compatibility [15].

Therefore, the development of multimedia telecommunications services in DCOM benefits greatly from the use of the proposed API, because it isolates the application domain semantics from the complexities of multimedia devices and continuous media communications, by providing services based on abstract data type interfaces. Additionally, it reduces and simplifies the required programming effort, by locating all the code related with the handling of streams inside easily extensible reusable components, preventing thus developers from "reinventing the wheel" using elementary capabilities and functionality.

3.3   Important Implementation Considerations

There are a few DCOM related issues that affect considerably the implementation of the proposed multimedia support services. These issues, which will be examined briefly in this section, include class factories, access to remote COM objects, and the available threading models.

In order to be able to use a (device or a StreamBinder) COM object, an instance must be created. This is done through a special COM object called a class factory, which implements the IClassFactory interface. This functionality is based on the design pattern of a "Factory Method", according to which, when a client wishes to instantiate a server object, a request is sent to a "Factory Object" for the corresponding class [5]. In complying, a class factory has to be created for every server component specified by the proposed multimedia support services. It has to be noted, that for optimisation reasons in some (not very common) cases (e.g. when a device has a large number of device specific interfaces, and depending also on their intended use), a custom implementation of the IClassFactory interface is allowed, but caution is needed to avoid possible compatibility conflicts / problems.

After performing all the necessary instantiations, a client that wishes to call operations on a (device or a StreamBinder) COM object has to obtain a pointer to a suitable interface of that object. When the desired COM object is remote, the CoCreateInstanceEx() function has to be used in a suitable manner to locate the server machine, create (an instance of) the appropriate COM object on that machine, and finally return the desired interface pointer. This function is called with an array of MULTI_QI structures as one of its parameters:

```
typedef struct _MULTI_QI {
const IID* pIID;    // pointer to an interface identifier
IUnknown * pItf;    // returned interface pointer
HRESULT hr;         // result of the operation
} MULTI_QI;
```

As it can be seen in Fig. 7, each pIID member of this array is given an IID of an interface of the remote COM object. If the CoCreateInstanceEx() succeeds, the desired interfaces can be obtained through the pointers in the pItf members. If there is an error, the hr member will receive the error code. Thus, except from the status of CoCreateInstanceEx(), the status of each element in the MULTI_QI array should also be checked, before a (valid) interface pointer can be extracted from the array.

```
// initialise the MULTI_QI structure
MULTI_QI qi[2]; // create an array of e.g. 2 structures
memset(&qi, 0, sizeof(qi)); //prepare the array for use
qi[0].pIID = &IID_IChain;    // add the 1st interface
qi[1].pIID = &IID_IEndpoint; // add the 2nd interface
// create a server COM object on the server machine
HRESULT hr=CoCreateInstanceEx(
          CLSID_CMyServer, // COM class id
          NULL,            // outer unknown
          CLSCTX_SERVER,   // server object scope
          &ServerInfo,     // name of the server machine
          2,               // length of the MULTI_QI array
          qi);             // pointer to this array.
if (SUCCEEDED(hr)) {hr=qi[0].hr;} // check the qi codes
if (SUCCEEDED(hr))
{ // extract interface pointers from MULTI_QI structure
  m_pComServer=(ICpServer*)qi[0].pItf; };
```

Fig. 7. Using an interface of a remote server object in DCOM

When a client requires access to a particular remote COM object, and this object has more than one interface to which the client needs pointers, an array of MULTI_QI structures should be created, containing as many pIIDs as necessary to keep all the IIDs of the interfaces that the client will (or intends to) use on the COM object. In that way, the CoCreateInstanceEx() will be called only once and multiple calls to it, due to an incomplete (in terms of requested IIDs) MULTI_QI array, will be avoided. This tactic reduces the number of necessary RPC calls across the network. This also improves the efficiency of the code especially when remote COM objects exhibit more than one interface (e.g. as in the case of COM objects used to model continuous media devices), and / or the network performance is or becomes slow.

Finally, shifting the focus to the internal structure of COM objects, the way that threading is performed needs to be examined. In DCOM, threads are established to improve performance (minimise execution time), to simplify the code, and to avoid the blocking of COM objects (e.g. to prevent the blocking of the StreamBinder when executing an operation on a device). Therefore, in the proposed multimedia support services threading is used in the implementation of the StreamManager and the IChain interface of the COM objects used to represent devices, in the interfacing with physical devices, and in the realisation of stream connections using transport protocols. When programming using DCOM, and therefore in the proposed API, except from the thread handling functions of Win32 (e.g. CreateThread(), ExitThread(), etc.), the COM threading models, i.e. the simple single threaded model, the Single Threaded Apartment (STA) model and the MultiThreaded Apartment (MTA) model can be applied [3][7].

4   Validation and Experimentation

The proposed multimedia support infrastructure and the related API have been tested in several simple scenarios (like the one depicted in Fig. 6) involving different configurations of source and sink devices associated by various stream connections. It has been found that they constitute a viable, flexible, consistent, coherent and relatively intuitive way of building multimedia telecommunications services in DCOM.

To verify and reinforce these results under realistic conditions, and to determine also the true practical value and applicability of the proposed API, an extended prototype of a MultiMedia Conferencing Service for Education and Training (MMCS-ET) has been developed [14]. This service is implemented using MS Visual C++ 6.0 and DCOM on MS Windows NT 4.0, and is executed on a number of workstations connected via a 10 Mbit/s Ethernet LAN. All the interconnected work-stations belong to the same (MS Windows NT) domain and one of them functions as a primary domain controller.

The main objective of the MMCS-ET is to facilitate the establishment of an educational / training session between one teacher / trainer and a number of remote students / trainees, which is equivalent to the educational / training session that would have been established between the same people in a traditional classroom. For this reason, the MMCS-ET implements a variety of scenarios supporting session manage-ment requirements (session establishment, modification, suspension, resumption, and shutdown), interaction requirements (audio/video, text, and file communication), and collaboration support requirements (chat facility, file exchange facility, and voting).

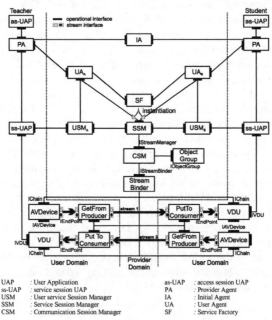

Fig. 8. The main computational objects of the MMCS-ET and the position of the interfaces of the proposed API

The computational view of the MMCS-ET in the simple case where one teacher interacts with only one student can be seen in Fig. 8. From the computational objects that appear in this figure, it is evident that the MMCS-ET is designed according to the TINA-C service architecture (version 5.0) [16]. Fig. 8 emphasises additionally that A/V communication is achieved between the teacher and the student by the establishment of two streams of opposite directions, presenting the position of the interfaces of the proposed multimedia support services for DCOM. It has to be noted that the Communication Session Manager (CSM), which is at the boundary with the resource layer, incorporates the functionality of the StreamManager, and that the GetFromProducer and PutToConsumer COM objects are used for the realisation of the stream connections.

The MMCS-ET validated the proposed API and confirmed the results of the initial tests with the simple scenarios, but also gave an insight, through several experiments, for the optimisation of the proposed API in terms of its use and its more efficient implementation in DCOM. More specifically, two types of experiments were conducted. The first type involved the application of object groups in the stream communication algorithm. This examined the complexity of the resulting code (which is actually the main piece of code written by the service developer when using the proposed API) in terms of the number of necessary calls of operations to other COM objects. The number of such calls (with and without the use of object groups) for an increa-sing number of stream connections can be seen in Fig. 9. It has to be noted that different stream connections are established between different source and sink devices. Thus, for example, 4 stream connections imply the existence of 4 sources and 4 sinks. From Fig. 9 it is evident that the use of object groups, as the number of stream connections is increasing, reduces considerably the number of operation calls that have to be made, and therefore simplifies the code of the stream communication algorithm (together with the task of the service developer) and increases its efficiency.

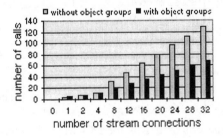

Fig. 9. Experimenting with the use (or not) of object groups in the stream communication algorithm

The second type of experiment involved the examination of the performance (in terms of execution time) of the different COM threading models when applied to the proposed API. A choice between these models becomes especially important when the StreamManager creates separate threads for the instantiation of the (COM objects representing the) devices and the execution of device specific actions, for the instantiation and initialisation of the GetFromProducer and PutToConsumer COM objects, and for the instantiation of the StreamBinder and the execution of the stream communication algorithm. It has to be noted that only the STA and MTA models are considered because the single threaded model is really a special type of the STA model.

Table 1. Comparison of COM threading models using the proposed API

COM Threading Models	Time 1	Time 2
STA	19.6 ms	12.6 ms
MTA	18.19 ms	11.66 ms

When each of the STA and MTA models are applied to all of the (COM objects) of the proposed API the time needed (in ms) to start (Time 1) and stop (Time 2) a stream connection between one source and one sink device is measured for each of them. Time 1 corresponds to steps 4 and 5 of the stream communication algorithm, while Time 2 corresponds to steps 6 and 7. The results of the measurements can be seen at Table 1. From this table is evident that the MTA model has a better performance (which becomes even better as the number of stream connections increases). Therefore, for the proposed API, taking also into account that there are no synchronisation issues, the MTA model is the preferred choice. Its performance superiority is mainly due to the fact that inter-thread access is direct (as all the treads are in the same apartment), requiring no proxy intervention as in the STA model.

Table 2. Comparison of modelling approaches for handling continuous media in CORBA and DCOM

Important Concepts	Modelling in OMG A/V Spec. (CORBA)	Modelling in the proposed API (DCOM)
Multimedia device	MMDevice interface/object	Device COM object
Device specific aspects	Vdev interface/object	Device dependent interface
Device control aspects	StreamEndpoint interface/object	IChain interface
Stream endpoint	StreamEndpoint interface/object	Iendpoint interface
Stream binding	StreamCtrl interface/object	StreamBinder COM object
Stream	StreamCtrl interface/object	Connect&Transfer operation (StreamBinder COM object)
Stream flows	FlowConnection interface/object	A stream has only one flow

Finally, to place the proposed API for DCOM in a more general context, and increase in that way the confidence in its use, a (high level) comparison with the approach followed by OMG for the handling of continuous media [13] is attempted. This is because OMG's CORBA is considered to be the main alternative to DCOM. The results of this comparison, which focuses on how continuous media communication is modelled, can be seen in Table 2.

From this table it can be easily deduced that the proposed API and the OMG A/V streams specification have the same scope as they are modelling the same basic concepts (due to their common influence by the RM-ODP), albeit in different ways. Therefore, they are "conceptually compatible", although their target technological domains are divergent, facilitating thus service developers to map their designs regarding continuous media interactions easily to either a DCOM or a CORBA DPE. It has to be noted that the comparison of Table 2 wasn't extended to cover performance issues, because performance depends greatly on the actual implementation of the OMG A/V streams specification by the different ORB vendors. Futher, the result of such a comparison wouldn't be very important as the decision for the adoption of one of the approaches depends almost entirely on the choice of the base DPE technology e.g. DCOM or CORBA; a choice which is relatively difficult as detailed in [2].

5   Conclusions

There is a technology push in the area of multimedia communications, which is acting as a catalyst for the specification and development of new multimedia telecommunications services. These services will be deployed in a distributed object environment. Therefore, there is an increasingly important need for distributed object platforms to support continuous media interactions in a flexible manner.

In this paper, we have proposed a number of RM-ODP compliant multimedia support services together with a related API in order to enhance DCOM with continuous media support. These extensions, which offer an abstraction over stream communications and multimedia devices, do not affect the core DCOM architecture, but only add the necessary functionality in terms of additional services (DPE services).

The viability of the proposed approach was evaluated by the implementation of the MMCS-ET, which demonstrated that DCOM's features can be successfully extended to address multimedia requirements in such a way that a substantial amount of software reuse can be achieved, which is the target of flexible DPEs. It remains to be seen if DCOM will form the basis of telecommunication services in the future.

References

1. Adamopoulos, D.X., Papandreou, C.A.: Distributed Processing Support for New Telecommunications Services. Proceedings of ICT '98, Vol. III, IEEE (1998) 306-310
2. Adamopoulos, D.X., Pavlou, G., Papandreou, C.A.: Distributed Object Platforms in Telecommunications: A Comparison Between DCOM and CORBA. British Telecom. Eng. 18 (1999) 43-49
3. Brown, N., Kindel, C.: Distributed Component Object Model Protocol DCOM. Microsoft Corporation (1998)
4. Coulson, G., Blair, G.S., Davies, N., Williams, N.: Extensions to ANSA for Multimedia Computing. Computer Networks and ISDN Systems 25 (1992) 305-323
5. Gamma, E., Helm, R., Johnson, R., Vlissides, J.: Design Patterns. Elements of Reusable Object-Oriented Software. Addison-Wesley (1995)
6. Gay, V., Leydekkers, P.: Multimedia in the ODP-RM Standard. IEEE Mult. 4 (1997) 68-73
7. Grimes, R.: Professional DCOM Programming. Wrox Press (1997)
8. IONA Technologies: Orbix MX: A Distributed Object Framework for Telecommunication Service Development and Deployment. (1998)
9. ITU-T, ISO/IEC Recommendation X.902, International Standard 10746-2: ODP Reference Model: Descriptive Model. (1995)
10. Kinane, B., Muldowney, D.: Distributed Broadband Multimedia Systems Using CORBA. Computer Communications 19 (1996) 13-21

11.Martinez, J.M., Bescós, J., Cisneros, G.: Developing Multimedia Applications: System Modelling and Implementation. Multimedia Tools and Applications 6 (1998) 239-262
12.Mühlhäuser, M., Gecsei, J.: Services, Frameworks, and Paradigms for Distributed Multimedia Applications. IEEE Mult. (1996) 48-61
13.Object Management Group: Control and Management of Audio / Video Streams. OMG document telecom/98-06-05 (1998)
14.Papandreou, C.A., Adamopoulos, D.X.: Design of an Interactive Teletraining System. British Telecom. Eng. 17 (1998) 175-181
15.Schmidt, D.C., Cleeland, C.: Applying Patterns to Develop Extensible ORB Middleware. IEEE Com. Mag. 37 (1999) 54-63
16.TINA-C: Definition of Service Architecture. Version 5.0 (1997)
17.TINA-C: Engineering Modelling Concepts - DPE Architecture. Version 2.0 (1994)
18.Waddington, D., Coulson, G.: A Distributed Multimedia Component Architecture. Proceedings of EDOC '97 (1997)

Near Client Caching in Large Scale iVoD Systems

P.Hoener, U.Killat

Technical University of Hamburg-Harburg
Denickestr 17, D-21071 Hamburg, Germany
phone: +49 40 42878 3249
{p.hoener,killat}@tu-harburg.de

Abstract. The efficiency of interactive Video on Demand systems (iVoD) heavily relies on the performance of the video servers. This paper presents methods to increase the iVoD quality of service and to simultaneously increase of the number of video clients by the addition of simple distributed Near-Client-Caches. This paper discusses intentionally cheap and simple algorithms and asserts their performance for systems in the transient state because this better reflects real-life situations than the usual steady state assumptions.

The examined algorithms permit an increase of the number of clients at a video server by a factor of 3 compared with conventional iVoD systems.

1   Introduction

The current developments in the computer and telecommunication technology pave the way for fast innovations in the entertainment world. Pay-TV-providers allow near Video on Demand (nVoD) by transmitting videos in fixed time slots of e.g.15 minutes length and increase the number of channels by using MPEG-technology. The movie industry produces more movies than can be shown in the cinemas. The first manufacturers therefore build large Video on Demand (VoD) and interactive Video on Demand (iVoD) systems, which do not scale very well in terms of server and network requirements: A 90 minutes MPEG-II video with a transfer rate of 6MBit/sec needs approximately 4 GByte storage space on a video server. The collection of typical video stores contains more than 10000 videos. A large scale iVoD system should offer at least 1000 videos to at least 10000 users. The needed storage capacity of 6TBytes is nowadays achievable. However, the server bandwidth for iVoD of 60GBit/sec and the distribution network dimensioned for 10000 simultaneous users is rather expensive. In addition, the real time requirements of iVoD are high to get a high quality of service (QoS). The system has to play back any video and allow interactivity like rewinding/fast forwarding with very small delays (t_d<10sec).

The restrictions for iVoD-systems are given by the bandwidth of the network and the number of video streams the server can transmit. There are methods to increase the possible server load like disc striping-strategies [1] in disc-farms, hierarchical storage [6] [7] and load balancing [2]. They permit higher server bandwidths, but are limited by the mechanical characteristics of the discs. RAM caches in the server increase the server bandwidth. Preloading of video information to the client's "preloading cache" may relax the real-time requirements. This assumes, however, complex hardware in all clients, which is undesirable for the given large number of clients. Moreover, this does not allow interactivity on videos. Rewinding or fast-forwarding within the video is an unexpected event for the preloading cache for which the system cannot be prepared. Complex caching techniques using video preloading on the basis of interaction probabilities [5] try to solve this problem. The interaction probabilities are based on user models and depend on the interaction scenario.

All these methods use complex hardware and complex algorithms. This paper will present methods to reduce the needed network and server bandwidths using only simple algorithms. The paper assumes simple clients connected to a Near-Client-Cache (NCC) by a direct link e.g. ADSL (see fig. 1). The connection between video server and NCC is limited in bandwidth and incurs a certain propagation delay.

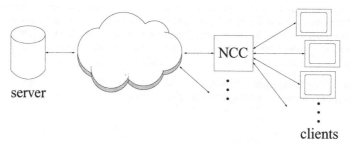

server

clients

Fig. 1. Network for iVoD-System with NCC

We can assume a correlation in the user's daily routine, which will result in a correlation in the start-up times of the clients. Normally user behaviour and their access to the videos is correlated. Therefore the system will never reach the steady-state. This paper discusses new algorithms, that will profit from the transient-state of the iVoD systems and the aforementioned correlation in user behaviour.

This paper addresses the design of simple NCC algorithms in section 2, describes 2 scenarios in section 3 and presents simulation results in section 4.

2   Near-Client-Caching

We insert an NCC in the last node terminating the 'last mile' (see fig. 1). The last mile is assumed to produce no further delay variations, so that the cache size in the client can be reduced to zero. The cache memory is moved from the client to the last node in the network, the NCC, to keep the clients cheap and simple. It is now used by several clients simultaneously. Thus the total cache size can be dimensioned smaller than the sum of all client caches. Moreover, the caching reduces the necessary bandwidth to the video server. Due to the fact that all clients use one NCC, a direct connection between server and client is no more necessary and the transmission technique between server and client can be chosen independent from the client-NCC connection.

In the sequel we discuss four different caching strategies.

2.1   Multi-Casting

The idea of multicasting is similar to current nVoD pay-TV-providers to send out all video data in a constant time pattern independent of the users' behaviour. Every video is started in fixed time slots of constant length ℓ_s (e.g. 5 min), where ℓ_s is much smaller than the length of a video ℓ_v. Thus the load of the video server is only depending on the number of available videos, n_v, the length of the videos, ℓ_v, and the slot length ℓ_s. Contrary to broadcasting the video data are only routed to NCCs, who request the data.

In this paper the video data transmitted between two multicasts will be called a segment. The video of length ℓ_v is divided in n_s segments of constant length ℓ_s, $n_s = \lceil \ell_v / \ell_s \rceil$ (only the last segment might not be full of video data, because of the ceiling operator).

The iVoD system should be able to start any video at any time with only a small delay which is implied by the time for initialisation of the transmission to the clients. The delay is bounded to a maximum of e.g. $t_d = 10$ sec to give a high iVoD quality of service. This means

- the segment length must be less than t_d or
- a copy of the first segment of each video must be available in every NCC at any time.

The first approach leads to a length of segment $\ell_s < 10$ sec and a high number of necessary multicasts. One video of $\ell_v = 90$ min will require up to $\ell_v / t_d = 540$ multicasts to guarantee $t_d = 10$ sec. If the number of clients requesting the video is smaller than 540 (in most scenarios), the number of multicasts will exceed the number of clients and the number of connections between server and clients in a conventional iVoD-system with individual transmissions for every client.

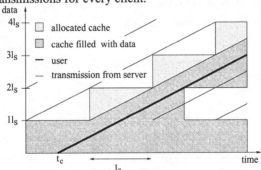

Fig. 2. Content of NCC at multicasting (t_c: startup time of client, ℓ_s: segment size)

The second approach is the recommended one because the number of multicasts is smaller, depending on ℓ_s, but the needed cache memory in the NCC increases rapidly. The first segment of every movie will be stored in the NCC permanently to allow every client to start every movie within the delay of t_d. The NCC will start storing a multicast of the segment i whenever a user is watching the segment i-1 and segment i is not stored in the NCC. Needed video data will be transmitted at fixed time slots as presented in figure 2. The thin rising lines symbolise the multicasts of video data from the server (not every multicast will reach every NCC). They have a time distance ℓ_s. The hatched area symbolises the data contained in the NCC, the light grey area the allocated cache. The user moves along the thick line through the video. Huge memory in the NCC will be needed for the first segment of every movie. The memory size for the first segments will rise linear with the number of videos on the server and the segment size ℓ_s. Additional memory is needed for caching the following video segments, to fulfil the requested QoS.

To reduce the memory size it is possible to transmit video data on request to the NCCs whenever the NCCs needs new video data for their clients. We call this block-caching.

2.2   Block-Caching

The block-caching technique reduces the cache size by storing the needed video data only. This happens only at the time when a client is requesting a new video not contained in the NCC.

Similar to the multicast technique a video is divided into n_s segments of fixed lengths ℓ_s. The NCC is divided likewise into blocks of the length ℓ_s. A table indicates which block in the NCC contains which video segment. Because blocks of the NCC are filled continuously the table for each NCC block has three entries: video id V, beginning, t_{sb}, and end of the already loaded data of the segment, t_{se}; $t_{sb}=\ell_s \cdot x$, $x \in \{0 \ldots n_s-1\}$, $t_{sb} \leq t_{se} < t_{sb} + \ell_s$.

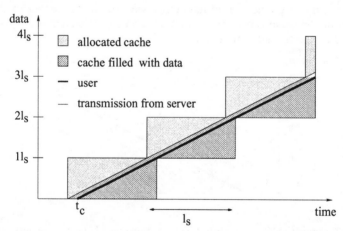

Fig. 3. Content of NCC in block-caching (t_c: startup time of client, ℓ_s: block size)

On a video request, the first segment of the video is loaded into the NCC. Play back at the client will be delayed by a preloading time t_p, which is needed to smooth delay variations during the loading from the server to the NCC (see fig. 3). If the client plays back a video at the pointer position t_c, due to continuous loading, the NCC contains the data interval $[\ell_s \cdot \lfloor t_c/\ell_s \rfloor, t_c+t_p]$. As long as a client needs data from this segment, the segment remains in the NCC, otherwise its loading is abandoned and the block is deleted from the NCC. If the client is at the transition between two segments, both are remaining in the NCC, as far as they are already loaded. Thus rewinding is possible within an interval of t_{rewind} without time delay, if abandoned segments remain in the NCC as long as $t_c > \ell_s \cdot t_c/\ell_s + t_{rewind}$.

Using the NCC fast-forwarding is only possible by maximum distance t_p within the loaded block, because the NCC block is filled continuously. If the client fast-forwards further, the data must be transmitted from the video server. Caching in the NCC nevertheless will go on, so that two video streams from the video server will be requested. If no more clients are active in the time interval of the cached segment the block will be deleted from the cache.

In comparison to multicasting the beginning of each video is not kept permanently in the cache but an equivalent iVoD quality of service, in terms of delay of play back and possibilities of rewinding and forwarding, are still ensured up to time t_p.

The block-caching technique uses a rigid block position, so that whenever a client's watching pointer is at the transition of a segment to the next, two segments of video data each occupy one block in the NCC.

The idea of the dynamic block position caching is to move a cache block of fixed size continuously along the video to reduce cache size and server load (the server load is equivalent to the network load).

2.3   Dynamic-Block-Position-Caching

In the previously selected notation the cache can store video segments of fixed length ℓ_s that contain the interval $[\max(0, t_c + t_p - \ell_s), t_c + t_p]$. This interval will move with the client's watching pointer over the video data (see fig. 4). The client can rewind and fast-forward in the bounds of the cached segment. If the client pauses (break-state), the loading into the cache is interrupted. If several clients use the same block (this should be the normal case to profit from caching) and one of them goes into the break-state, then he can run out of the block at the lower edge. If this happens, a new block must be created which runs behind the first, because the block length is limited to ℓ_s. The video data are then copied from a block in the NCC to the new created one without loading it again from the video server.

In case the leading client in the block is going to break-state he can be overhauled by another client and the now leading client will control the loading process into the NCC. The same procedure is necessary if the leading client changes the video or requests no more video. The algorithm needs a table with one entry per client and a table of all blocks in the NCC to administer the whole process. The first table contains for each client the relevant block and its actual position in the block. The second table contains the load-state of the segments. With each load step of the clients the client table must be updated and checked whether the distance between t_c and t_{sb} is smaller than t_p, to determine whether preloading to the NCC is needed.

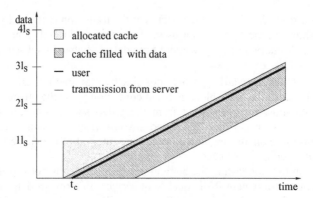

Fig. 4. Content of NCC in dynamic-position-block-caching (t_c: startup time of client, ℓ_s: block size)

The algorithm must assure that in case of shifting the block forward no client will refer to data behind t_{sb}. Therefore the algorithm has to scan the whole table for clients with a watching pointer t_c moving out of this segment, to create a new segment for them, and to move data to this new segment. Also upon loading of data into the NCC it must be checked whether the data are already in the NCC in another segment. Nevertheless a cache segmentation problem does not occur, because segments always have a fixed length ℓ_s.

The algorithm uses the same block length for blocks used by one client or by more than one client, although in the former case the cached data will not be used by more than one client.

The idea of dynamic-block-length-caching is to override this restriction.

2.4   Dynamic-Block-Length-Caching

In this algorithm video data requested from the client are loaded into the cache in blocks of dynamic length (limited to ℓ_s). The blocks are moving over the video like in dynamic-block-position-caching.

If more than one client is in an interval $[t_c+t_p-\ell_s, t_c+t_p]$ of length ℓ_s it is clever to create a block which contains all these clients. If there is only one client in the time interval and no other client is coming into that interval soon, it is not necessary to keep a block of size ℓ_s in the NCC. So this algorithm will reduce the size of a block $[t_c+t_p-\ell_m, t_c+t_p]$ to a minimum size ℓ_m whenever the video data will not be needed in time ℓ_s-t_p (see fig. 5). If the gap between two clients is getting bigger than ℓ_s-t_p (e.g. due to break-states of one client) the block will be split up. If on the other hand clients get close to each other their blocks will be melted together. When starting a video not contained in the NCC, the first block will be loaded to its maximum size ℓ_s because the decision whether another client is following the first one or not can be made only after ℓ_s data have been loaded.

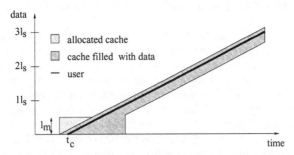

Fig. 5. Content of NCC in dynamic-length-block-caching (t_c: startup time of client, ℓ_s: maximum block size, ℓ_m: minimum block size)

For this algorithm a minimum block size ℓ_m and a maximum block size ℓ_s have to be defined. The minimum block size ℓ_m is taken as large as needed for rewinding a few scenes. The maximum block size is chosen larger than in the other algorithms (depending on the scenario by a factor of 2 to 5). The total cache size will not increase, because most blocks will shrink to minimum block size ℓ_m.

From increasing and shrinking the length of the NCC blocks, a fragmentation problem in the NCC arises. Additionally the administration of these blocks becomes more complex than in the algorithms working with a constant blocksize. For each client it has to be controlled whether he runs out of or into another block and which new suitable blocks have to be created in the memory. A micro segmentation of the videos in segments a of small fixed length (e.g. 1 minute) simplifies the algorithm. A table stores for each micro segment of each video whether it is in the NCC, in which NCC block and whether a client is using this micro segment. Additionally, for every client his pointer in the video is stored.

3   Simulation Scenario

An analytic investigation of the algorithms with the methods of traffic theory is only possible for the steady-state of the systems. The following scenarios will never reach this state; so simulations of the transient state are made by starting the simulation in the initial zero state (see section. 1). Two scenarios will be discussed:

Local Area iVoD: The first scenario is e.g. the cabin of an aeroplane. An aeroplane's starting and landing activities lead to a zero-state of the system, because no video transmission will then take place. During the flight the watching behaviour is strongly correlated in time. The flight duration is so small that no steady-state will be reached. Furthermore, the passengers will mostly watch block buster videos. The approach in general used in literature to model the viewers' preferences is to use the Zipf distribution [4] limited to the number of videos. This truncated Zipf distribution produces only one extensively used video. More-class distributions are probably more realistic. We assume here for simplicity reason 15 videos with probability 5%, 10 more videos with 0.5% probability and 975 with 0.02% probability of access. The collection of videos in the library should cover 1000 videos. The assumed number of users is 100 per NCC.

Home iVoD: Also within the home area, it has to be assumed that a correlation in time is given due to social events, like lunch time, beginning of work, end of work etc. The system will not fall back into the zero-state, however, steady-state will also never be achieved. The watching behaviour is not as strongly correlated as in an aeroplane. Due to the daily use of the iVoD-system the number of often watched block busters is rising and the hit rate on each of them is going down. A Zipf distribution describes this user behaviour well. We assume that the video library for a home iVoD-scenario will cover 1000 videos for 1000 clients per NCC.

The user behaviour will be described by the transition probabilities in a Markov model. Transition-probabilities are estimated, because no real user profiles are available. The probabilities that clients change or terminate a video are independent from the chosen video and the duration of watching. The transition to the break-state will also be described through the Markov chain. Duration of break-state is assumed to be 1 minute. This model leads to scenarios in which nearly 80% of the clients are active at the same time.

Based on this Markov chain model the performance of the algorithms has to be evaluated.

4   Performance

The evaluation of the algorithms must be based on the complexity of the algorithms, the used system resources and the offered QoS. The QoS criteria of an iVoD system are fulfilled only if the iVoD system operates with a time delay of at most t_d. Fast-forwarding and rewinding has to be possible within the bounds of the cached video data without delay and out of this bounds with delay t_d. In the following analysis only scenarios are considered in which the QoS criteria are fulfilled with a probability of 99.99%. Two scenarios are analysed: An NCC with 100 and 1000 clients and a video server with 1000 videos are used. We assume that both scenarios will return in zero state after maximum 5 hours. Each measurement is repeated at least 20 times to get a small confidence interval. Network technology is not of consideration for the evaluation of the algorithms.

4.1   Multi-Casting

The video server must be able to transmit each of the n_v videos to the NCCs in time interval of length ℓ_s, if requested. Since each video consists of n_s segments the server must transmit up to $n_v \cdot n_s$ video streams to the NCCs depending on the number of clients and the distribution function of the accesses to the videos. The needed memory size in every NCC is mainly given by the needed size for the fixed stored first segments of every video. The figure 6(b) and 7(b) show in the upper left corner the needed memory in each NCC. It is linearly growing with the segment size plus a comparatively small component for additional amount of cache due to the caching algorithm. The algorithm cannot profit from the correlation of the user behaviour.

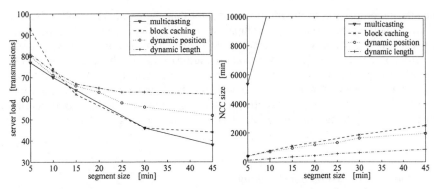

Fig. 6. Server load and cache size for multiclass-distributed access to 1000 videos by 100 clients (local area iVoD scenario)

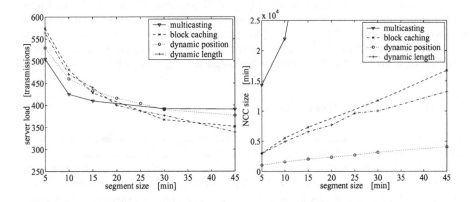

Fig. 7. Server load and cache size for Zipf-distributed access to 1000 videos by 1000 clients (home iVOD scenario)

The figures 6(a) and 7(a) show the server/network load in number of transmissions depending on the scenario. In the local area iVoD scenario (see fig. 6(a)) the server load for 80 (see section 3) active clients will be reduced to less then 50% by using a segment size of 45 min. A reasonable explanation for this load reduction are the fixed stored data in the NCCs. In home iVoD a further reduction in server load is reached due to the correlation in accessing of videos. The needed cache size to get the reduction of 50% in server load is enormous in both scenarios. Compared to the following strategies it is at least 1000 times higher.

4.2   Block-Caching

The block-caching algorithm reduces the needed cache size, because only used segments are loaded into the NCC. The needed cache size is, depending on the scenario, growing slower than the cache size for multicasting (see fig. 6(b) and 7(b)). In the local area iVoD scenario it takes even for a small segment size only 10% of cache size in comparison to the multicasting algorithm. A reasonable explanation for this is the small number of clients in comparison to the number of videos. In the home iVoD scenario is takes only 30% of cache size at a segment size of 5 min. The server load will not be reduced in comparison to multicasting. In the local area iVoD scenario the load is always higher than in multicasting (see fig. 6(a)). In the home iVoD-scenario the large number of clients and the high correlation of the Zipf-distributed access to the videos allows this algorithm to be as good as the multicasting, even though this algorithm has no fixed stored video data in the NCC (see fig. 7(a)).

4.3   Dynamic-Block-Position-Caching

The expected positive effect on load and NCC size has not been observed. The simulation shows a small reduction of cache size for both scenarios but an increase of server load by 10% to 15% for the local area iVoD scenario. The hit rate in the cache for a small number of clients is going down in comparison to block-caching. In the home iVoD scenario we achieve an improvement of only 5% at a cache size of 12500 minutes and block sizes of 35 and 45 minutes. The improvements in terms of server load and NCC size are not outweighing the complexity of the algorithm.

4.4   Dynamic-Length-Block-Caching

In figures 7(a) and 7(b) the influence of segmenting is well recognisable. Small segments reduce the necessary NCC memory, large segments reduce the server load. In this algorithm both characteristics are combined. In a home iVoD scenario with n_v=1000 videos on the server and n_c=1000 clients (Zipf-distributed access to the videos), the NCC can use blocks of a size ℓ_s of 45 minutes and thereby reducing the server load by about 20% and at the same time the storage requirement by 80% in comparison to block-caching (see fig. 8). In comparison to multicast-caching the reduction for the local area iVoD scenario reaches a factor of 45 in cache size at a small segment size of 5 minutes and a factor of 85 at large segment sizes of 45 minutes for equivalent server loads. In the home iVoD scenario we reach a reduction of server load of 20% at (comparing multicasting with ℓ_s = 5 min and dynamic-length-block-caching with ℓ_s = 45 min (see fig. 9)). At the same time the cache size will be reduced by 50%.

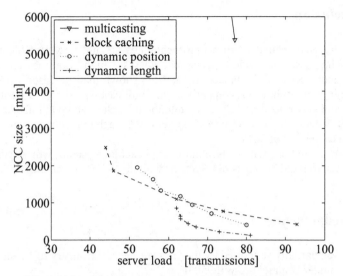

Fig. 8. NCC size over server load on for the home iVoD system

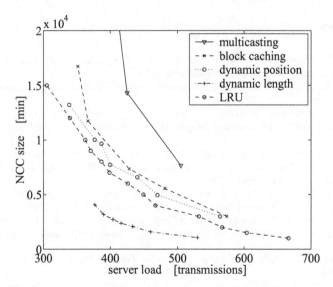

Fig. 9. NCC size over server load on for the local area iVoD system

One reason for this win in server load and cache size is the adaptation of the algorithm to inhomogeneous access rates to the videos as expressed by the Zipf distribution. If many clients do use the same video, then it remains in the cache. A rarely used video is only loaded to the extent to which it is needed for preloading. The reduction of the block length reduces the cache size and allows a higher upper bound for ℓ_s at equal NCC size especially for highly correlated access.

In comparison to iVoD-systems without caching the server load is reduced in home iVoD-systems by almost a factor of 3.

5   Conclusion

The paper has shown that multicasting techniques are not very suitable for iVoD systems. The assumed video scenarios possess high time correlation and inhomogeneous access to the videos. It has been shown that the described dynamic-length-caching algorithm, in comparison to multicasting, shows a reduction in bandwidth and server load of 20% or a reduction of cache size up to a factor of 45 for small iVoD systems. In very large iVoD systems we reach reductions in server load of 20% and in cache size of 50%.

Further investigations will analyse multi-level caching systems, which will connect several NCCs, to build bigger iVoD-Systems with up to 100,000 clients.

Acknowledgements

This work was partially supported by the German Federal Ministry of Economics and Technology under contract no. WIF9503P.

References

1. Tat S. Chua, Jiandong Li, Bend C. Ooi, Kian-Lee Tan, Disk striping strategies for large Video-on-Demand Servers, ACM Multimedia 1996.
2. Yitzhak Birk, Deterministic load-balancing schemes for disk-based Video-on-Demand storage servers,
3. Miranda Ko, Irene Koo, An overview of interactive Video on Demand System,
4. D.E. Knuth, The art of computer programming, Addison Wesley Publishing Company, 1973.
5. Frank Moser, Achim Kraiss, Wolfgang Klas, L/MRP: A buffer management strategy for interactive continuous data flows in a multimedia DBMS, Proceedings of the 21th VLDB Conference, Zurich Switzerland, 1995.
6. David W. Brubeck, Lawrence A. Rose, Hierarchical Storage in a Distributed Video-on-Demand System, http://www.bmrc.berkeley.edu, Berkeley, CA, USA, 1996.
7. M.Blaze, R. Alonso,Dynamic Hierarchical Caching in Large-Scale Distributed File Systems, Proceedings of 12th International Conference on Distributed Computing Systems, June 1992.
8. D.J.Gemmell et.al. Multimedia Storage Server: A Tutorial, IEEE Computer, Vol.28 No.5, May 1995.

Integrated Trouble Management to Support Service Quality Assurance in a Multi-provider Context

Dan Dragan, Thomas Gringel, Jane Hall, Richard Sinnott, Michael Tschichholz,
Wilhelm Vortisch

GMD FOKUS,
Kaiserin-Augusta-Allee 31,
10589 Berlin, Germany
flowthru@fokus.gmd.de

Abstract. Liberalisation of telecommunications encourages competition between the various actors in the Open Service Market (OSM). In this highly competitive context, Connectivity Service Providers (CSPs) and Value Added Service Providers (VASPs) are investigating opportunities to provide differentiated Service Quality related Service Layer Agreements (SLAs) to their customers. The services provided will span several administrative domains which makes their management complex. The key element for end users when choosing a particular service is the guarantee of support to be provided when using the service and the desire to interact with as few actors as possible. On the other hand, key issues for network operators and service providers are the cost-effective maintenance of equipment and services. The aim of this paper is to present a novel architecture that provides the necessary infrastructure, models and mechanisms to help VASPs and CSPs to rapidly introduce customer care services for user quality assurance in a Multi-Domain environment. The architecture aims at integrating TINA, TMF and TMN concepts as well as established *legacy* in-house customer care and help desk systems. This work is being undertaken within the Assurance part of the CEC ACTS project FlowThru.

1   Introduction

The ongoing liberalisation of the telecommunication market is making the industry in this area very active and subject to many changes. These changes are breaking down the traditional barriers between public and private domains and encouraging more relationships to be matde between the various actors in telecommunications for the purpose of end-to-end service provisioning. This is permitting the emergence of what is now called an Open Service Market (OSM). This liberated market provides an open area where network and service providers can co-operate and compete to improve their business.

To be efficient in such a market, operator and service providers need to develop novel solutions that permit the rapid introduction and maintenance of new telecommunication services [33]. These solutions should permit the rapid deployment of co-operative policies by means of interoperable interfaces at the network and service levels so that new services can be set up quickly and efficiently. Furthermore, if customers have accepted a service offer, the overall end-to-end service quality has to be assured by the service provider. Service management systems are required to help reduce operating costs and to interact efficiently with customers and suppliers [1].

From this perspective, the provisioning of a particular service can necessitate complex configuration and maintenance that involves a number of actors: service providers, connectivity providers[1] and also service brokers. The TINA Business Model [19] gives a good model of such actors and their interactions. However, this multiplicity of intermediate actors makes the process complex and should be hidden, e.g. by the retailer business role, from the customer. Hence, differentiation between service providers will be based on the capability of the provider to offer the customer flexible differentiating or customised SLAs (Service Level Agreements) and related facilities that permit the customer to monitor and control the QoS provisioning.

In this paper, we present an integrated framework for service quality assurance in an OSM supporting various sets of customer and service providers. This framework is based on TINA-C, TMF, OMG, TMN, and Internet management concepts and solution sets.

The concepts outlined below are addressed in the context of the European ACTS project *FlowThru* [9]. This project focuses on the information "flow through" between customers and multiple service providers in a multi-domain OSM environment. It covers the overall service life cycle of service provisioning, including fulfilment, assurance and billing. Furthermore, it defines guidelines for the development of distributed management systems [11]. The problem handling process is treated in the network and service quality assurance part, which is described in this paper. The objective of the assurance part of the project is to automate the interaction between customers, service providers and network operators for the purpose of problem identification, awareness distribution and resolution, SLA production, assurance, SLA fulfilment assessment and discounting in case of failure.

This paper is organised in five sections. Section 1 provides a short description of the problem and the general approach. It also outlines major recent initiatives in the area. Section 2 describes the overall goal of the project and the general component architecture of the solution. Section 3 is an internal description of the TINA Trouble Report System and Section 4 introduces details of the TMN based TTS. Finally a conclusion and perspectives are outlined in Section 5.

1.1   Problem Description

Problem handling is an already known aspect in telecommunication network management but will become increasingly important in the near future. This problem is also more and more concerning the service level where differentiation between service providers is moving towards Quality of Service (QoS) and discount policy competition.

Connectivity Service Providers (CSP, also referred to as a NO Network Operators) and Value Added Service Providers (VASPs) are investigating opportunities to provide differentiated Service Quality related Service Layer Agreements (SLAs) to their customers [13]. When a Service Provider offers a telecommunications service,

---

[1] In this paper we use the TINA term connectivity service provider. In other contexts, terms such as Public Network Operator (PNO) are used.

there is always a possibility that there will be a partial or total failure of that service. Such a problem is known as a 'trouble' and the process of trouble administration is concerned with identifying and resolving that trouble. The existence of a trouble has an adverse effect on the quality of the service as perceived by the Customer.

Violation of SLAs by 'troubles' and the production of evidence of the performance provided (or more generally the QoS) is still a major issue. Furthermore, the impact of SLA violations on tariffs and accounting should be taken into account to satisfy customer expectations. Because of QoS dependencies between service and networks levels, these issues require a more integrated network and service management environment. However, this process is made complex by the deregulated market as the service can span a number of service provider domains. Thus the processes of detection, localisation and resolution are becoming more difficult in such an OSM environment. Furthermore, it is quite complex to associate network related faults with specific services, customers or users. Therefore, the provisioning of QoS based SLAs and Service Quality assurance requires a more general Service Quality Management Framework, as being developed in the Eurescom project P806 [13] and as specified in the TMF [TMF-501, TMF-503].

1.2   General concepts for Inter-Domain Problem Management

To support the needs of the various actors in the growing OSM a "multi-domain problem management support system" is required. Such a system should enhance the functionality of established *customer care centres* and of *in-house trouble ticketing systems*, not changing the current often phone based service but enabling the additional exchange[2] of Trouble Tickets (TTs) via standardised interfaces. This would require "embedding" *legacy trouble ticketing systems* such as the well-established Remedy ARS product [14], as depicted in Figure 1. This figure depicts the general principles of multi-domain problem management.

For connectivity service providers a TMN based solution [34] might be more suitable. The same principle of "embedding" the legacy *in-house TT-System* can be used. To exchange TTs between the customer and connectivity provider domains (i/f type 4) or between connectivity provider domains, standardised interfaces (i/f type 5) based on GDMO / CMIP specifications of Eurescom project P612 [8] will be used.

The exchange of TTs between connectivity provider and value added service provider domains requires a mapping between TMN and Corba technologies. This can be done by a gateway which maps CMIS to IDL interfaces, e.g. based on the JIDM standards [35], [36].

---

[2] More generally: Remote Trouble Handling enables TT/TRs to be created, cancelled, monitored, updated, escalated and closed as well as the verification of problem resolution.

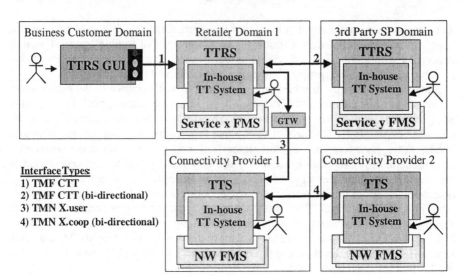

Fig. 1. Principles of multi-domain Problem Management

1.3   Recent Initiatives

Different consortia have undertaken initial work to handle problem management processes, up to now more or less independently at the network or the service level. Furthermore, today's TT-System products are mainly designed to support the internal needs of a service provider, enabling in some cases customer access (e.g. ARWeb). The concepts introduced below have been selected pragmatically as contributions for our architectural design and the implementation of the FlowThru Quality Assurance Trial System.

1.3.1   ITU-T Contribution

Quoting from [34], *"this Recommendation is concerned with the management of malfunction in systems and communications networks from the perspective of a provider of service and user of that service. In the Recommendation these malfunctions are referred to as "troubles". A report format is defined to allow a user to report a trouble, which will then be progressed to resolution by a provider. During problem resolution by the service provider, the service user may determine the current state of resolution by issuing a request for this information. When a trouble has been cleared the provider may notify the user."*

Defined in 1995, this Recommendation specifies the Trouble Management functionality for:

• Reporting of troubles on services or resources on a managed network or system;

• Tracking the progress of the trouble to resolution;

• Clearing and closure of the trouble.

The recommendation follows the TMN X series approach by using the GDMO interface specification language as a vehicle for defining the management functionality.

To support the largest possible deployment, the format of the trouble report is made up, with few exceptions, only from (a large number of) conditional packages. That implies for a particular implementation project a preliminary effort of 'profiling' the X.790 GDMO for its specific purposes.

Both, Eurescom Project P612 and the TMF CTT specifications, as introduced below, are based on the principles as defined in this ITU-T Standard.

1.3.2   Eurescom P612 Contribution

The aim of Eurescom [8] is to carry out pre-competitive R&D projects in order to support the Shareholders (European Telcos) in establishing future-oriented telecom networks and services. The Eurescom P612 project was a purely TMN project, focused on taking the International Recommendation ITU-T X.790 Trouble Management [34], developing it and validating practical implementations of this standard in the EURESCOM Pan European TMN Laboratory environment.
The main objectives of this project were to:

- Carry out a requirement analysis for trouble management functions in a number of operational telecommunications service environments, and to capture a set of operational scenarios to guide and test the technical work. The final aim was the development of a generic, interoperable trouble ticketing (TT) process.

- Profile the management functionality of the base ITU-T Rec. X.790 and its associated information model to match the functional requirements and the operational scenarios. This should be done on the essential X interfaces involved in trouble management [P612-D2]:

  - X.user interface between a CSP Management Domain and a Customer Network Management Domain and

  - X.coop interface between two peer CSP Management Domains (that have to cooperate in order to resolve some Customer problem by exchanging information between them).

- Carry out interoperability tests between different implementations in different laboratories based on previously developed test suites.

One important conclusion of this project was that the flow of messages on the X.coop interface is similar to that on the X.user, adding the fact that the same message can flow in both directions depending on the role taken by the CSP with respect to a specific Trouble Report. The specifications of P612 have been used to design and implement the IMA TTS (see section 4).

1.3.3   TeleManagement Forum Contribution

TeleManagement Forum (TMF) [24] is a non-profit, global organisation that provides the telecom industry with leadership on the most effective ways to streamline the management of communications networks and services. Membership includes Network Operators, Telecommunication Service Providers, Telecommunication Systems Vendors, etc.

TMF's principal mission is to enable the development of 'standardised' Management System solutions. In order to move standardised Telecommunications Management Services forward, the TMF uses as a focal point a framework of agreed business processes. The current focus is upon the integration of all these processes into process "flow-through" services built around three high level processes of Fulfilment, Assurance and Billing of Telecommunication services. Collectively, the TMF calls this set of agreed business processes an "Operations Map" [31]. Use of the agreed business processes makes it considerably easier for service providers to work together to deliver global services, enable customer access and control of services, etc.

Within the scope of the TOM framework the TMF has defined a set of detailed specifications to support important customer-to-business and business-to business management processes. Within the scope of multi-domain problem handling the following documents have been taken into account:

- Service Provider to Customer Performance Reporting Business Agreement [26];

- Performance Reporting Definitions Document [28].

In both documents, requirements, concepts and terms have been defined for service level agreements, QoS measurement and performance reporting. These concepts have been used to define SLAs in the context of the Assurance Trial System.

- Trouble Administration Business Agreement [25];

- Customer to Service Provider (SP) Trouble Administration Information Agreement [27];

- Customer to Service Provider Trouble Administration Analysis Specification [29];

- Corba Interface Specification for Customer to SP Trouble Administration [30].

The first two documents listed above define requirements, concepts and terms for problem management between customer and service provider. The latter two documents specify the interfaces for Corba based systems. They have been used for the TTRS design (see section 4).

1.3.4   TINA-C Contribution

Over 40 of the world's leading network operators, telecommunications equipment and computer equipment manufacturers have formed the Telecommunications Information Networking Architecture Consortium [18] to define and validate a common and open software architecture for the provision of telecommunication and information services, known as TINA.

TINA defines a set of concepts, principles, rules and guidelines for constructing, deploying, and operating TINA services. The major principles are based on the Reference Model for Open Distributed Processing [15]. The purpose of these principles is to insure interoperability, portability and reusability of software components and independence from specific technologies, and to share the burden of creating and managing a complex system among different business stakeholders, such as consumers, service providers, and connectivity providers [19]. Reference Points are defined to specify conformance requirements for TINA products [22].

TINA provides a set of specifications, e.g. Computing Architecture, Distributed Processing Environment Architecture, Service Architecture and Network Resource Architecture [TINA-CA,TINA-DPE,TINA-SA,TINA-NRA], which formed the basis for the development of the PLATIN TINA Service Platform [37] and used in the FlowThru Service Quality Assurance System Trials.

Although TINA covers the major market requirements caused by deregulation and globalisation (multi-provider environment, need for flexibility, customisability, etc) the service quality assurance and fault management issues are not sufficiently covered. While fault management in the context of network resource management is covered by the Network Resource Architecture, the Service Architecture stresses, but does not define, these issues in the service management context. The concepts described below could be used to enhance useful TINA concepts with problem management related solutions.

2   Service Quality Assurance Proposal in the OSM

The availability and quality of communications services[3] are of increasing importance as businesses automate and rely heavily on computer-based applications. The actual objective of all telecommunication actors is the rapid, accurate and reliable exchange of trouble information between the customer and its service providers to minimise the trouble resolution time and to optimise customer satisfaction in case of SLA violations. In anticipation of an OSM environment, the customer has only to interact with the retailer (one stop shopping) and does not want to be concerned with the various supporting actors/providers. For the Assurance scenario, subscription of the customers and the necessary network and service configurations have already been specified for the fulfilment phase.

The service level problem management service, implemented by the TTRS, will be offered to customers as a TINA service. It makes use of TINA SA principles and enhances the current definitions and reference point specifications to support distributed problem management business processes. The TTRS also makes use of enhanced definitions of reference points between retailers and 3rd Party service providers [3].

---

[3] Covering connectivity as well as value added services.

2.1   Scenario Description

To validate the general concepts for a multi-domain service quality assurance framework as introduced in Section 1.2, an integrated Network and Service Quality Assurance Trial System is being developed in the EC/ACTS project FlowThru. The objective is to evaluate, according to business related use cases, how service quality assurance can be improved by multi-domain problem management. This will cover the exchange of the service level trouble reports (TRs) and the network level trouble tickets (TTs) event correlation as well as partially automated problem resolution.

The trial scenario focuses on the information flow between the components of the inter-domain problem management system distributed in the various autonomous administrative domains of the OSM environment. The TINA Trouble Report System (TTRS) concentrates on the service level management issues (see section 3) whereas the TMN Trouble Ticketing system (TTS) focuses on the issues at the connectivity service management level and the network and network element levels below (see section 4).

The assurance trial system is being developed to demonstrate the following aspects:

- service offers to customers based on differentiated (Gold, Silver) or customer specific (customisable) Service Level Agreements (SLAs);
- customer controlled problem handling, i.e. to enable customers to generate TRs and to monitor problem resolution in a multi-provider service environment;
- exchange of Trouble Tickets (TTs) between customer and connectivity service provider domains and connectivity-connectivity provider domains based on X.790, EURESCOM P612 (TMN-based) specifications;
- exchange of Trouble Reports (TRs) according to TMF specifications (Corba-based);
- exchange of TTs between TMN and Corba technology by JIDM based TMN/Corba gateways;
- correlation of problem events, e.g. where the same trouble is identified by different sources;
- discounting if a QoS / SLA violation has occurred for a user, customer or set of customers depending on specific SLAs.

The assurance business process of the FlowThru system shows the interaction between these different levels of problem management, based on the various defined use cases. All elements of the Assurance System scenarios will be demonstrated to show how to improve significantly the distributed problem handling process [31] and the information *Flow Through*. This would increase customer satisfaction and can reduce the overall maintenance costs and SLA violation related penalties.

2.2   The Business Model

Figure 2 describes the various domains involved in the Business Model of the FlowThru Assurance Scenario. It identifies the customer domain and a set of co-operating service provider domains (based on the TINA Business Model) [19] and major personnel roles. The connectivity service being offered is a premium IP service over an ATM network infrastructure. The value-added service being offered,

MusicShop, is a Web based public (but secure) file system allowing up- and downloading of multimedia documents for individuals or globally distributed user groups.

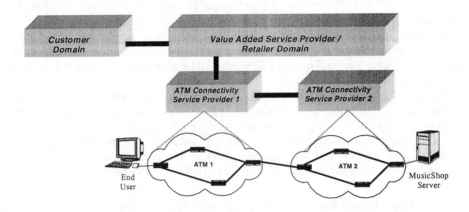

Fig. 2. Business Model

For the Assurance Trial scenarios, two different CSPs were introduced to reflect the possibility that the connectivity service may span a number of network infrastructures. Furthermore, it was decided to not differentiate the service retailer from the 3rd Party service provider domain, but to avoid additional complexity we have considered that both business roles are performed by the same actor.

2.3 The Scenario Use Cases

The functional overview for the Assurance Trial System is given in terms of use cases and their relations to external actors. These are based around:

- the occurrence of troubles in the different systems, e.g. in the network or the service;

- the persons or components that identify those troubles, e.g. the service provider, local/remote connectivity providers or the customers themselves;

- how information related to these troubles navigates through the system and is subsequently used to resolve the troubles by appropriate fault management systems;

- the subsequent impact of the troubles once they have been resolved, e.g. discounts given to affected customers should the troubles violate customer SLAs.

2.4   The Technical Approach

All service providers (including the CSPs) will offer their services as TINA services. Information will be exchanged at TINA reference points [3] making use of Corba [4] or TMN [12] technology at the network level. The use of enhanced TINA subscription concepts allows connectivity or value added service troubles to be associated with affected services, customers and/or user sessions. An integration of the TINA Trouble Report System (TTRS) and TINA accounting systems will enable discounts in case of SLA violations.

Figure 3 depicts a simplified computational model of the Assurance Trial System configuration at the TINA service level. The TINA Service Environment provides the infrastructure to run the FlowThru service quality assurance trial. A precondition for the scenarios is that the customers have already subscribed to the MusicShop, the assurance trial service, and the required premium IP connectivity services, which enable authorised end users to use the MusicShop service at any time.

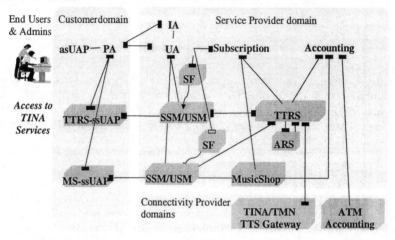

Fig. 3. Description of the TINA based Assurance System Components

The MusicShop service and the TINA TR management service are embedded within the PLATIN TINA Platform Y.TSP [37] which provides an implementation of the major TINA service architecture components. They are configured for the assurance trial (service templates, tariffs, customer and user profiles) to enable the services to be offered to the end users.

- The *access session components* (asUap, PA, IA, UA) enable service selection and secure service access in a TINA environment. The interface between the customer and service provider domain is based on the TINA Retailer reference point definition [22].
- The *Subscription Component* contains information related to registered customers, the existing services and SLAs between the service provider and customers on the use of services subscribed to, e.g. the negotiated QoS. It also contains information on customer connectivity access points.

- The *Accounting Component* is responsible for the accounting of the service and network usage. The TINA Accounting System is connected with the *ATM Accounting System* to exchange network usage charge records and with the *MusicShop* to exchange service usage charge records. The interface to the TTRS allows granting of discounts in case of SLA violations.
- The Assurance Trial service, i.e. the *MusicShop* (MS-ssUap, USM/SSM, MusicShop) is a TINA Service offered to customers. Authorised users can access the service from different network access points to up/download documents. This service is used to demonstrate problem handling and discounting at the TINA service level.
- The *TINA Trouble Report System* (TTRS) implements a management process that permits the handling of the quality assurance process at the TINA service level of the FlowThru trial system, including SLA management. The TTRS is considered as a specific TINA management service embedded in the TINA environment through the required components (SSM/USM, SF, ssUAp). The SSM/USM components are required in TINA environments to maintain user specific service sessions and the interface to the ssUAP in the customer domain.

3   The Service Level TINA Trouble Report System (TTRS)

As mentioned above, the TTRS could be considered as an additional component for the TINA Service Architecture for Problem Management, supporting service quality assurance and customer care. It is designed to exchange Trouble Reports between the customer, retailer, 3rd party service provider and connectivity provider domains. It also supports SLA handling as well as initiating discounts in case of SLA violations. The TTRS encapsulates an in-house trouble ticketing system which handles trouble tickets between the help desk and the 1st and 2nd level support. Customers interact with the TTRS using the TINA compliant TR management service (ssUAP-SSM/USM). These are started automatically within the user access session to enable problem notifications to be delivered.

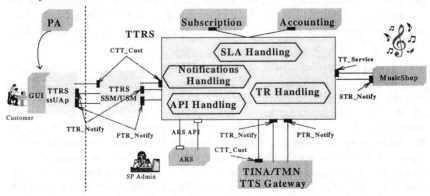

Fig. 4. TTRS Interfaces and Component Architecture

3.1 Description of the Interfaces

The following interfaces are associated with the TINA Trouble Reporting System (see Figure 4):

- The *CTT_Cust* interface is used by a customer to create, modify, track the status of, view, verify, delete, and cancel trouble tickets; to grant an authorisation for repair activities or to escalate a trouble ticket. The service provider exports this interface so that its customers can manage the lifecycle of trouble tickets effectively.

- The *TTR_Notify* interface is used by a customer to receive notifications from the service provider about trouble tickets, for example that a trouble ticket has been created due to a problem that another user, the service or connectivity provider or some remote service or connectivity provider has identified. This interface also allows customers to be notified about modifications to existing trouble tickets, status updates of trouble tickets, deletion notifications of existing trouble tickets and cancellation notifications of trouble tickets.

- The *PTR_Notify* interface is used by customers to receive notifications from service providers about trouble tickets associated with scheduled maintenance.

The IDL for these interfaces is provided in the TM Forum interface specification document [30]. The TTRS component provides a CORBA based wrapper around the Remedy ARS component so that it can be incorporated into the Service Quality Assurance System using open CORBA interfaces.

- The *TT_Service and STR_Notify* interfaces between the TTRS and MusicShop components correspond to a subset of the functionality of the CTT_Cust, PTR_Notify and TTR_Notify interfaces. More precisely, the TT_Service interface is used by the MusicShop service to inform the TTRS of intended maintenance periods for the service and to request that trouble tickets are created for problems identified by the MusicShop service trouble management system (and their states changed when necessary, etc.). The STR_Notify interface is used for receiving notifications from the TTRS component related to trouble tickets, e.g. the TTRS might inform the MusicShop service trouble management system that it has created a trouble ticket when the problem lies with the MusicShop service itself.

The TTRS also possesses interfaces to the TINA accounting and subscription components. The interface to the accounting component is used primarily for issuing discounts to users who have incurred some form of SLA violation for one of the services they use. The interface to the subscription component is used for several purposes. First, it is used for querying whether users who complain about services are actually subscribers to those services. If so, the subscription component returns subscription information about the user and any other customers affected. This includes both the normal subscription information and service properties that they might have as well as the SLA that they agreed to when subscribing to those services. These are subsequently used by the TTRS when deciding whether a SLA violation has taken place and if so, how much the discount should be.

3.2   Underlying Technology Aspects

The TTRS prototype is being implemented as a TINA service in the PLATIN TINA platform Y.TSP. It is being developed using Java to enable easy installation and porting over various operating systems, such as UNIX and Windows/NT. To enable an integration with the customer care or help desk related business process, the trouble report database is based on a well-established product for trouble ticket management, the Remedy ARS product. This use of existing technologies permits an enhancement of the normal business process to support automated exchange of trouble reports over administrative domain boundaries making use of well established technologies, i.e. Remedy, Corba/IDL, Java, JIDM based Corba/TMN Gateway, etc.

4   The Network Level Trouble Ticketing System (TTS)

At the connectivity provider level of the multi-domain problem management system, the Trouble Ticketing System (TTS) component is used. It is based on a TMN compliant product [6] that implemented the P612 specifications and was extended for the FlowThru service quality assurance system. Its function is to implement the Trouble Ticketing Service for connectivity service providers or Network Operators. The overall model is depicted in Figure 5.

4.1   Description of Interfaces

The system presents 4 interfaces:

- a *TMN X.user* interface that allows the Value Added Service Provider, e.g. the MusicShop provider, using a CMIP/Corba Gateway, to act as a Customer to the Network Operator for the Trouble Ticketing Service requirements [5];

- a *TMN X.coop* interface that allows two Network Operators to co-operate in deploying ATM VP connections for end users to exchange trouble tickets relating to these connections;

- a *TT.Q* interface allows a connectivity provider call centre to dispatch a trouble report to one of its regional areas for resolution. This dispatching is operationally identical to the referral of a trouble report between two co-operating network operators. The only difference is that this dispatching may be executed in a single direction. From the point of view of the TTS agent it is identical with the one used for the X.coop interface – apart from the configuration parameter.

Besides these interfaces, a local interface should exist for carrying out some parts of the trouble resolution process. Depending on the pre-existing SLAs and the complexity of the legacy systems, the ratio between human and automatic activities involved in the trouble resolution process may vary. Even if this ratio is zero for common daily activities, it may still be necessary to resort to human intervention for unforeseen cases. So this local interface has to be preserved.

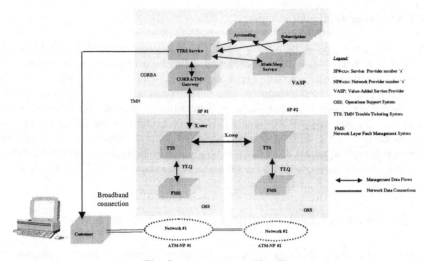

Fig. 5. TTS position in FlowThru

For brevity we do not discuss the details of the connectivity and fault management systems of the ATM based network providers. We note however that they consist of simulated network element management systems, network management systems and fault management systems based on Q3ADE technologies. In addition we note that the fault management system itself supports trouble ticket management functionality where trouble tickets are created based upon notifications from the network management system, or from requests from the TTS component.

4.2   Underlying Technology Aspects

Internally the TTS product is based on the Hewlett-Packard Open View Distributed Management (HP [10] platform supporting the OSI protocol stack for CMIP and the HP OV Managed Object Toolkit for rapid generation/implementation of a TMN agent starting from its GDMO specification.

It is to be mentioned that the C++ classes have a stub automatically generated by the MOT generation utility from the GDMO specification. These stubs are subsequently extended by manually writing the implementation of behavioural statements of the GDMO specification and the functionality not exposed at the manager interface.

5   Conclusion

This work has shown how existing but independently developed concepts from different problem domains have been integrated to design and develop a trial system to enable multi-domain service quality assurance through an automation of the problem handling business process. Through the architecture proposed here, generic and reusable components have been designed and implemented which can be used to support the differentiation of service provision through a QoS assurance process. These include TINA based components (TTRS) that allow users to be informed (or themselves be the informers) of problems with offered services or the networks those services use. Similarly, generic TMN based network provider trouble ticketing systems (TTS) have been developed which support trouble management and administration between the service and network management domains, as well as network-network management domains.

To support the integration between TMN and CORBA based technological domains, gateways have been developed. These support the flow through of trouble information necessary to provide an integrated solution to trouble management in a multi-provider domain.

We conclude by emphasising that the FlowThru assurance trial is an open solution and not some overly prescriptive, non-reusable "one off" scenario. The components themselves can be applied to a myriad of CORBA based services and TMN based networks. Similarly the gateways are generic and largely independent of the specific instances of services and networks that they are associated with.

References

1.   Adams, E.; Willetts, K.; The Lean Communications Provider: Surviving the Shakeout though Service Management Excellence; McGraw-Hill; 1996.
2.   N. Agoulmine, D. Dragan, T. Gringel, J. Hall, E. Rosa, M. Tschichholz, Trouble Management for Multimedia Services in Multi-Provider Environments, to appear in Journal of Network and Systems Management, March 2000.
3.   Schoo, P; Egelhaaf, C; Eckardt, T.; Agoulmine, N.; Tschichholz, M.; Modularization of TINA Reference Points for Information Networking, Proceedings of IS&N'99, 27.-29. April 99, Barcelona, Spain.
4.   The Common Object Request Broker: Architecture and Specification. OMG Document Number 92.12.1, Revision 1.1. Object Management Group, 1992. And: Object Request Broker 2.0. Object Management Group, OMG Publication, John Wiley & Sons, 1995.
5.   Covaci, S.; Dragan, D."Customer Care Solutions: Interoperable Trouble Ticketing Management Service" in: Mullery, A., Besson, M.; Campolargo, M.; Gobbi, R.; Reed, R. (eds.), "IS&N'97: Technology for Co-operative Competition", Lecture Notes in Computer Science 1238, Springer Verlag, Berlin (D), pp. 255 - 262, IS&N'97, Cernobbio (I), 27 - 29 May 1997.
6.   Covaci, S.; Marchisio, L.; Milham, D. J., "Trouble Ticketing X Interfaces for International Private Leased Data Circuits and International Freephone Service" in Proceedings of the 6th NOMS '98 - Management for the New Millenium, pages 342 - 353, New Orleans, USA, 15 - 20 February 1998.
7.   www.commerce.ie/difference/
8.   www.eurescom.de/
9.   www.cs.ucl.ac.uk/research/flowthru/
10.   HP Corp - Part No. J1064-90023 "HPOV: Distributed Management - Developer's Guide".
11.   D.Lewis, A Development Framework for Open Management Systems; To appear in the Interoperable Communication network Journal; Special issue on Management, Baltzar Science Publishers, Q1, 1999.

12. Principles for Telecommunications Management Network, ITU-T Rec M.3010, Geneva, May 1996.
13. Eurescom P806, Deliverable 1, A framework for a common and harmonised service quality, March 1999. http://www.eurescom.de/public/deliverables/dfp.htm
14. Remedy Corp - Part Number GSG-320-001, "Getting Started Guide"
15. Reference Model of Open Distributed Processing, ITU-T Recommendations X.901..4 / ISO International Standard (DIS) ISO/IEC 10746-1..4, 1995.
16. www.tinac.com/TINA2000/workgroups/sarp.html
17. www.tinac.com/TINA2000/workgroups/sm.html
18. www.tinac.com
19. H. Mulder (Ed.), TINA Business Model and Reference Points, Version 4.0, May 22, 1997
20. A. Gavras, W. Takita (Eds), TINA DPE Architecture, Version 2.0b0, November 24, 1997
21. F. Steegmans, TINA Network Resource Architecture, Version 3.0, February 10, 1997.
22. P. Farley, R. Minetti (Eds.), TINA-C Ret Reference Point Specifications, V. 1.0, Jan. 27, 1998
23. L. Kristiansen (Ed.), TINA-C Service Architecture, Version 5.0, June 16, 1997
24. www.TMForum.com
25. TMF, Trouble Administration Business Agreement, Issue 1.0, August 29, 1996
26. TMF, Service Provider to Customer Performance Reporting Business Agreement, Issue 1.0, March 1997
27. TMF, Customer-Service Provider Trouble Administration Information Agreement, Issue 1, March 1997
28. TMF, Performance Reporting Definitions Document, Issue 1.0, April 1997
29. "Customer to Service Provider Trouble Administration Analysis Specification", Version 7.0, NMF March 23, 1997
30. Corba Interface Specification for Customer to Service Provider Trouble Administration" revision 0.2, TM Forum (formerly NM Forum) SMART Group, November 4, 1997
31. Telecoms Operations Map, TeleManagement Forum GB910, Evaluation Version Release 1.0, Oct 1998
32. M. Tschichholz, D. Lewis, W. Donnelly, Use of TMF Business Model Concepts for TINA, Berlin, October 15, 1998
33. V. Wade. Three Keys to Developing and Integrating Telecommunications Service management Systems; IEEE Communications Journal; To be published in 1999.)
34. Trouble Management function for ITU-T Applications – ITU-T Recommendation X.790
35. Inter-domain Management: Preliminary Corba/CMISE Interaction Translation Architecture. Technical Report, Joint Inter Domain Working Group, X/Open and Network Management Forum, April 1995.
36. Inter-domain Management: Specification Translation, Technical Report, Joint Inter Domain Working Group, X/Open and Network Management Forum, 1997.
37. Eckert, K.-P.; Moeller, E.; Schoo, P.; Schürmann, G.; Tschichholz, M., Verteilte Objekt-technologie in der Telekommunikation, Meuer, H.W. (ed.), Praxis der Informations-verarbeitung und Kommunikation, PIK 21 3/98, K.G. Saur Verlag, München, September 1998, pp. 149 - 154, ISSN: 0930-5157. *Further information on Y.TSP is available at:* www.fokus.gmd.de/research/cc/platin/

Service Creation Techniques
for Software Development and Deployment

Peter Schoo

GMD National Research Center for Information Technology,
Research Institute for Open Communication Systems (FOKUS), Germany
schoo@fokus.gmd.de

Service creation spans a whole lot of different activities and tasks, when producing software to provide the functionality of a service. Technology selection is often approached rather early in the creation process and, based on available requirements, design activities are carried out that result in software development. Within this field of service creation various decisions have to be taken. They range from the selection of basic technology up to the re-use of available solutions, the adoption of pre-products or the integration of legacy elements. Furthermore, methodological issues and guidelines for use of the technology come into play. Finally, once the software development activities come to an end, the software has to be deployed and made available for use. In this group of papers a small slice of all of these tasks is captured.

Whenever a new technology becomes available, the question arises: Is it worthwhile to be used for developing services and what aspects of service creation will become easier with the new technology? Nowadays, the *eXtensible Markup Language* (XML) is such a candidate. On the other hand, the telecommunications community considers CORBA for quite a while as the solution to integrate existing systems and applications making use of the object-oriented paradigm. It is thus only natural that the question *"XML or CORBA, synergistic or competitive?"* is raised. This paper describes the XML and CORBA 'approaches', and the synergies and/or competition between these technologies. Considering the intersection of XML and CORBA, different criteria are identified, where comparison is possible and relevant, such as: specification, deployment, and tool support. The conclusion is drawn that XML and CORBA are complementary technologies; XML can, in conjunction with other protocols, provide the same functionality as CORBA, but will only replace CORBA in those cases where using CORBA is undesirable.

When the fundamental decision is taken on which technology to be used, the software design requires a variety of decisions regarding which options to be taken and adopted. In the intersection of the areas of human-computer interaction and mobile agent technology the paper *"Web-based Agent Applications: User Interfaces and Mobile Agents"* discusses solutions how to develop user interfaces for agents to enable end-users to interact directly with them. While one way is based on HTML and the provisioning of applets for each agent, the authors present a more sophisticated approach: the mobile agent encompasses its own classes to be rendered when interaction with human users takes place. It is argued that the management and maintenance of code is thus simplified and additionally the agent can bring different types of user interfaces for different users.

Often service creation comes along with a specific methodology, which enables a set of development processes to be used. Such methodologies typically address questions of efficient and effective re-use of existing solutions and the identification of any remaining (and hopefully minimised) development effort. The paper "*Multi-level Component Oriented Methodology for Service Creation*" is a perfect example. It presents a multi-level Component Oriented methodology for Service Creation, which could provide for both rapid service provisioning and high service quality. Especially interesting is the combination of means to develop components as software developers typically do, and the means to compose services by users that are not software developers and do not write code.

In the overall service creation process the code production will eventually be finalised and the targeting and deployment of applications will be the subsequent burning issues. "*Cooling the Hell of Distributed Applications' Deployment*" is a paper that certainly will give the reader some insights in the specific deployment requirements, in particular when the target environment is CORBA. The authors describe the need for abstract models, expressive notations and supportive tools, which are still missing when it comes to the deployment of distributed applications. Also, from a user's point of view, some shortcomings of the CORBA3.0 specifications are identified.

In summary, a slice of all of the tasks that need to be fulfilled during service creation in the development and deployment of software are presented in this set of papers.

Web-Based Agent Applications: User Interfaces and Mobile Agents

Alberto Rodrigues da Silva, Miguel Mira da Silva, Artur Romão

IST / INESC, Rua Alves Redol, 9, 1000 Lisboa, Portugal
University of Évora, 7000 Évora, Portugal
alberto.silva@acm.org
mira-da-silva@ip.pt
a.romao@computer.org

Abstract. The process of developing agent-based applications requires at least two tasks that are usually tackled separately by programmers. On one hand, programmers need to develop business rules and other support tasks for agents. On the other hand, programmers need to develop user interfaces (UI) for agents in order to enable end-users (not only owners but also other third parties) to interact directly with them. This paper focuses on this second task (developing user interfaces) and describes the solutions offered by the AgentSpace mobile agent system. Basically, we show and discuss two complementary ways to gather user-interfaces with mobile agents. On one hand, mobile agents don't provide any UIs. This situation promotes the separation of the UI and the backend (i.e., the agent) which allows flexibility and reuse. On the other hand, mobile agents provide by default UI components, which consequently promotes agents as better units of development and management. This situation can be very suitable in the context of dynamic and large-scale applications such as those found in electronic commerce domains. This paper also shows the relationship between the application of these mechanisms and the model-view-controller architecture used currently to build user interfaces in modern object-oriented frameworks. Finally, this paper presents some concrete examples with source code based on the AgentSpace system in order to validate and clarify the discussed mechanisms.

1 Introduction

Software agents and in particular mobile agents are an interesting paradigm recently proposed to support a new kind of application, mainly in Internet and Intranet contexts, which are large, dynamic and unstructured spaces of resources.

The main idea of the *software agent* paradigm is to provide indirect (delegated) human-computer interaction (HCI) as well as a model to better solve many complex problems [GK94, JSW98, Sil99]. On the other hand, *mobile agents* [Whi94, Whi97, RP97, RH98] provide another new and interesting paradigm particularly because they may be suitable to: reduce the network load, overcome network latency, encapsulate protocols, execute asynchronously and autonomously, adapt dynamically, be naturally heterogeneous, and be robust and fault-tolerant [LO98].

However, the process of developing agent-based applications for the Web context requires at least two tasks that are usually tackled separately by programmers. Firstly, they develop agent's business rules and support tasks. Secondly, they develop agent's user interfaces in order for end-users (the agents owners as well as other third parties) to interact directly with them.

The first task (business rules and support tasks) is usually achieved using the supported language (in this paper we assume Java [Sun98] as the main language for mobile agent programming) as well as an API provided by the supported mobile agent system. Aglets [IBM97, LO98], Grasshopper [IKV98], CyberAgents [FTP96], Odyssey [GM97], AgentSpace [SMD98], Mole [SBH96], MAO [MLC98], Jumping Beans [AdA99], Voyager [Obj99] are some examples of these systems, each providing its own API for programming agents.

The second task (developing user interfaces) is achieved traditionally using a combination of Java, usually through sub-classing the Java `Applet` class, and HTML pages. In some situations, for each agent there are as many as three complementary files: the first with the Java agent class, the second file with an HTML page referencing the applet, and the third file with the Java applet class. This approach for developing agent's interfaces can become impractical, from the management and development point of view, in concrete, large and/or dynamic applications. However, on the other hand, the separation of UI and the backend (i.e., the agent) allows flexibility, reuse, and innovation to occur independently at either end. As a complement, this paper advocates the principle of integrating the business rules and support task components together with the UIs in the same agent component. As a consequence of this principle, the agent would be a better unit of development and management in context of dynamic and large-scale agent-based applications. In the rest of the paper, we will show how this can be achieved using the AgentSpace mobile agent system.

The paper is organized in six sections, including this one. In Section 2 we introduce the AgentSpace system, its main goals, architecture, and conceptual object model. Section 3 discusses the generic communication mechanism between applets and agents, namely from applets to agents and from agents to applets. Section 4 discusses the communication models between end-users and agents, namely the conventional model (each agent has a corresponding applet) and our proposed integrated model (all HCI components are integrated with the rest of the agent's code). Section 5 describes some concrete examples developed for the AgentSpace framework. Finally, Section 6 summarizes the main conclusions and contributions of this paper.

2   Overview of the AgentSpace Framework

AgentSpace [SMD98] is a Java mobile agent framework being developed at the Technical University of Lisbon (IST) and INESC in Lisbon, Portugal. In this section we introduce its main goals, architecture, object models, and the notion of AgentSpace-based applets.

2.1   Architecture

The main goals of AgentSpace are the support, development and management of dynamic and distributed agent-based applications. These goals are provided by three separated but well-integrated components as depicted in Figure 1.

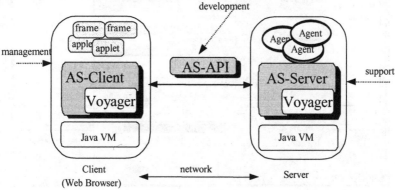

Fig. 1. The AgentSpace Architecture

Both the server and client components run on top of Voyager [Obj97] and the Java Virtual Machine (JVM). They can execute in the same or in different machines. Agents run always on a AgentSpace server's context (AS-Server) never in the client which is a Web browser such as Netscape Navigator or Internet Explorer. On the other hand, agents interact with their "master" (end-user) through applets/frames or by means of generic applications (AS-Client) directly on the Java VM.

The *AgentSpace server* (*AS-Server*) is a Java multithreaded process in which agents are executed. The AS-Server provides several services: (1) agent and place creation; (2) agent execution; (3) access control; (4) agent persistency; (5) agent mobility; (6) generation of unique identities (UID); (7) support for agent communication; and optionally (8) a simple command-line interface to manage/monitor the server and its agents.

The *AgentSpace client* (*AS-Client*) supports – depending on the corresponding user access level – the management and monitoring of agents and related resources. The AS-Client is a Java applet/application (stored on the AS-Server's host machine) offering all Internet users the possibility to easily manage their own agents remotely. Furthermore, the AS-Client is able to access several AS-Servers at the same time, providing a convenient trade-off between integration and independence between client and server. Figure 2 shows the AS-Client being used to create a new agent.

The *AgentSpace application programming interface* (*AS-API*) is a package of Java interfaces and classes that defines the rules to interact with the AgentSpace framework at the Java level. In particular, the AS-API supports the programmer in the following ways: (1) building agent classes and their instances (agents) that are created and stored in the AS-Server's database; and (2) building client applets/frames in order to provide an interface to agents.

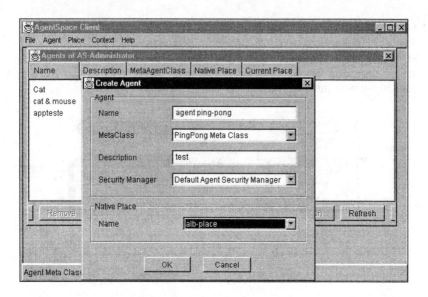

Fig. 2. The AS-Client tool – the "create agent" frame

The AgentSpace framework follows the Aglets approach, and as such it provides a complete environment to develop Java agent-based applications. However, AgentSpace provides a more complete object model regarding security mechanisms, integration with end-users, and location transparency [SMD98, SD98]. Below we briefly describe the main features offered by AgentSpace. The interested reader is referred to our previous papers for details about the system.

2.2   Object Model

AgentSpace involves the support, development and management of several related objects: contexts, places, agents, users, groups of users, permissions, ACLs (access control lists), security managers, tickets, messages, and identities. Figure 3 shows the relationships between these objects through a UML class diagram.

The *context* is the most important and critical object of the AS-Server, as each AS-Server is represented by one context. The context contains the major data structures and code to support the AS-Server, such as lists of places, users, groups of users, meta-agent classes and access control lists.

Each context has a number of places. The *execution place*, or simply *place*, has mainly two objectives. First, to provide a conceptual and programming metaphor where agents are executed and meet other agents. Second, to provide a consistent way to define and control access levels, and to control computational resources.

The place has a unique global identity and knows the identification of its owner/manager. It also maintains a keyword/value list that allows an informal characterization. Optionally, places can be hierarchically organized. The place can also contain the maximum and current number of agents allowed in order to support some resource management.

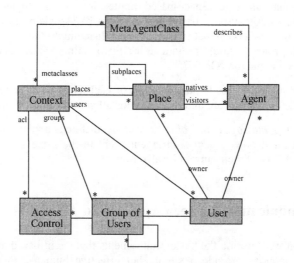

Fig. 3. UML class diagram of AgentSpace

In order to keep track of its agents, the place keeps a list containing its visiting agents and another with its native agents. The place also knows in which place its native agents are executing at a given point of time.

The *agent* is the basic element of the system. Agents are active objects that execute in some AS-Server, but from a conceptual perspective, they are currently in some place. Agents can navigate to other (local or remote) place if they have permission to do it. Just one user owns an agent. Nevertheless, other users (or even agents from other users) can interact with this agent, if interaction is granted by the agent' security policy.

The AS-Server also maintains lists of *users, groups of users* and *ACLs* to implement the permission and access control mechanism. A user may belong to one or more groups. Groups may be hierarchically organized to simplify permission management. This means that all users of some specialized group have implicitly all the permissions they inherit from the more general groups. By default, every AS-Server defines four groups of users and establishes a convenient security access policy, based on them: anonymous group; end-users group; place owners group; and AS-Server's administrators group.

2.3   AgentSpace-based Applets

The applets supported by AgentSpace have the capability to access AS-Servers and consequently to access their respective resources. In particular we describe in Section 3 the general interaction model between applets and agents, and indirectly the interaction model between agents and end-users.

The ability to support interaction between agents (executed in some AS-Server context) and applets (executed in some Web browser and/or Java Virtual Machine context) is mainly due to the features provided by the Voyager object request broker. This means that an AgentSpace-based applet is also a Voyager-based applet. Consequently, an AgentSpace-based applet has to be associated to a given program (in our case, the AS-Server) that runs as a server, listening to the network.

This server program should provide at least two main services, as described by the Voyager's documentation [Obj97]:

- Remote, network-based, *class loading* of each class used or referenced by the applet. These classes are automatically installed in the applet's codebase.

- *Redirecting messages and objects* sent by different applets, as well as those messages and objects sent from agents kept in the same or other AS-Servers, independently of their current localization.

3   Communication Between Applets and Agents

In this section we describe the basic requirements that agent-based applications need to fulfill in order to provide a simple but effective human-computer interaction. Namely, we will present two complementary communication mechanisms: *from applets to agents*, and from *agents to applets*. It should be noted that applets and agents usually run in different machines because applets are executed in the context of web browsers while agents are executed in the context of AgentSpace servers.

3.1   Communication from Applets to Agents

The code below is an example of a simple applet that shows how an AgentSpace-based applet can access and interact with an agent running in some AgentSpace server when the initiative is taken by the applets.

```
public class SimpleApplet extends Applet   {
   Label label = new Label( "Power3 of " );
   TextField operand = new TextField( "", 5 );
   Button operation = new Button( "=" );
   TextField value = new TextField( "", 3 );
   String userLogin = "admin";
   String userPwd = "adminpwd";
   InternalUser user = null;
   ContextView cv = null;
```

```
public void init() {
    try {
        Voyager.startup(this); // startup in applet mode
        String asAddress = Voyager.getServerAddress();
        user = AgentSpace.getUserByLogin( asAddress,
         userLogin, userPwd);
        cv = AgentSpace.getContextView(asAddress, user);
        setLayout(new FlowLayout());
        add(label); add(operand); add(operation);
        add(value);
        add.addActionListener( new ActionListener()
            { public void actionPerformed( ActionEvent
event )                          { power3(); } } );
        } catch(Exception ex ) {...}
    }
  void power3() {
    String aidStr = ...  // get remote agent identifier
    try {
        AgentView av = cv.getAgentOf(aidStr);
        Message m = new Message(getId().toString(),
        "power3");
        int n = Integer.parseInt( operand.getText() );
        m.putContent(new Integer(n));
        Result result = av.doOperation(m);

value.setText(Integer.toString(result.readInt()));
        } catch (Exception ex) {...}
    }
    ... }
```

This applet invokes a service (called "power3") provided by a public agent. The service executed by the remote agent is the calculus of the cube of an integer number. The code of this agent is not shown here due to space restrictions and because it is not relevant for the explanation regarding the user interface aspects.

In the example above, the reader should note the method init in which the Voyager distributed system is setup (startup) and the address of the corresponding AgentSpace server from which the applet comes from is retrieved (getServerAddress). The port number in which the AgentSpace server is listening to the network is also retrieved, because there may be many AgentSpace servers executing in the same host machine.

Based on the server's address and information about the user authentication (login and password), a ContextView object is then created which represents a view to the context in the remote server machine. The contextView object then offers a gateway to easily access all the functionalities and resources maintained by the AgentSpace server.

After the setup operations just described, the applet can then interact with the AgentSpace server. In this example, the applet uses the method power3 to obtain a reference (i.e., an object of class AgentView) to the agent that provides the desired functionality. Because AgentSpace provides location transparency, the desired agent can be anywhere on the Internet – as long as the AgentSpace server in which it was created is still running. In particular, the agent does not need to be in the same

AgentSpace server from which the applet comes from. AgentSpace uses the agent's identifier (the `aidStr` string) to look up the agent's current place.

After retrieving a view to the agent (a handler to talk to the agent) the applet builds a message of type "power3" and sends this message to the agent. Of course the message should be something the agent will understand, otherwise the agent would not know what to do after receiving the message. In particular, "power3" is the name of the method that will be called on this agent when the message is received by the AgentSpace in which the agent is currently executing.

The `doOperation` method is synchronous by default, that is, execution in the applet stops until the answer arrives from the agent. AgentSpace provides other communication protocols that will not be further elaborated in this paper.

3.2   Communication from Agents to Applets

Although communication from applets to agents solves most of the interaction needs between applets and agents, there are applications in which the agent should take the initiative. This applies between agents and applets, but also between agents and objects in general such as other agents or frames.

This kind of agent-initiated interaction can be useful when agents are executing time-consuming tasks that cannot be completed in real time. In this case, synchronous communication is not desirable, and instead the applet should send the data to the agent and continue doing something else. Later on (maybe days or weeks after the request) the agent has an answer and needs to send this answer to the applet without having the applet necessarily waiting for a reply.

AgentSpace provides a number of communication models between agents and applets. For example, an agent may send a message to a specific applet or instead send this message to all applets currently being executed against an AgentSpace server. Examples of this kind of communication include: when something interesting has happened such as sport news, when something will happen such as a machine crash in a nuclear power station, or just periodical control and management information.

In order to facilitate the interaction between agents and applets, AgentSpace offers a distributed event notification mechanism compatible with the AWT event model, and, as a consequence, to JavaBeans [Sun99].

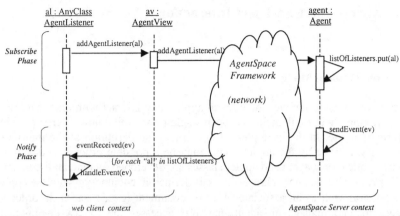

Fig. 4. UML sequence diagram of the event notification mechanism in AgentSpace

The event mechanism is illustrated in Fig. 4. The object `al` belonging to class `AnyClass` represents an applet/frame.

In order to receive events from agents, this object should implement the interface `AgentListener` that includes the method `eventReceived`.

```
public interface AgentListener extends EventListener {
public void eventReceived(AgentEvent e);            }
```

The abstract class `AgentEvent` represents an event and has to be implemented for each application depending on the events used on that application. Each agent class may receive one or many objects of class `AgentEvent`. The same happens the other way around, an `AgentEvent` object can be used by any number of agents.

```
public abstract class AgentEvent extends EventObject
             {
   public AgentEvent(String agentId) { super(agentId);
}
   public String getAgentId() { return (String)
super.getSource(); }        }
```

In addition, the Java interface `AgentView` provides the following methods:

```
public void addAgentListener(AgentListener listener)
public void removeAgentListener(AgentListener listener)
```

These methods help the programmer to register any objects implementing the interface `AgentListener` to receive events generated by agents with an agent view. For example, when `addAgentListener` is called, the object passed as an argument is added to the list of listeners and will be notified each time an event is sent to this agent. On the other hand, when the agent calls `sendEvent`, the event generated is sent to all listeners previously registered – that is, for each object registered as a listener, a method `eventReceiver` will be called. Section 5 will give a concrete application of this mechanism.

4   Agents and End-Users Interaction

4.1 Conventional Model

In the previous section we saw how an applet may interact with one or more agents, by invoking their final methods when asking for well known services, obtaining information, etc. In the examples we presented, the definition of the interfaces with the user was done by the applets. This may be a possible process to develop agent-based applications: agent and applet classes are developed incrementally, in parallel.

For example, the agent class AgentA is defined in conjunction with the applet class AppletA as well as an associated HTML document (e.g., HtmlA). In order to have access to, and manage an agent, the user accesses the HtmlA document, which references AppletA. Finally, AppletA obtains a reference to an instance of AgentA, which is used to interact with the agent. (see Figure 5).

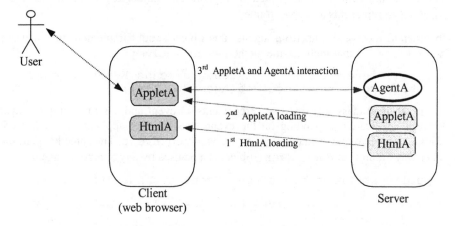

Fig. 5. Conventional model of the user-agent interaction

4.2   Integrated Model

As an alternative to this mechanism, AgentSpace offers a more integrated and easy to use model, based on the agents' ability to provide their own interfaces. In this "integrated interaction model" between agents and users, there is no need for specific applets for each agent class (which makes the management of applications more difficult). It is only necessary to develop "relatively generic" applets, such as the Client-AS shown in Figure 2, supporting management and presentation of the interfaces defined specifically by each agent.

The graphical interfaces of an agent are defined in two callbacks, doInitializationInterface and doManagementInterface, of its class. These methods are obtained from the AgentView's getInitializationInterface and `getManagementInterface` methods, respectively (see focus on the Table 1). Both methods return an instance of `Frame`, which is a component of type container of components defined in the `java.awt` package.

The main difference between the initialization interface (`InitializationInterface`) and the management interface (`ManagementInterface`) is that the former is obtained automatically after the creation of an agent, and the latter has to be requested explicitly from some other object. Thus, the initialization interface is obtained only once, by the agent owner. On the other hand, the management interface may be obtained many times by different users.

Table 1. Events and methods associated to the agent execution

Event	Final Method	Respective Callback
Life cycle:		
- Create	*createAgent*[3]	*onCreation*
- Activate	*start*[1]	*run*
- Clone	*clone*	*beforeCloning; run*
- Memory Management	*flush*	*beforeFlush; afterLoaded*
- Persistence	*save*	*-*
- Remove	*die*	*beforeDie*
End-user interaction:	*getInitializationInterface*[1]	*doInitializationInterface*
	getManagementInterface[1]	*doManagementInterface*
Navigation:	*moveTo*	*run; ou atPlaceX*[2]
	backHome	*afterBackHome*
Communication:		
- Asynchronous (by default)	*sendMessage*[1]	*HandleMessage*
- Synchronous (by default)	*doOperation*[1]	*HandleOperation*

(1) Methods only invoked through the `AgentView` interface.
(2) Callback not predefined – defined by the programmer in concrete classes.
(3) Methods only invoked through the `PlaceView` interface.

The code below illustrates the definition of graphical interfaces (i.e., instances of `Frame`) for agents, for the specific case of the management interface.

```
public class AgentWithFrame extends Agent
{   ...
  public Frame doManagementInterface(InternalUser user)
  {
    Frame frame;
    if(user.equals(getOwner())
      frame= new MyPrivateFrame();
    else
      frame= new AuthenticationFrame();
    return frame;
  }
... }
...
public class AuthenticationFrame extends Frame {...}
public class MyPrivateFrame extends Frame {...}
```

The doManagementInterface method creates and returns one out of two possible instances of Frame, depending on the user that invoked the method. If it is the agent's owner, then an instance of MyPrivateFrame is returned. Otherwise the method returns an instance of AuthenticationFrame, which is defined internally in the agent's class. This interface management policy could be extended in order to integrate other criteria: for example, depending on the agent's state, user groups defined by the agent itself, etc.

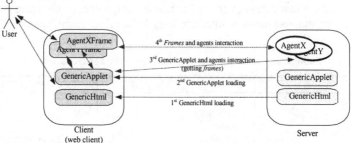

Fig. 6. Integrated model of the user-agent interaction

Figure 6 illustrates the integrated model for interaction between users and agents. The main advantages of this model over the conventional model illustrated in Figure 5 are: (1) simplicity of management and maintenance (e.g., there is no need to keep multiple applet files associated to HTML documents); and (2) possibility of defining different types of interfaces to the same agent. On the other hand, this model may reduce the flexibility, reuse and innovation that can occur independently from both ends at any time. This trade-off should be conveniently analyzed in every particular situation and application.

It is worth noting that the issues related to the interaction between agents, or between agents and other objects (in particular, Applet objects) are similar. The mechanism of these interactions is composed of the following main steps: (1) publishing, sharing and knowledge of agent identifiers; (2) acquisition of references to agents; and (3) invocation of agents' final methods, some of which have a higher level of variability [JGJ97], such as sendMessage, doOperation getManagementInterface, or moveTo.

4.3   Relationship with the MVC Architecture

The model-view-controller (MVC) architecture was originally used to build flexible user interfaces in Smalltalk-80 [KP88]. It defines a trend on modern object-oriented frameworks. For example, this architecture can be found on the Java Foundation Class (JFC) included with the Java 1.2.

The MVC architecture involves the interoperation of three different classes as showed in Figure 7.

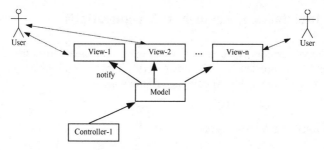

Fig. 7. The MVC architecture.

The model consists of the application data as well as application business rules classes. The view consists of the application presentation. The controller classes define how user interface is handled in the application. Views and models usually communicate through some kind of subscribe/notify protocol.

The MVC architecture decouples these different components promoting easier reuse of code and better flexibility. Figure 7 shows that views have subscribed to the model data source. A source of input or some other event causes the first controller to send a message that changes the state of the model. Consequently, it publishes a change message to each of the views that have previously subscribed with the model.

Figure 8 shows the application of the MVC architecture to the AgentSpace's user interface mechanisms.

The model is the agent object, which keeps the data and logic of the application. AgentListner instances are the views that should subscribe previously a specific set of events and are notified when they occur. Controllers are objects of any type that can send messages to the agent. Controllers and AgentListners can be user interfaces (such as applets or frames) or any other type of objects (e.g., other agents). Nevertheless, any object to communicate with an agent need to have an AgentView reference.

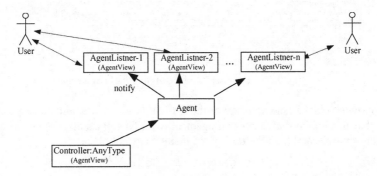

Fig. 8. The AgentSpace's user interface mechanisms based on the MVC architecture

It is important to refer that the adoption of the MVC architecture is somehow orthogonal to the conventional or integrated models discussed above because this architecture is present in both situations.

5   User Interfaces of Agent-based Applications

In this section we explain some issues discussed in this paper, from the user interface point of view. We present a real, albeit very simple, example of an agent-based application, built on top of the AgentSpace framework.

5.1   The PingPong Agent Class

The "PingPong" application demonstrates the classical example of an agent moving between two different places several times, as specified by the user. After arriving at a place, the agent creates and sends the PingPongMoveEvent event, which notifies all the listener objects. These objects are monitoring frames, associated to the agent (see Figure 9).

This example also illustrates the integrated fashion of using frames with agents. (The complete code of this example application can be downloaded from the AgentSpace web site [AS99]).

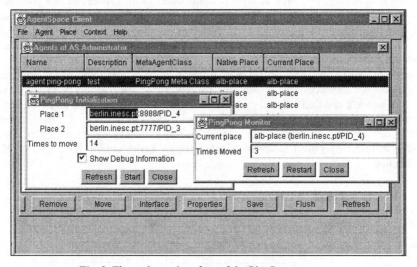

Fig. 9. The end-user interface of the PingPong agent

The following class is an extension of the AgentEvent class, and corresponds to an event that is triggered whenever an agent moves. This class also keeps track of the number of moves and the identifier of the agent's current place.

```
public class PingPongMoveEvent extends AgentEvent {
   private String place;
   private int times;

   public PingPongMoveEvent(String AgentId, String
place, int times) {
      super(AgentId); this.place = place; this.times =
times;
```

```
      }
   public String getCurrentPlace(){ return this.place; }
   public int getTimes(){ return this.times; }
}
```

The PingPong agent class, shown below, uses the sendEvent method to send a PingPongMoveEvent event. This is done at each place that the agent visits.

```
public class PingPong extends Agent {
   private int ntimes = 0;
   private void startMoving() {
      try {
            sendEvent(new PingPongMoveEvent(getId(),
"Home", ntimes));
         Thread.sleep(1000);
         moveTo(new PlaceId(getProperty("place1")), new
Ticket(this),
                 "atPlace1");
      }
      catch (Exception e) {...}
   }
   public void atPlace1() {
      try {
            sendEvent(new PingPongMoveEvent(getId(),
getCurrentPlace(), ntimes));
         Thread.sleep(1000);
         moveTo(new PlaceId(getProperty("place2")), new
Ticket(this),
                 "atPlace2");
      }
      catch (Exception e) {...]
   }
   public void atPlace2() {...}

   public Frame doInitializationInterface(InternalUser
user) {
      try {
         AgentView av =
getCurrentContext().getAgentOf(this.getId());
         return new PingPongInitializationFrame(av);
      }
      catch (Exception e) {...]
   }
   public Frame doManagementInterface(InternalUser user)
{
      try {
         AgentView av =
getCurrentContext().getAgentOf(this.getId());
         return new PingPongManagementFrame(av);
      }
      catch (Exception e) {...]
   }
   ...    }
```

The `PingPongManagementFrame` class, which is a `Frame`, illustrates the event subscription mechanism. In this example, the frame object associated to the agent registers, at startup, its willingness to receive events from the agent (referenced by av). Upon termination, it removes this subscription (method `finalizeListener`).

```
public class PingPongManagementFrame extends Frame
                    implements ActionListener,
Serializable, AgentListener {

   public PingPongManagementFrame(AgentView av) {
      this.av = av;
      try { ... } catch(Exception e) {...}
      initializeListener();
      setTitle("PingPong Monitor");
      pack();
   }
public void initializeListener() {
      try { av.addAgentListener(this); } catch(Exception
e) {...}
      refresh();
   }
private void finalizeListener() {
      try    { av.removeAgentListener(this); }
catch(Exception e) {...}
   }
   public void eventReceived(AgentEvent e) {
      if(e instanceof PingPongMoveEvent)
         updateFields((PingPongMoveEvent) e);
      if(e instanceof PingPongActivateEvent)
         rsb.setEnabled(((PingPongActivateEvent)
e).isActive());
   }
   ...
}
```

The events received by the frame object are processed by the `eventReceived` method, which is transparently invoked by the AgentSpace system.

5.2   Other Examples

Other agent-based applications and specific agent classes can be found in the AgentSpace web site [AS99] such as the Book Market Simulator or the Virtual Stock Market examples. We encourage the reader to download the main components of the AgentSpace system as well as the code source of several complete examples.

6   Conclusions and Future Work

In this paper we presented and discussed several issues regarding user interface in the context of Java agent-based applications, and in particular how to solve several problems using the AgentSpace framework.

First, we discussed the *conventional model* of user-agent interaction, which is mainly based on HTML and Java applets, and concluded that in several situations it is not suitable since it suffers from management, maintenance, and eventually scalability problems. In order to cope with these limitations, we proposed the *integrated model*, in which the end-user interface belonging to each agent is developed and kept together with the rest of the agent's code.

It should be emphasized that AgentSpace supports both models. Furthermore, concrete agent-based applications have been developed on top of AgentSpace and are available for download from our web site [AS99]. The applications we built suggest that the agent paradigm is suitable to design and develop many applications, namely in dynamic and distributed environments.

We are currently preparing a paper that will describe the most important applications we have built or are still building, as agent applications are still unusual. Among these, perhaps the most interesting are the following:

In the COSMOS ESPRIT Project [COS97] we are using AgentSpace to carry business contracts that are being negotiated between a number of parties. The agent paradigm is especially suited to this kind of application because it can carry not only the data (the contract) but also code (i.e. the contract editor, signing facility, etc.). Furthermore, agents are autonomous entities and may operate while the user is disconnected from the COSMOS server. We are also investigating how to send agents by e-mail.

A final year project is working on a project called Personalized News that aims at building a system based on AgentSpace to collect news from national newspapers on the Web and deliver news personalized for each user. Each newspaper will have an agent that is responsible for collecting the news every day at a certain hour. Each person will also have an agent to manipulate and store his or her profile, as well as sending the news to the user. The agent paradigm adapts quite well because each agent represents a real-world entity, so it becomes easier to design and build the application, not to mention to evolve and change.

A third project is investigating how to build an Electronic Commerce Mall using AgentSpace. In this application, electronic shops are represented by agents, as well as shoppers and the mall itself. The main idea is to create each shop autonomously from all other shops such that, at least in theory, anyone can write a new shop and install that new shop in the mall without rebooting the system. Representing shoppers as agents will also make it easier to manage the interaction between shoppers and shops.

Apart from developing applications to get feedback from the AgentSpace framework, there are many technical research topics that are still open and deserve to be addressed individually. For example, if it is possible to encapsulate an agent as a Java bean. If so, an agent could be used easily and more efficiently in the context of visual tools. Other also topics arise here, such as those regarding security (protecting AgentSpace servers from agents, and agents from malicious AgentSpace servers) and writing general-purpose agents.

References

[AdA99] Ad Astra Engineering Inc. Jumping Beans White Paper, October 1999.
http://www.JumpingBeans.com/
[AS99]   AgentSpace Web Site. 1997-1999.
http://berlin.inesc.pt/agentspace/
[BTV96] J. Baumann, C. Tschudin, J. Vitek (editors). *Proceedings of the 2nd ECOOP Workshop on Mobile Object Systems*. Dpunkt, 1996.
[COS97] Ponton, COGEFO/CEFRIEL, Hamburg University, INESC, Interzone Music Publishing, Oracle UK, and SIA. *COSMOS – Common Open Service Market for SMEs, ESPRIT Research Project Proposal*, 1997.
[FTP96] FTP Software. *CyberAgents*. 1996.
http://www.ftp.com/cyberagents
[GM97] General Magic, Inc. *Odyssey Product Information*. 1997.
http://www.genmagic.com/agents/odyssey.html
[GK94]   M. Genesereth, S. Ketchpel. Software Agents. In [*Rie94*].
[IBM97] IBM Research. *The Aglets-based e-Marketplace: Concept, Architecture, and Applications*. Research Report RT-0253, Tokyo Research Laboratory, Japan, 1997.
http://www.ibm.co.jp/trl/aglets
[IKV98] IKV++ GmbH. *Grasshopper, An Intelligent Mobile Agent Platform written in 100% Pure Java,* 1998.
[JGJ97] Ivar Jacobson, Martin Griss, Patrik Jonsson. *Software Reuse – Architecture, Process and Organization for Business Success*. Addison Wesley, 1997.
[JSW98] N. Jennings, K. Sycara, M. Wooldridge. A Roadmap of Agent Research and Development. *Journal of Autonomous Agents and Multi-Agent Systems*, 1(1), Kluwer Academic Press, 1998.
[KP88]   G. Grasner, S. Pope. A cookbook for using the model-view-controller user interface paradigm in Smalltalk-80. *Journal of Object-Oriented Programming*, 1 (3), 1988.
[LO98]   D. Lange, M. Oshima. *Programming and Deploying Java Mobile Agents with Aglets*. Addison-Wesley. 1998.
[MLC98] D. Milojicic, W. LaForge, D. Chauhan. Mobile objects and agents (MAO). In *Proceedings of the USENIX Connference on Object-Oriented Technologies and Systems (COOTS)*, April 1998.
[NC98]   P. Nixon, V. Cahill (editors). Special Issue on Mobile Computing. *IEEE Internet Computing*, 2(1), 1998.
[Obj99] ObjectSpace Inc. The ObjectSpace Voyager Universal ORB. 1999.
http://www.objectspace.com/
[PL97]   A. Park, S. Leuker. A Multi-Agent Architecture Supporting Services Accesses. In [*RP97*].
[Rie94]   D. Riecken (editor). Special Issue: Intelligent Agents. *Communications of the ACM*, 37(7), July 1994.
[RP97]   K. Rothermel, R. Popescu-Zeletin (editors). *Lecture Notes in Computer Science 1219 (Mobile Agents '97)* Springer, 1997.
[RH98]   K. Rothermel, F. Hohl (editors). *Lecture Notes in Computer Science 1477 (Mobile Agents '98)* Springer, 1998.
[SBH96] M. Strasser, J. Baumann and F. Hohl. Mole: A Java-Based Mobile Object System. In [*BTV96*].
[Sil99]   A. Rodrigues da Silva. *Software Agents on the Internet* (in Portuguese). Edições Centro Atlântico. March 1999.
[SD98]   A. Rodrigues da Silva, J. Delgado. AgentSpace versus Aglets: Infraestruturas de Agentes para as Futuras Aplicações da Internet. (SBC – SEMISH'98, Brasil, Belo Horizonte), Anais do 18º Congresso da Sociedade Brasileira de Computação – Rumo à Sociedade do Conhecimento, August 1998.

[SMD98] A. Rodrigues da Silva, M. Mira da Silva, J. Delgado. AgentSpace: An Implementation of a Next-Generation Mobile Agent System. In [*RH98*]

[Sun98] Sun Microsystems, Inc., The Java Development Kit (JDK), 1998. http://www.javasoft.com/products/jdk/

[Sun99] Sun Microsystems, Inc., The Java Beans, 1999. http://www.javasoft.com/beans/index.html

[Whi94] J. White. General Magic, Inc. *Telescript Technology: The Foundation for the Electronic Marketplace*. General Magic. 1994.

[Whi97] J. White. Telescript Technology: An Introduction to the Language. *White Paper*. General Magic, Inc. Appeared in J. Bradshaw, Software Agents, AAAI/MIT Press. 1997.

XML and CORBA, Synergistic or Competitive?

Steven Vermeulen, Bart Bauwens, Frans Westerhuis, Rudi Broos

Alcatel,
Francis Wellesplein 1,
2018 Antwerpen, Belgium
Steven.Vermeulen@alcatel.be
Bart.Bauwens@alcatel.be
Frans.Westerhuis@alcatel.be
Rudi.Broos@alcatel.be

Abstract. The eXtensible Markup Language (XML) is gaining a lot of attention in the Internet world and is adopted by major companies (IBM, Microsoft, Oracle). XML, the open-standards child of SGML, promises to provide platform- and language neutral data encapsulation and separates application logic from application data. Meanwhile, various object-oriented technologies and standards such as Java and CORBA have also progressed rapidly in the past few years. Java is being presented as the perfect partner for XML. Java supports the development of Web-aware, platform-neutral applications, and XML is a platform-neutral document description meta-language. But doesn't CORBA promise exactly the same? This paper describes the XML and CORBA 'approaches', and the synergies and/or competition between these technologies, taking into account the "philosophy" of each approach. Different criteria are identified where comparison is possible and relevant, such as: specification (e.g. expressive power), deployment (parsing, marshalling, scalability), and tools. XML is the next step in Web-protocols (after IP, HTML). It is concluded that XML can in conjunction with a multitude of other protocols provide the same functionality as CORBA, but will only replace CORBA in those cases where using CORBA is undesirable.

1   Introduction

The Internet world is being transformed before our eyes as open standards such as XML. The XML technologies are being seen as harbinger of various new functionality's in numerous domains ranging from electronic commerce to electronic publishing to healthcare delivery to manufacturing to telecommunications. Various object-oriented technologies and standards such as Java, CORBA and DCOM have also progressed rapidly in the past few years. With the surging interest in XML, there is abundant speculation as to what its role will be in the emerging Web and computing landscape. Over the past year, a number of views emerge on the critical role that XML will play in transforming the Internet and data exchange standards. For instance, being a platform-neutral document markup language, one potential role could be to use it to describe data that must be shared and exchanged between distributed applications. XML in cooperation with the appropriate application and transport protocols can provide an underlying glue to application-to-application data exchange, in essence become a Web-native environment for distributed applications. In this role XML, could be used for cases where using CORBA is not possible, eventually it could become a potential competitor of CORBA. Furthermore, XML presents a possible medium for interchange of data between CORBA based systems and other systems. This paper is looking at the intersection of the XML and CORBA, it tries to describe the synergies and/or competition between these technologies in a number of areas (section 4). Before comparing both approaches, let's have a closer look at XML (section 2) and CORBA (section 3) and their different uses.

2   Extensible Markup Language (XML)

XML, eXtensible Markup Language, became a W3C Recommendation [7] in the beginning of 1998, and was announced as a much "better" solution than HTML, enabling advanced Web applications. Both XML and HTML use "tags" to delimit information. XML and HTML have also similar roots: XML can be seen as a successor of SGML, (ISO8879), the Standard Generalized Markup Language, and HTML was also derived from SGML. Nevertheless, XML and HTML are fundamentally different.

First of all, XML is a meta-language, which means that you can create new "mini-languages" with it, which are called DTDs (Document Type Definitions). A DTD is a means to define different structures for specific types of information, for example, a DTD for books, invoices, or even cars. The DTD defines a number of elements (e.g. <title> and <author> for the book DTD, <price> and <address> for the invoice DTD, or <color>, <model> for the car DTD) and also specifies which elements can or should contain other elements, the number and sequence of those elements, the attributes you can associate with them, etc.

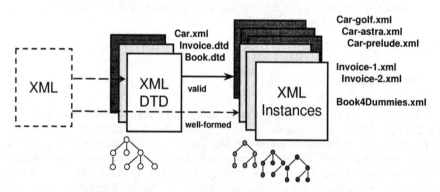

Fig. 1. Relation between XML, XML DTDs and XML instances

As a second step, one will typically generate XML "Instances" (i.e. the actual "documents", or "data"), which satisfy the requirements of the DTD. The DTD acts here as some kind of template, which guides the completion of the data. The

relationships between XML, DTDs and Instances are further illustrated in fig 1. This figure also distinguishes between two important types of XML instances:
1. *Well-formed instances* which follow all rules XML prescribes for "tagging" your data as a tree structure, but do not necessarily follow the rules of an XML DTD.
2. *Valid instances* which are well-formed and also conforming to a DTD.

Another important difference with HTML, is that XML strictly separates data from presentation. In HTML, you can place text in bold by surrounding it with tags (<B>), while in XML you will put this type of formatting info in a separate style-sheet. This separation allows you to generate a variety of formats originating from the same source of data, for example HTML, RTF, or a database format.

Tagging your data with XML, offers you a number of additional advantages. Being a purely textual format, it is independent from any platform, operating system or transport mechanism. It is often called "self-documenting", as it is easily readable by humans. XML is clearly compatible with today's Web technologies (HTML alike tags, HTTP transport, …). Because of its extensibility and flexibility, it was rapidly supported by major IT companies as Microsoft, IBM, Sun and Oracle.

In the following sections, we will categorize the possible scenarios and different uses of XML, which is needed before we can compare the XML and CORBA approaches.

2.1   XML as data exchange language

Since XML's inception, many different vocabularies have emerged on the Internet in very different industry domains, such as e-commerce, mathematical sciences, meta-data representation and so on. Different companies are now sitting together to develop open common XML-based formats, which can be used to exchange data across organizations. In this view, XML becomes the de-facto language for representing almost all data: XML is here the logical data structure of all Web applications, while HTTP servers offer a valid alternative to physically store all resources in native XML format.

2.2   Towards XML interfacing

Besides using XML to exchange data between disparate clients and servers, some companies are starting to use it also for defining application-layer protocols, establishing interfaces between different peers on the Internet. Different scenarios are possible when using XML as language to define interfaces between distributed applications, ranging from simple loose approaches, to more elaborated and strict models.
• In its loosest form, one can define a set of XML tags, and agree on the meaning of each separate tag, without writing a DTD. Then, it is up to the developers to write a piece of code which interprets each tag. Well-formed XML messages can be exchanged (e.g. over HTTP) and processed by an interpreting application. Fig. 2 gives an example in the domain of e-commerce which illustrates this approach.

```
<?xml version='1.0'?>
  <Order xmlns='http://ecommerce.org/schema/'>

            <Item Price="540.000"
  xmlns:C='http://www.cars.org/schema/'>

              <C:car id="car-astra-508087" year="1998">

               <C:trademark>Opel</C:trademark>

               <C:model color="oceanblue">Astra</C:model>

               <C:picture scr="astra-blue.jpg"/>

               <C:optiongr type="safety">

                 <C:option>ABS</C:option>

               </C:optiongr>

              </C:car>

            </Item>

  </Order>
```

Fig. 2. XML interfaces in its loosest form: agree on a set of XML tags

- A second approach is to use valid XML data, i.e. which conforms to an XML DTD. All semantics defined on a high level, implied in documentation (tailored DTD for one application; including an application protocol elements). When you want to reuse an existing DTD, but have a strong idea about how the internal representation of your objects should look like, you need to establish a mapping (by building translators) between your XML data structure and your object models.
- One can avoid this case-by-case translation mapping, by using the DTD itself as the object model to navigate through your data and make changes. One can do this via the DOM, the Document Object Model [17], a W3C specification that provides standard functions for navigating and manipulating the elements in the XML tree structure. This is an interesting approach, certainly when a 'standard' DTD is already available, which can be used as-it-is.
- Instead of using the DTD itself as your object model, one can also agree on a set of rules to generate from an existing object model (based on UML for example) new XML DTDs. In this way one creates a standard way to serialize your objects in an XML encoding. This approach is useful when no suitable DTDs exist which can be reused and/or you need to start from an existing object model. This is the approach taken by the OMG's specification XMI (the XML Metadata Interchange). It results in a separate class of XMI compliant DTDs.
- A more "extreme" solution is to define also tags for a complete distributed object model, including method invocations, object references etc. Efforts such as XML-RPC [8] and WDDX [9] are good examples of this approach. See Fig. 3.

```
POST /RPC2 HTTP/1.0

User-Agent: Frontier/5.1.2 (WinNT)

Host: betty.userland.com

Content-Type: text/xml

Content-length: 181

<?xml version="1.0"?>

<methodCall>

        <methodName>examples.getStateName</methodName>

        <params>

          <param>

            <value><i4>41</i4></value>

          </param>

        </params>

</methodCall>
```

Fig. 3. XML-RPC example (including HTTP header)

2.3   Typical building blocks of an XML-based interface architecture

In what follows, we will describe the different software modules which can be combined to build an architecture where XML messages are exchanged between different peer applications (see Fig.4). We will call here the sending application the "client" and the receiving application the "server", *not* to be confused with Web client (browser), resp. HTTP server.

1. As a first step, the client application will need to create XML messages via an *XML Generator Module*. This module can create the tagged instances "manually", or use the DOM interfaces. The creation process can be "guided" by a DTD (but this is not required), or also by an XSL style-sheet used in an XML editor.
2. Secondly, the message needs to be sent via some transport protocol. A logical choice will be a TCP-IP based protocol such as HTTP or SMTP. However, one can in principle select any transport protocol (including WAP, or even IIOP).
3. When the server application receives the XML message, it will first parse the message. The *XML Parser Module* can choose to use the DTD or only check on well-formedness. Another important design choice for parsing is the option between event-based or a tree-based parsing. An *event-based* parser is typically based on SAX, the Simple API for XML [18]. The SAX API notifies you when certain events such as start- or end-tags occur during parsing of your document. The process is linear which means that if you do not act upon an event, this data is

completely discarded and you can not go back to it, unless you start again parsing from the beginning. The advantage of the SAX approach is that it is fast and saves a lot of memory if your messages are long and you only need to pass a few elements of those messages. A *tree-based parser* is typically based on the *DOM*. The advantage of the DOM approach is that it offers a very rich interface, but the construction of the tree may require a lot of memory.

4. As explained in section 2.2, one can also decide to transform the parsed XML data first into an internal object representation, which is better suited to your server application. This is done in an *XML Translation Module*.

5. All actions taken on the output of the XML parser (or possibly the output of the Transformation Module), constitute the actual "application logic", the *XML Interpreting Module* of the Server Application. As XML separates content from presentation, one will typically find a separate *Visualisation Module*. This module can use an XSL style-sheet to automatically display the message in a browser such as Internet Explorer 5, or can transform it into another viewing format. Also the DOM can be used to send a tailored version of the message to your browser.

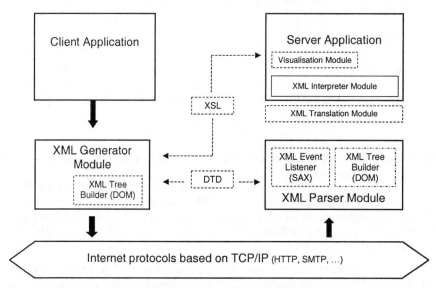

Fig. 4. A common architecture for XML-based interfacing

3   CORBA

Fundamentally, CORBA is an application integration technology. Since its inception in 1991, CORBA has provided abstractions for distributed object-oriented programming that have allowed developers to seamlessly integrate diverse applications into heterogeneous distributed systems [2]. The core of any CORBA based system, the Object Request Broker (ORB), hides the low-level details of platform specific networking interfaces, allowing developers to focus on solving the problems specific to their application domain than having to build their own distributed computing infrastructure [3].

3.1 Architecture

CORBA is the middle-ware that establishes the client/server relationship between objects. Using an ORB, a client object can transparently invoke a method on a server object, that can be on the same machine or across a network. The ORB intercepts the call and is responsible for finding an object that can implement the request, pass it the parameters, invoke its method, and return the results. The client does not have to be aware of where the object is located, its programming language, its operating system, or any other system aspects that are not part of the object's interface. It is very important to note that the client/server roles are only used to coordinate the interactions between two objects. Objects on the ORB can act either as a client or server, depending on the occasion [1].

In the following we will present a brief panoramic overview of the ORB components and their interfaces. For a complete description of the CORBA standard the reader is referred to the Object Management Group.

Fig. 5. shows the client and server sides of a CORBA ORB. The following elements are provided by CORBA at the client side:

- The *Client IDL stubs* provide the static interfaces to object services. These precompiled stubs define how clients invoke corresponding services on the servers. From the client's perspective, the stub acts like a local call - it is a local proxy for a remote server object. The services are defined using the Interface Definition Language (IDL), and both client and server stubs are generated by the IDL compiler. A client must have an IDL stub for each interface it uses on the server. The stub includes code to perform marshalling. This means that it encodes and decodes the operation and its parameters into flattened message formats that it can send to the server. It also includes header files that enable you to invoke the method on the server from a higher-level language like C++ or JAVA without worrying about the underlying protocols or issues such as data marshalling. The program invokes a language method to obtain a remote service.

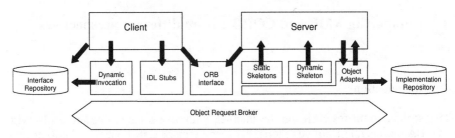

Fig. 5. The Structure of CORBA

- The *Dynamic Invocation Interface* (DII) are used to discover methods to be invoked at run time.

- The *Interface Repository* APIs are used to obtain and modify the descriptions of all the registered component interfaces, the methods they support and the parameters they require.
- The *ORB interface* consists of a few APIs to local services that may be of interest to an Application. For example, CORBA provides APIs to convert an object reference to a string, and visa versa.

Static invocations are easier to program, faster and self documenting. Dynamic invocations provide maximum flexibility, but they are difficult to program; they are very useful for tools that discover services at runtime.

- The *Server IDL Skeletons* provide static interfaces to each service exported by the server. The skeletons, like the ones on the client, are created using an IDL compiler.
- *Dynamic Skeleton Interface* (DSI)
- *Object Adapter* provides the runtime environment for instantiating server objects, passing request to them and assigning them object Ids.
- *Implementation Repository* provides a run-time repository of information about the classes the server supports, the objects that are instantiated, and their IDs

3.2   CORBA Services

OMG defines a set of standard services in the CORBA services specification. These services provide a standardized solution to some common problems. The most well know and used services are:

- *CORBA Naming Service.* The naming service allows associating abstract names with objects and allows clients to find those objects by looking up the names. This one sentence explains both the simplicity and the power of the naming service. It just provides a standardized way to name and find an object.
- *CORBA event services.* The event service provides a way to de-couple the communication between objects. It's based on a supplier-consumer model and allows suppliers to send data asynchronously to some consumers through what is called an event channel.

4   Comparing XML and CORBA from different perspectives

4.1   Introduction

This section compares XML on the Internet, with all its supporting protocols, with CORBA on several identified criteria. A general remark is that one can almost always find a way to do everything with both approaches but this section looks at what is typically in line with the "philosophy" of each approach.

To limit the scope of the comparison, two widely deployed architectures were chosen, an architecture with applications that make use of XML to express data and HTTP to communicate with each other, and on the other hand a CORBA based architecture.

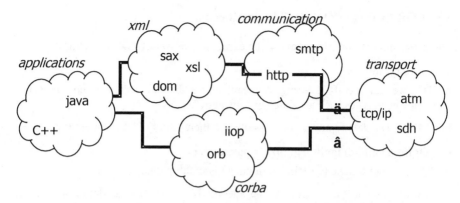

Fig. 6. CORBA based and XML/HTTP based architecture

Even with this limited scope of two architectures, a general comparison is very difficult and depends very much on the type of application. Different criteria were identified where comparison is possible and relevant.

4.2 CORBA versus XML/HTTP: General

- **Standardization process**

Although OMG and W3C are obviously different standard bodies with different standardization processes, both will standardize on generic technologies, while application specific interfaces are typically specified outside those bodies. Typically, a number of companies or organizations will sit together to define a common XML DTD or a common IDL in a specific domain.

- **Political arguments**

Some companies are pushing XML on the Web into the arena of distributed computing out of a political desire to support open standards without threatening own developments. For instance Microsoft is embracing XML because the alternative CORBA is a threat to their DCOM.

- **Open interfaces**

XML interfaces are often created with an open-source attitude and published on the Web. This lets the XML DTD evolve to a de-facto standard, used outside the consortium which created it, providing overall reusability. In the case of IDL, the interfaces are often kept private within the consortium.

- **Protocol fan**

OMG aims to provide, as far as possible, an integrated solution. While in the IETF a fan of protocols are developed. The "IETF philosophy" is to use protocol on protocol on protocol. IETF offers a fan of protocols, so it is up to the developer to choose the appropriate protocol based on his requirements. The advantage of this approach is that each protocol can be developed with few and small constraints. The disadvantage is that stacking protocols creates overhead.

4.3   CORBA versus XML/HTTP: Specification

There are some major differences in the expressive power of IDL vs. XML.

- **Data typing**

In IDL, we can specify a String, boolean, int, … However there is no support for more enhanced types such as "date" or "currency".

In XML DTDs, data typing is not possible yet, there is only PCDATA (strings).

- **Optional and default arguments/parameters**

In IDL it is not possible to define optional and default parameters.

However, optional parameters can be specified in a DTD: e.g. define an element substract as (x, y?, z), where y is optional. So it is now possible to specify x-y-z or x-z.

Default attribute values can be specified in XML (e.g <!attlist car color NMTOKEN (green)>). So in case you don't specify the color of the car, in your message arguments, it is assumed to be green by default.

- **Synchronous vs. asynchronous interaction.**

CORBA operations are synchronous. This has some important implications. An operation can return a result, both as a function result and as parameter. Its also possible to do exception handling. To achieve this in an asynchronous communication, return-messages have to be specified and the application has to keep some sort of state information about its requests. CORBA also allows asynchronous operations (one-ways).

XML/HTTP only allows asynchronous operations.

- **Naming Service**

CORBA has a standardized way to find an object (or server) by using the naming service.
When using XML/HTTP you are limited to using URI's. In essence both an URI and a CORBA object-reference contain the same information namely host and port, but using the naming service allows to make an abstraction of the actual location of the server. However there are protocols which allow to achieve location transparency on the internet using URI's.

4.4   CORBA versus XML/HTTP: Deployment

- **Heavyweight vs. lightweight?**

CORBA is considered to be more heavyweight, while the XML/HTTP approach is more lightweight. This has to be put into perspective, as XML/HTTP is lacking some of the features provided by CORBA When needed these features have to be added in the application (by using other protocols), thus making it more "heavy". On the other hand, there are also lightweight ORB's available. These are ORB's with reduced functionality and features so that they can be deployed on systems where a small footprint is required.

- **Marshalling**

CORBA uses marshalling to go to and from the flat binary encoding to language specific constructs, thus ensuring a correct passing of all parameters.

With XML you can check a XML message against its DTD, ensuring compliance with this DTD, but it's not required. This is called parsing. If marshalling is required, it should be implemented at the level of the application. See Fig. 4.

- **ASCII vs. Binary**

CORBA uses a binary message format, whereas XML uses plain old ASCII. This means that XML is human-readable. On the other hand this kind of encoding is not optimal, costing both processing and bandwidth.

This does not mean that every XML-message is understandable. Much depends on the names used for the tags and how this correlates with the meaning. E.g. using tags like "a", "b", "c", does not carry much meaning to the casual reader.

A binary format has its disadvantages. When using it to represent a serialized java class for instance both sides of the communication have to use the exact same version of this class for the de-serialization to work.

- **Statefull vs. Stateless**

CORBA uses IIOP, a protocol which keeps the state of the communication, thereby supporting return values and exceptions.
HTTP is a simple Web-protocol without any state. So if state is required in a bi-directional communication this should be added on the level of application. Return values and error conditions can be implemented by defining a return message in the DTD.

- **Performance**

CORBA/IIOP is optimized to make the most use of the transport protocol. It will for instance keep a connection between two objects open for a certain amount of time to minimize set-up delays.

HTTP on the other hand will open and close a connection for each request. This is slower but result in less used sockets on a server with many clients. CORBA can be used without this feature, losing performance but making it more scalable.

- **Firewalls**

IIOP causes problems on most firewalls, which can be solved with tunneling but remains tricky.

HTTP on the other hand doesn't have these problems.

- **Scalability**

CORBA is set to scale well, but not yet to the scale of the Internet. A CORBA-network requires some management, therefore it will only be deployed within a closed administrative domain.

HTTP is the protocol of the WWW. As such it has proven its scalability.

4.5   CORBA versus XML/HTTP: Tools

There are many different ORB's available, commercial and freeware. But the promise of overall interoperability is not completely fulfilled.

Also for XML there are many tools (parsers, browsers, editors, translators, …) available (free and commercial), but with varying quality and conformance. There are no de-facto standards yet for a generic XML interface mechanism.

4.6   CORBA versus XML/HTTP: Summary of features table

Feature	CORBA	XML/HTTP
OS independent	Yes	Yes
Language independent	Yes	Yes
Data types	Most used types like integer floating point, … supported	Only strings supported
Synchronous invocation	Yes	No
Asynchronous invocation	Yes	Yes
Return values	Yes	No
Exception handling	Yes	No
Location transparency	Yes using naming services	Yes, using other Web-protocols
Marshalling	Yes	Only possible to check data against DTD
Human readable	No	Yes
Network performance	optimized	New connection for each message
Scalability	Yes, but no to the internet	To the internet
Firewalls	Problem, use tunnels	No problem yet

4.7   XML on top of an ORB

XML is designed to encode human-oriented documents, as such it is a very convenient way to transfer such data between applications. CORBA provides a way to reliably transfer strings back and forth between applications, these strings can of course also contain XML.

4.8   XML on top of other protocols

XML is just describing the data, it can be transported by any means possible. Some of the possibilities are SMTP, WAP, FTP etc.. A discussion of these possibilities would be out of the scope of this paper.

4.9   Future Evolutions:

- **CORBA standards**

CORBA, like any technology or standard, has had to continually evolve since its original publication in order to remain viable as a basis for distributed applications. As part of its continuing evolution, several significant new features are being added to CORBA as it approaches version 3.0: Internet Integration, Quality of Service Control, and the CORBA component architecture. The new Messaging Specification as part of the Quality of Service Control defines a number of asynchronous and time-independent invocation modes for CORBA, and allows both static and dynamic invocations to use every mode. Asynchronous invocations results may be retrieved by either polling or callback, with the choice made by the form used by the client in the original invocation. Policies allow control of Quality of Service of invocations. Clients and objects may control ordering (by time, priority, or deadline); set priority, deadlines, and time-to-live; set a start time and end time for time-sensitive invocations, and control routing policy and network routing hop count [4].

- **XML Standards**

W3C is now working on the successor of DTDs: XML Schema [10], which will offer a solution for some important shortcomings of DTDs. Schemas will become more reusable and better aligned with OO technology as they will support inheritance mechanisms. Also, data typing will become possible with XML Schema. Finally, the syntax of the schemas will become the same as for the instances, i.e. XML, which allows reuse of all tools on the metadata level. Furthermore, standards for Xlink [11], XPath [12] and XPointer [13] will be soon finalized, so that more standard support will become available in browsers, to inter-link XML data/objects over the Web. Another concern is that highly structured XML data may become very verbose. In this context, standards are in progress, developing more compact binary versions of XML (e.g. WBXML [17]). Finally, the ultimate dream of a semantically enhanced Web, where one can find information in a much more automated and intelligent way, may become closer, when the XML schema concepts will be merged with real metadata models, similar to those already defined in RDF.

5   Conclusions

XML and CORBA are complementary technologies when used in line with the "philosophy" of each approach. XML is being used in many applications to represent various types of data structures, intended for human readable documents, while solutions like CORBA tie together cooperating applications exchanging data that will probably never be directly read by anyone. Because of XML's intensive use via the Web there is also a growing tendency to use it as glue in application-to-application communications. On this field it competes with CORBA, DCOM and Java RMI. XML does have some strong points, in its flexibility and simplicity. It also works on top of standard Web-protocols, thus inheriting their robustness and scalability. Although good in representing data in a human readable format, it cannot represent data in a machine-optimized format. To implement all features CORBA does require coding several protocols into the application. On the other hand CORBA is designed for application to application communication, but has a problem scaling to the internet. It's use will therefore be limited to distributed applications within one administrative domain. Because of its close relation with real programming languages IDL is less suited than XML when transmitting across domain messages with very flexible content requiring interpretation at application level.

There are many application areas where XML provides advantages, especially in areas requiring human interaction and representation of human readable data. As inter-application communication protocol it can, in combination with application and transport protocols, be used in cases where an orb is undesirable, such as on the internet. When the flexibility of XML is needed to transport data, XML can be used in a CORBA based system.

References

1. Client/Server Programming with JAVA and CORBA second edition, Robert Orfali & Dan Harkey, Wiley Computer Publishing, 1998
2. Object Management Group, The Common Object Request Broker Architecture and Specification
3. New Features for CORBA 3.0, Steve Vinoski, Communications of the ACM, October 1998, Volume 41, N0 10.
4. What's coming in CORBA 3.0, http://www.omg.org/news/pr98/compnent.html
5. CORBA and XML: Conflict or Cooperation?, Andrew Watson, Java Developers Journal, volume 4, issue 9, September 1999.
6. The Emerging Distributed Web, Jeremy Allaire, Allaire Corp. http://www1.allaire.com-/developer/technologyreference/columnsarticlesarchive.cfm
7. XML (W3C) http://www.w3c.org/xml
8. XML-RPC http://www.xmlrpc.com/
9. WDDX http://www.wddx.org/
10. XMLSchema http://www.w3.org/TR/xmlschema-1/
11. XLink http://www.w3.org/TR/NOTE-xlink-req/
12. XPath http://www.w3.org/TR/xpath
13. XPointer http://www.w3.org/TR/WD-xptr
14. WAP Binary XML Content Format (WBXML), on-line available at http://www.w3.org-/TR/wbxml/
15. Resource Description Framework (RDF), specifications on-line available at http://www-.w3.org/RDF
16. WBXML http://www.w3.org/TR/wbxml/
17. DOM http://www.w3.org/DOM/
18. SAX http://www.megginson.com/SAX/index.html

Multi-level Component Oriented Methodology for Service Creation

Kurt Verschaeve[1], Bart Wydaeghe[1],
Frans Westerhuis[2], and Jan De Moerloose[2]

[1] SSEL - VUB, Pleinlaan 2,
1050 Brussel, Belgium
{kaversch, bwydaegh}@info.vub.ac.be
[2] Alcatel Bell N.V., Francis Wellesplein 1,
2018 Antwerpen, Belgium
{Frans.Westerhuis, Jan.de_Moerloose}@alcatel.be

Abstract. This paper presents the COSEC methodology, a component oriented methodology for service creation. Important are the different abstraction levels, targeting a wide range of users, going from software engineers to high-end users. On a low level we create reusable developer components that groups consistent fragments of UML, SDL and Java. These developer components are composed and augmented with user-interface aspects, documentation and customization parameters to form a customer component. On a higher level, these customer components are customized and composed within a customer framework, yielding in an executable service. This combination forms an answer to the challenge of organizational and technological problems in the creation of telecom services.

1   Introduction

One of the key issues in the development of services is rapid service provisioning through re-use. Recently, interest in software reuse has shifted from the reuse of single components (procedures, functions, classes), to whole abstract architectures [1]. A software architecture that may be reused for creating complete services is called a framework. A framework is an extensible template for applications within a specific domain [2] and is usually build according to a certain architecture. The idea is that it should be relatively easy to produce a range of specific services within a certain domain starting from the framework software.

This paper proposes a multi-level Component Oriented methodology for Service Creation (COSEC), which could provide for both rapid service provisioning and high service quality. The approach assumes that for a certain class of services, a flexible and reliable software *framework* will be developed. The framework is a "nearly ready" service set of software components that can be used to build a large number of standard and customized services. The architecture of the frameworks developed here is based on TINA principles. A framework is instantiated and specialized by means of graphical service creation tools which allow service designers, business consultants, marketing people and also service end-users to create/customize services at a high level of abstraction.

The approach proposed here has two main objectives. The first is to minimize time-to-market for new services by simplifying the service creation process; and the second is to enable service creation by less–technically qualified and more market-oriented people as well as by the service end-user themselves. The first objective is achieved by doing as much development work as is possible prior to the actual service creation, i.e. shifting the bulk of development and validation effort from the service creation process to the framework creation process. The second objective is achieved by providing service creators with appropriate graphical service creation tools which allow them to design services on a relevant level of abstraction, i.e. without getting into each and every implementation detail, but rather keeping the user's view in mind.

The COSEC methodology illustrated in this paper is being applied and validated in the Advanced Internet Access (AIA) [3] project funded by the Flemish government's Information Technology Action (ITA) [4] program. The COSEC methodology takes into account the results from ACTS projects SCREEN [5] and TOSCA [6] and integrates both approaches, and extends it when necessary. COSEC and SCREEN have in common that they are both methodologies with two main players (customer and developer) and are based on components being developed by combining UML (or OMT) and SDL. These components are then incorporated into a framework. COSEC and TOSCA have in common that they provide a rapid service provisioning based on the specialization of a framework with a nearly ready service set of software components. The framework can be used to build a large number of standard and customized services. More details about the alignment of the TOSCA and SCREEN approaches can be found in [7].

The structure of the paper is as follows. Section 2 gives an overview of the methodology and the different abstraction levels. It also describes the role of the main players in the service creation process. Section 3 gives more details about the low-level service creation environment and defines developer components and customer components. Section 4 shows how customer components are customized and composed into the framework in order to build a service. Section 5 considers the current state of work and tool supports. Section 6 gives some conclusions.

2   Integrated Methodology

The methodology proposed in this paper is an integration of two abstraction levels, each aimed at different players in the service creation process. The high-level service creation environment targets end-users, service consultants and service developers. The low-level service environment targets component builders. Fig. 1 gives an overview of the important phases in the service creation life cycle and the major business roles as we see them today in the telecom world.

We distinguish four business roles and two major abstraction levels. The component builder works on the low level service creation. The end-user, service consultant and service developer works on the high-level service creation. The high-level can be divided again in several abstraction levels, as described below. The Customer Component (CC) ties the low-level and high-level together. Customer components are build on the low-level by composing Developer Components (DC) and adding extra information. On the high-level, customer components are composed and customized to build services. Both kind of components are described in detail in section 3. Here we give a profile of each business role involved in the process.

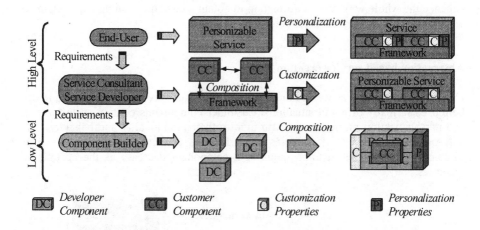

Fig. 1. Major business roles at different abstraction levels

Business Role Profiles

End-users are the ultimate target public of almost any telecom service. They distinguish themselves from the other roles by the fact that they are actually using the service in a runtime environment. They provide their requirements, directly or indirectly, to the service provider. The end-user's main activity in the service creation is to personalize a service, although more technically qualified users can perform full customization of user profiles and GUIs.

Service consultants capture service requirements from service end-users. They can do this the old way using natural language, before passing them through to technically qualified consultants or service developers. But our goal is to make it possible to build a service on the spot by searching and composing high-level components. Service consultants can easily acquire those skills and become actual participators in the service creation process. In that respect, the role of the service consultant may overlap that of the service developer.

Service developers are involved in setting up and commercially exploiting services. They build services and, if necessary, provide requirements for new components to the service builders. There are two scenarios for creating new services, depending on the technical skills of service developer. He can either build a new service by customizing existing services or he can be involved in the entire service life cycle. Basically they build a new service by searching and composing customer components. The customer components fit in a larger framework and are built to work together. In a second step, the customizable properties of the customer components are assigned a value.

Component Builders are responsible for creating components based on higher level requirements. They use the low-level service creation environment to build developer components and customer components. Developer components form the basic reusable building block. Component builders compose developer components and enhance the whole with more information to obtain a customer component. Section 3 elucidates on the low level service creation.

Abstraction based service creation

Customer components as well as the customer framework are highly versatile and can be reused in many different situations. Customer components contain customization and personalization properties that can modify their functionality and behavior. The TINA customer framework contains many features that can be enabled, tuned or disabled. The possible customizations of a component decrease as the abstraction increases.

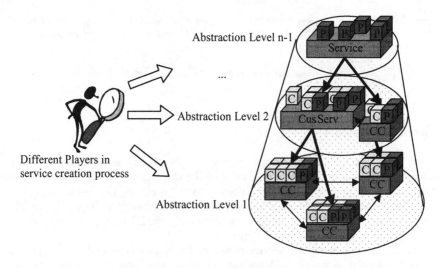

Fig. 2. Abstraction levels in the High-Level service creation.

The low-level and the high-level are the two main abstraction levels. If we focus on the high level abstraction level, we see multiple levels as depicted in Fig. 2. On the lowest abstraction level, all the customization properties are available and the customer components are not composed yet. On the highest abstraction level, only a set of selected personalization properties remains and the customer components are invisibly wrapped into the service. The abstractions of customer components are determined by the *granularity* and *grid* of the components.

The granularity of a component is determined by the functionality offered by the component. A finely grained Customer Component can e.g. be a distributed service feature. Service features cannot work standalone, they depend on the presence of other service features in order to be able to work properly as a service. A coarsely grained Customer Component would be a complete distributed service. In contrast to service features, services can work stand-alone.

The grid of a component determines the access capabilities of the customization and personalization properties of a service. Certain properties will be customizable, while other remain on their initial (default) values (and will not be visible to the service developer at this particular abstraction level).

Abstraction levels can use the same core functionality of the high-level service creation tool and inherently share the same look and feel concerning the way services are developed. The difference between both abstraction levels will be reflected by means of the options and functionality that are available in the high-level service creation tool. For example, views that are available at one abstraction level are not available on the other.

3   Low Level Service Creation Methodology

The low level of the COSEC methodology is targeted to component builders. The process of creating new components is triggered by the requirements of a service provider that does not find the right customer components to build his service. The new requirements can vary from an extra customization property to a completely new customer component in a new application domain. The requirements are described in the form of structured text, sequence charts and collaboration diagram. They should describe the purpose of the component, the difference with existing components, the way in which the component will work together with other components and outline the expected interface and behavior.

Developer Components

Developer components are the basic building blocks in our methodology. They are reusable components with a specific functionality and a clearly defined interface. They can be relatively small size, e.g. a few classes, and typically depend explicitly on a number of other developer components. So they are not really stand alone, but rather a part of a collaboration. However, several developer components can be composed into another developer component and in this way they can form larger, possibly stand-alone entities.

We use state-of-the-art object-oriented technology to build and document developer components. Requirement analysis and system analysis is done in UML, typically using the use-case diagrams, sequence diagrams, collaboration diagrams and class diagrams. These UML diagrams also serve as documentation for the components. Detailed design and implementation is done in a combination of UML, SDL and Java. The UML class and state diagrams are used to improve the system design, while at the same time, SDL is used to fill in al the details after the UML has been translated to SDL. The UML to SDL translator translates the UML class diagrams as well as state diagrams into a high quality, readable SDL package or system. An automatic round-trip process makes sure that both models keep being synchronized [8]. The user-interface aspects are written in Java. The GUI communicates with the SDL through signals and remote procedure calls. Java is also used as target implementation code. The code is generated from the SDL specification without modifications. The IDL interface of the developer component is written based on the documentation, UML models and interfaces of the individual classes.

Developer Component

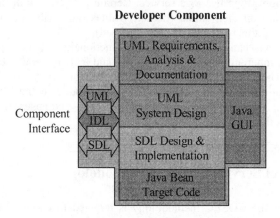

Fig. 3. Parts of a developer component. UML for analysis and system design, SDL for design and implementation & Java for the graphical user interface. The interface is described in UML, IDL and SDL.

Customer Components

Customer Components are the coupling between low-level and high-level service methodology. They are developed by the component builders and are deployed by service builders as basic building blocks for building their services. In order to make the customer component user-friendly and allow a wizard like composition, a customer component must contain a lot of information that can be used by the service building tools.

The core of a customer component is a composition of developer components. The documentation and the clear interface of the developer components ease this composition. The developer components are linked with each other by either a simple association or by using a composition pattern. A composition pattern is a sequence of abstract messages that describes how components should work together. Furthermore, small changes can be made to the developer components, and their composition, to write glue code or to overcome interface incompatibility. Because of the iterative support, these changes can be made on any level: UML, SDL and/or Java. The composition forms a solid entity on its own and can be browsed as such on all levels and can be simulated in SDL.

The developer component composition is then extended with extra information that is essential for using customer components during service creation. In the HLSCE, it should be possible to compose and customize customer components in a wizard like way. Therefore we need an explicit declaration of the CC's interface to other components and of the modifiable parts in the component. A customer component consists of the following parts:

• *Composition* of developer components, see above.

- *Documentation* is an abstract description of functionality in a free text format. It eases the search for suitable customer components in the HLSCE. It also hints at how this component can be integrated with others. This documentation is partially based on the documentation of the developer components.
- *Customization properties* allow service providers to tune the service for their customers. These properties relate to existing properties in the developer components. Typical examples include company logos and icons, menus, quality of service parameters and homepage address.
- *Personalization properties* are aimed at the end user. It allows him to customize this component with user specific data. These properties relate to existing properties in the developer components. Typical examples include setting the name of the user, user preferences and user priority.
- *GUI* aspects include a set of partially customizable user interface elements. They are taken directly from the developer components, although some of them can be removed because of an overlap or irrelevance for the customer component.

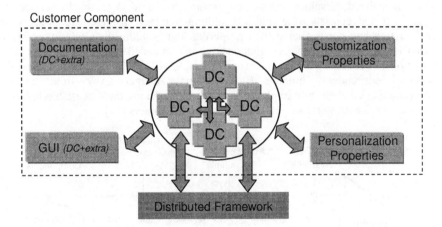

Fig. 4. Parts of a customer component.

Customer Framework

A customer framework provides a customer component with the necessary environment to work properly and to communicate with other components and the outside world. Although a customer component could be standalone on a conceptual level, it always needs the framework to become alive. In this project we use the TINA framework, for which we developed a default implementation in SDL. The TINA standard provides a lot of facilities to the customer components: connections, security, broadcast, invitation, sessions, streams, etc. At the same time it makes different customer components more compatible with each other as they are implemented according to the same standard.

The implementation in SDL makes it possible to simulate incomplete specifications or prototypes [9]. The parts of the prototype that have not been implemented yet can be filled-in by manually sending signals.

A customer framework is built upon a service platform and provides the architecture and functionality for a certain class of services. For instance the TINA service session components are regarded as part of the TINA Customer Framework. The actual customer framework is built by composing and modifying developer components as described in this section.

4   High Level Service Creation Environment

The customer components developed in the LLSCE, are deployed in the high-level service creation environment (HLSCE). A customer component is first fitted into the framework, then composed with other customer components and finally customized and personalized. Graphical editors and wizards are used to create the service at a high level and consists of graphically putting together customer components and customizing and personalizing their properties and building the GUIs in a fast and user-friendly way. Customer components are intended to be high-level DPE (Distributed Processing Environment) counterparts of IN SIBs (Intelligent Network – Service Independent Building Block). They abstract away from complex programming concerns and implementation details, and are thus accessible to a wide range of service creators, not just experienced programmers [11].

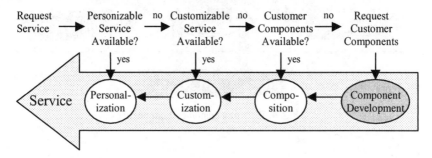

Fig. 5. High-level service creation process.

An overview of the high-level service creation process is shown in Fig. 5. For example, an end-user would like to use a chat/shared white board service. If his provider has this service available, he enters his preferences and his name in the wizard and he can start chatting/drawing. Otherwise he sends a request to his provider for a new service. Presuming that the service provider wants to provide this service, he looks for an existing service that he can customize. If not, he can build a new customizable service by composing the chat and shared white board components. If one of those customer components is missing or does not fit entirely, the service provider requests the necessary customer components to be developed.

The requirements for a service flow top-down, but the development happens bottom-up. The first step in high-level service creation is to find the right customer components and make a composition. Then the customizable service must consecutively be customized and personalized. We will focus on these this process and illustrate it by means of an example. Suppose we want to develop a Desktop Video-Conference service that also supports chat, shared white board and digital library functionality.

Selecting Customer Components

The service builder can use a standard component browser with search facilities to choose the right customer component. The browser shows the documentation, the properties and the relations with other components.

When a particular Customer Component is not available or properties should be set that cannot be set at this abstraction level, a request for development/adaptation of a customer component is made. Several techniques can be used to capture the requirements of the missing service (for example: text, UML use cases, GUI design etc). This request is then forwarded to the component developers to build a new component.

Composing Customer Components

The graphical service creation tool, which could be a distributed service, provides a means of composing these services represented as customer components. Composition can be wizard-based, by guiding the composer through a series of questions related to frequently used Customer Components. Based on these answers, the service is now automatically generated by tying together the underlying logic of all the developer components and plugging it into the service framework. Default GUI components are generated, which can be visually composed and edited by the composer. Composition can also be drag-drop based. Customer components, represented as icons, can be visually plugged into the framework on a user-friendly way.

Each customer component has an explicit list of all properties that can be modified within the component. The personalization and customization properties of an entire service are the sum of the properties of all customer components in the service, minus the duplicated and overlapping properties.

Interestingly, the information provided by the composer can also be used to generate a wizard for the customization stage. A similar relationship exists between the customization and personalization processes, i.e. the options available for personalization are a subset of the options for customization and can be extracted from the latter in a fully automated way.

Personalization and Customization

When the composed service is ready we need to do the customization. This encompasses the customization of the selected generic Java GUIs and behavior of the service. For example do the participants in the service have the ability to personalize the service? Can people join the service without an invitation? Can the participants change the Quality of Service of the audio video stream binding? The service creation tool detects customization conflicts when services are composed. For example when the chat service is customized, in such a way that all people can join the service while, within the video conference service there is a limit on the number of participants, then the service creation tool automatically resolves these conflicts (i.e. with minimal human interaction).

When the composed service has been customized we can proceed to personalize the service. Personalization consists of adapting the GUI and the behavior of this service for a particular end-user. The GUI of the service can be adapted by loading the customized GUIs and changing them according to the needs in a fast, reliable and user-friendly way. The behavior could perhaps be adapted using a visual scripting language. For example, when we start the service we want to invite automatically the same people continue the discussion on that particular subject, illustrated by a library of pictures and drawings.

5   Current State and Tool Support

At the current state of the project, we have a clear insight on the overall methodology and on many of the technical solutions. The methods to develop developer components using UML and SDL already have a solid foundation and we are currently building tool support for translating UML to SDL in a commercial product. Also many off-the-shelf UML, SDL and Java bean environments can be used and extended for our purpose.

Currently, our main research focuses on the documentation, customization and composition of components. Composition is the most difficult part, because it should take the structural as well as the behavioral aspect into account, without the user or the developer having to (re)write code. To achieve this, the changes made on a higher level must be translated to the lower levels without interfering with the changes made before, see [8].

It is still unclear yet how much tool support we will be able to offer, but our goal is to create an integrated environment for the composition of developer component and one for the composition of customer component. This encompasses a component browser and a graphical way to link the selected components. A first important feature of such a composition environment is adding a component to a composition on all levels at the same time: UML, SDL and Java. In other words, when adding a component to a composition, the documentation in UML, the design in SDL and the implementation and GUI in Java are all joined with the (possibly empty) models already available in the composition. A second feature will use the scenario documentation of the components to check whether a certain composition is valid or which conflicts need to be resolved. Finally, the round-trip engineering support will update the SDL specification after UML has been modified.

6   Conclusion

This paper presented a multi-level, component oriented methodology for service creation. The low-level methodology combines UML and SDL to create and compose developer components. A composition of DC's together with extra information forms a customer component. The customer components can be viewed on different abstraction levels, according to the level of experience of the user creating the service. The high-level methodology uses this characteristic to compose customer components and to customize and personalize services on different abstraction levels.

The work presented in this paper is the continuation of work done in other projects like SCREEN and TOSCA [1]. Except for updating the languages used (UML, SDL'96), we also opt resolutely for the component oriented approach, which allows more reusability and better documentation. The high-level environment of our methodology adopts many aspects of an IN service creation environment [10]. Our concept of customer component, however, is more flexible and forms part of a single methodology.

References

1.   Fiona Lodge, Kristofer Kimbler, Manuel Hubert. *Alignment of the TOSCA and SCREEN approaches to service creation*, ISN'99, Barcelona, April 1999
2.   Rational Unified Process 5.5, Rational Software Corporation, http://www.rational.com
3.   Advanced Internet Access project description, ITA-2 AIA consortium, Brussel, November 1998.
4.   ITA-2 - Information Technology Access program, second part. http://www.iwt.be
5.   SCREEN deliverable D28, *SCREEN Engineering Practices for Component-based Service Creation,* SCREEN/A21-D29, ACTS SCREEN consortium, December 1998.
6.   TOSCA deliverable D9, *Specification of the TOSCA Process Architecture for Service Creation*, ACTS TOSCA consortium, December 1998.
7.   Fiona Lodge, Kristofer Kimbler, Manuel Hubert (1999). *Alignment of the TOSCA and SCREEN Approaches to Service Creation.* IS&N'99, LNCS 1597, pp. 277-290.
8.   K. Verschaeve (1997). *Automated Iteration between OMT* and SDL.* Eighth SDL Forum, Evry, France.
9.   K. Verschaeve, A. Ek (1999). *Three Scenarios for Combining UML and SDL'96.* Eighth SDL Forum, Montréal, Canada.
10.   ITU-T Q.122x-series. *Recommendation for Intelligent Network*, CS-2, ITU-T, March 1993.
11.   L. Demounem, F. Westerhuis, *Fast and User-friendly Service Creation Methodology for TINA*, ICIN 98, May 1998, Bordeaux

Cooling the Hell of Distributed Applications' Deployment

José Bonnet[1], Fabrice Dubois[2], Sofoklis Efremidis[3], Pedro Leonardo[1], Nicholas Malavazos[4], Daniel Vincent[2]

[1] PT Inovação, Porto, Portugal
{jbonnet,leonardo}@ptinovacao.pt
[2] CNET, Lannion, France
{daniel.vincent,fabrice.dubois}@cnet.francetelecom.fr
[3] INTRACOM, Brussels, Belgium
efremidis@intracom.be
[4] OTE, Athens, Greece
nickmala@oteresearch.gr

Abstract. The deployment of distributed applications has been mainly a manual task, despite of the proprietary solutions that already exist in the market. Due mainly to the heterogeneity of the operational environment and the software to be deployed, deployment is something that is usually done by writing installation and configuration scripts in an appropriate scripting language. With the wide acceptance of Enterprise Java Beans and the arrival of the CORBA 3.0's Components, an extraordinary opportunity for automating the deployment process has emerged. This is the basis for this paper, in which we define the concept of deployment, present a list of assumptions that would enable the automation of the process and establish a list of requirements on the deployment process itself and on deployment means.

1   Introduction

Deploying a distributed application on the nodes of a network where it is supposed to work has undoubtedly been a daunting task. Experience shows today that software support is needed to manage software distribution in heterogeneous environments featuring different middleware products, development tools, and methods. Deployment relates to an efficient combination of service modularization and distribution, which can reduce the system development effort and the dependencies on software vendor's middleware products, and shorten the time to market. Deployment is considered as a software management task that will need to be handled by the network operators themselves in future. The deployment task effectively involves the following steps:

- Decision on an appropriate distribution pattern of the application (i.e., assignment of parts of the application to system nodes).
- Distribution of the application on the system nodes each of which hosts an instance of the necessary Distributed Processing Environment (DPE) functionality.
- Installation of application components.

- Configuration of each application component (i.e., relevant parameter setting).
- Configuration of the application (i.e., linking, registration, etc., of involved components).

Relying on the recently released OMG's CORBA 3.0 specification [6], namely on the Component specification, the first step in the endeavor of automating the deployment process is the definition of the requirements, the terminology to be used, and the relevant concepts. Thus, in the next section, we state our view of the current state of the art in deploying a distributed application with the technologies available today. Next, we briefly present how things might change when products compliant with the new CORBA 3.0 specification are used, and what parts of the deployment problem CORBA 3.0 is expected to solve, or at least ease. But CORBA 3.0 does not solve all the problems of deploying distributed applications. We, therefore, then propose a deployment environment model to serve as an intermediate layer between a deployment tool and the system nodes. A concise list of requirements on the deployment process and the means for it is presented and the paper is concluded with the identification of future work items and conclusive remarks. The work presented in this paper is preliminary, hence no experimental results are available.

2   Deployment of Distributed Applications Today

Proprietary solutions for this problem exist already in the tools market, such as IBM's Tivoli [1], Microsoft's Transaction Server [2], and Microsoft's Management Console [3]. These tools though relatively open for Independent Software Vendors' (ISV) plug-ins, are targeted to specific environments and, therefore, present limitations in heterogeneous environments such as the ones that are commonly encountered in Telecommunications companies. Another kind of proprietary solution for the problem of deploying distributed applications is represented by the tools coined as Application Servers, which are usually sold by the Java programming language Interactive Developing Environment's vendors. These kinds of tools were born as supporters for the deployment of Enterprise Java Beans (EJB), thus being restricted applications or components developed using Java and the EJB specification[1] [4]. Application Servers have also appeared from the side of Database, Transaction Monitors, and even ORB vendors.

There also exist languages, which provide capabilities for deployment support in terms of notations that express distribution policies of software components within a Distributed Processing Environment (DPE). Specifically, there have been a number of initiatives in projects with distinct kinds of results in the Configuration Languages (CL) levels, like Darwin [10]. The OMG has standardized the Unified Modeling Language (UML) [9] and the Object Constraint Language (OCL) [5], which is an extension of UML.

---

[1] The EJB specification has strongly influenced the specification of CORBA components, the later being considered a super-set of the former.

3   Deployment of Distributed Applications Using CORBA 3.0

The CORBA 3.0 specification [6] presents, among other things, a specification of the CORBA Component Architecture comprising:

- An Abstract Component Model,
- A Packaging and Deployment Model,
- A Container Model,
- A mapping to EJB, and
- An Integration Model for Persistence and Transaction.

By splitting an application into components, the ability to distribute these components is eased. This section presents only a brief description of the Packaging and Deployment Model, this being the most relevant for the deployment problem. A short overview is given for the purpose of completeness.

The CORBA 3.0 Component specification introduces to CORBA the notion of a component container, which already exists in EJB. The container takes the responsibility for the implementation of several infrastructure issues, like the transactional and security policies of the component. For this, and beyond the API needed to connect to the component, the container uses a CORBA Usage Model for its interaction with the Portable Object Adapter (POA), the Object Request Broker (ORB) and the needed CORBA Services (COS). These relationships are shown in Fig.1.

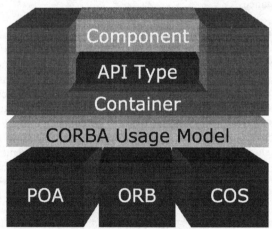

Fig. 1. The CORBA Container's associated API Type (for interaction with the containers) and CORBA Usage Model (for interaction with the POA, the ORB and the needed COS).

The Packaging and Deployment Model follows most of the Open Software Description (OSD) [11] proposed by Microsoft and Marimba to the World Wide Web Consortium (W3C) for standardization of software distribution. The extensions to this proposal mainly concern the broader scope of the whole CORBA technology, namely programming language and operating systems independence.

The main concept in packaging CORBA 3.0 components is to include a *component descriptor* together with the component itself, but in a different file, defined as an Extended Markup Language (XML) file. Components are then packaged with their descriptors in an archive file called *component package*. This package may then serve as input for a deployment tool or become part, together with other related component packages and an *assembly descriptor*, of a *component assembly package*. The assembly descriptor, also an XML file, specifies all the components that make up the assembly, the components' partitioning constraints and the connections between them. These connections are made between interface ports represented by the "provides" and "uses" features of each component definition and event ports (represented by the "emits", "produces", and "consumes" features of each component definition). The component assembly package may also be an input to the deployment tool. Each component of this assembly package may be installed one or more machines of the network.

The deployment tool can then get these packages and create in each node the appropriate environment for each component of the distributed application to be activated. From the component descriptor the deployment tool can automatically infer the following:

- The *component kind*: this can be session, service, process, or entity.
- The *transaction policy*: i.e., if the component implements all the transactional aspects it needs, or if it relies on the container for doing it.
- The *event policy*: used to indicate the quality of service desired at the event ports.
- The *threading policy*: if more than one thread can manipulate each single component instance.

The deployment of each component type implies the creation of at least one home (like in EJB) for managing its instances. More than one home can be declared and created for each component type, though each component instance can only be managed by one home. Homes have component factories (for component instance creation), finders (for methods finding), and primary keys (for instance finding).

All the components in a component assembly package may be interconnected. For example, a component that uses interface A may be connected to another component that provides interface A, based on the information provided by the component descriptor. Fig. 2 shows this kind of relationship between components.

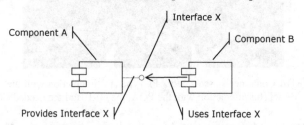

Fig. 2. The relationship "provides-uses" between two components.

Two or more components can also establish a relationship as to what kinds of events each one of those components emits, publishes, or consumes. This relationship, which is represented by a connection between those components, is established during the design of the components (see above). Such connections are only an initial configuration, and might be changed at deployment time.

The CORBA 3.0 Component specification defines a set of helper objects, which actually perform the installation and configuration of the deployed components. There are the following kinds of helper objects:

- *ComponentInstallation*: the one (and only) instance per node is used to install, query, and remove component implementations on that node.
- *AssemblyFactory*: the one (and only) instance per node is used to create assembly objects on that node.
- *Assembly*: used to instantiate all components in the assembly and create the connections between them, as specified by the assembly descriptor (building the assembly), and for removing all the connections between components and destroying all of them (tearing down the assembly).

4   An Environment Model

The CORBA Component Architecture mentioned in the previous section partially solves the deployment problem, by addressing the following issues:

- The heterogeneity of programming languages and operating systems;
- The modularization of the application by splitting it into co-operating pieces, or components;
- The connection of these different pieces before their installation (off-line) at a human specified processing node.

We consider that we can improve the issue raised in the last bullet above, by proposing the definition of a *deployment environment model*. This model can take into consideration all the restrictions of the dynamic environment and distributed application characteristics, on which a deployment tool might work upon, to deploy that application in a more efficient way. This increase in efficiency could be gained by pro-actively knowing almost in real time the most important characteristics and/or restrictions of the processing nodes where to deploy each component of the application.

The proposed environment model (Fig. 3) would be an intermediate layer between the deployment-helping tool and the nodes where the distributed application is to be deployed. This layer could get updated upon request by the deployment tool (e.g., right before the first installation) or by the nodes themselves (e.g., when some pre-defined condition, like the nodes workload, evaluates to true), thus allowing for an on-line configuration of the components.

The environment-model creation is quite eased by the structure envisaged for the CORBA 3.0 Deployment model: the objects implementing the functionality needed by the environment may be implemented using the model's helper objects. Each node can use the CORBA Event (Notification) service in order to alert on its changing environmental conditions.

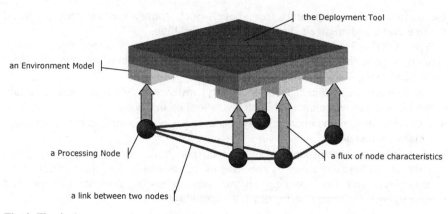

Fig. 3. The deployment environment model, between the nodes and the deployment tool.

5   Requirements

From what has been discussed above, and considering the steps involved in the deployment process, as they have been introduced in the introduction section, four main deployment requirements prevail:

- The ability to decide on an appropriate distribution pattern for the application.
- The ability to automatically distribute the application.
- The ability to automatically install the application components.
- The ability to configure the constituent components and the application itself, both on-line and off-line.

Specific, non-exhaustive means to address the requirements above are proposed in the following, together with identified requirements on the proposed means.

The problem of deciding on an appropriate distribution pattern for an application is defined as follows:- given a set of application components, a set of nodes on which the application will be distributed, a cost function (which captures computing and communications costs incurred by the application components), and a set of constraints imposed by the components, nodes, and environment - assign components to nodes so that they respect the constraints and minimize the cost function. Solving the deployment problem requires that:

- Constraints imposed by each component be properly expressed and associated with the component (e.g., in the component descriptor mentioned in section 3).
- Constraints imposed by each node be properly expressed and associated with the node in the respective environment model.
- A cost function be determined that accommodates several distribution-related concerns, such as resulting communication cost, fault tolerance, security, etc.
- An algorithm be defined that produces appropriate distribution patterns on the basis of the above.

From the known notations and the above requirements, we see that the automatic distribution, installation and configuration of the components of a distributed application requires the definition of a new notation that is able to express the distribution pattern. As the development of a notation is closely related to the development of supporting tools, a further requirement calls for the creation of appropriate tools that are able to process the notation and carry out the necessary tasks. A further requirement is an instance of the necessary DPE functionality to be installed and configured in each considered system node.

A non-exhaustive list of identified requirements for the needed notations and tools is the following:

- *Heterogeneity*: non-CORBA systems or non-CORBA Components are expected to co-exist with fully fledged CORBA Component distributed applications; the deployment characteristics of each node might also be very heterogeneous; the deployment notations and tools should therefore correctly cope with this.
- *Ease of use*: human resources scarcity and the high costs and time associated with their training, calls for notations and tools that allow for a default behavior that is quite acceptable on average; qualified users might override these defaults, improving the final result thereof.
- *Flexibility*: within this ever changing world, a flexible notation and tool, probably developed using CORBA Components in itself, is a must.
- *Openness*: is another must for such a notation and tool; the latter should have the ability to be extended by ISV' s plug-ins.
- *Robustness*: every aspect of the notation and the tool must gain be clearly and unambiguously defined.

From this last set of requirements it appears that using CORBA, and in particular CORBA Components, may ease the development of deployment-related processes, notations, and tools. For the moment, however, we do not impose the use of a CORBA infrastructure as a requirement for the deployment tool.

6   Future Work

The presented work will be extended and deepened to result in:

- A notation for expressing configuration information and distribution requirements of distributed applications.
- A methodology for distribution and configuration of distributed applications and systems particularly suited for use in the Telecommunications domain, including dynamic configuration services. Based on QoS constraints and performance analysis, special support will be provided for reconfiguration, migration, and replication of software components. The methodology will encompass mechanisms for the relocation of persistent data during runtime.
- The validation of the concepts with use of an appropriate middleware platform, like [7].
- Integration of tools for development and configuration of distributed applications and systems, which will provide network operators the means to deliver new services faster and with better quality.

7   Conclusions

This paper derives mainly from work performed in the EURESCOM project P924 that addresses Distribution and Configuration support for Distributed PNO (Public Network Operator) Applications. The authors as well as all members of the project are convinced that there is a need for more powerful and widely-applicable deployment-helping means (i.e., models, notations, and tools). In today's very complex, dynamic and demanding (24x7) environments, these tools should have the support of a deployment environment such that not only static configuration (i.e., one shot configuration) but also dynamic and on-the-fly configuration be possible.

Working with OMG's CORBA 3.0 as a base has been very promising, but it is not complete, namely in regards to aspects like dynamic configuration of a distributed application. Other approaches have to fill the existing gap between component and assembly descriptors and the engineering aspects of actually deploying every piece of a distributed application. In this paper, we have suggested that a deployment environment be created for each node where a component has to be deployed at. We feel that the CORBA infrastructure present at each node may dramatically improve the flexibility of the creation and the update of these environments.

Acknowledgements

The work discussed has been performed within the context of EURESCOM project P924-PF, partially funded by EURESCOM. The authors would like to thank participants of this project and colleagues within their companies for their collaboration in this work, and also for ideas, comments, and fruitful discussions.

References

1. IBM: A Project Guide for Deploying Tivoli Solutions, http://www.redbooks.ibm.com.
2. Jennings, R.: Microsoft Transaction Server 2.0, SAMS Publishing, 1997.
3. Microsoft Corp.: Microsoft Management Console Overview, 1999.
4. Sun Microsystems Inc.: Enterprise JavaBeans™ Specification, v1.1, Preliminary version, 1999, http://www.javasoft.com.
5. Warmer, J., Kleppe, A.: The Object Constraint Language, Precise Modeling with UML. Addison-Wesley Object Technology Series.
6. Object Management Group: CORBAtelecoms: Telecommunications Domain Specifications, OMG Document formal/98-07-12 ed., Version 1.0, June 1998.
7. EURESCOM Project P715 Report: Definition of Reference Points for Stream Management, 11 February 1998.
8. Frankel, D.: CORBA Components – alive and well, Java Report, October 1999.
9. Fowler, M., Scott, K.: UML Distilled, Applying the Standard Object Modeling Language. Addison-Wesley Object Technology Series.
10. http://www-dse.doc.ic.ac.uk/research/darwin/darwin-lang.html.
11. http://www.w3c.org/TR/NOTE-OSD.html.

Agent-Based Management

Andreas Kind

C&C Research Laboratories (Berlin)
NEC Europe Ltd.
ak@ccrle.nec.de

The management of computer networks becomes more and more difficult with an increasing variety of hardware devices, protocols and services. Agent technology has been proposed to support service and network management in several ways. The resulting area of agent-based management emerged as one of the few acknowledged application domains for intelligent and mobile software agents. The typical characteristics of intelligent agents, like autonomy and high-level communication, can for instance lead to more flexible and general network control structures. Furthermore, mobility of agents can result in reduced communication delay and bandwidth if computation involved with managing a network node can be performed closer to the node or even directly on the node. The autonomous character of agents can also give support for better fault tolerance in case of connectivity problems.

The papers presented in this section cover in various ways the area of agent-based management:

In the first paper *An Approach to Network Control and Resource Management Based on Intelligent Agents*, open programmable networks are described as a way to overcome the difficulties in customizing network resource control for the benefit of new services. The authors suggest a new network control architecture using intelligent agents. The architecture addresses the requirements for flexible network control by providing an open framework for the deployment of network control schemes.

The second paper in this section on agent-based management, entitled *Providing Customisable Network Management Services Through Mobile Agents*, shows how an initial set of management services could be customized and extended on demand. The units of extension are mobile software agents that are used as deployment vehicles carrying management functionality to specific servers. Agent mobility provides in this application context better performance and support for fault management.

The next paper in this section, *Accounting Architecture Involving Agent Capabilities*, presents an accounting management model that incorporates intelligent and mobile agents supporting user/provider negotiation, creation of service offers, event collection and service analysis. The author assumes a TINA-oriented service creation architecture into which software agents are carefully introduced.

The paper *IN Load Control Using a Competitive Market-Based Multi-Agent System* presents an agent-based load control mechanisms to prevent QoS degradation at Service Control Points in Intelligent Networks. The proposed mechanism uses a decentralized, market-based model for IN load balancing. The realization of the model as a multi-agent system is described and evaluated.

An Approach to Network Control and Resource Management Based on Intelligent Agents

Evangelos Vayias, John Soldatos, Nikolas Mitrou

Computer Science Division, National Technical University of Athens
9, Heroon Polytechneiou Street, Zografou 15773, Greece
{evayias;jsoldat}@telecom.ntua.gr, mitrou@cs.ntua.gr

Abstract. Network Control is currently carried out mainly by means of signalling protocols. Although these protocols are robust and facilitate standardisation, they present several drawbacks such as the inability of the Service Providers to adapt network resource control to the particular needs of their services, or the difficulty in deploying advanced traffic control schemes in order to achieve increased utilisation. An open distributed software architecture for Network Control based on concepts such as Intelligent Agents can provide a solution to this problem. This paper highlights the major issues of this problem and presents the architecture and implementation of a system that enables open and flexible network control and resource management using intelligent agents. The description of the architecture is complemented by experimental results as well as a generic implementation framework for deploying open programmable control capabilities in existing networks.

1 Introduction

Quality of Service (QoS) is widely acknowledged as a crucial feature of future multi-service networks. Commercial networking infrastructures are expected to make use of QoS features and it is anticipated that customers will be willing to pay for reliable and high performance networking services. Traffic control mechanisms are traditionally considered as a vehicle towards enabling QoS provision. The essentiality of such mechanisms is nowadays under question, since fibers and techniques such as Wavelength Division Multiplexing (WDM) promise to deliver QoS based on bandwidth profusion. However, the continuous increase in network traffic, which is provoked from the emergence of new bandwidth-consuming applications, keeps network control in the foreground. It is no accident that most manufacturers of networking equipment are adding some sort of QoS mechanisms in their products (e.g., [23]).

In networks that support QoS, traffic control aims at ensuring QoS for all connections (or flows) within the network, while maximizing the utilization of network resources. Specifically, network control decides whether newly offered traffic can be admitted into the network and configures the internal traffic control mechanisms of the network elements, so that the established connection gets the requested QoS while not affecting the QoS of other connections. Thus, the network control algorithms should take decisions that guarantee, on one hand, the QoS of connections and permit, on the other hand, an efficient utilisation of network resources. Another objective of network control is to achieve a specific connection-level performance, e.g. by trying to keep the ratio of blocked (rejected) connection requests or the connection set-up time under some predefined limits.

Note that by the term "network control", we usually refer to a set of mechanisms and procedures that should be performed in a short time scale (real-time), each time a new connection is requested (or new traffic is being injected into the network in the case of datagram networks). These procedures deal with the admission, the routing, the resource allocation and the establishment, maintenance and tear-down of connections/flows. Since a network is a distributed system, network control involves an interaction protocol called the "signalling" protocol for exchanging control information (connection requests, topology updates) among the network elements. It also involves certain algorithms for executing procedures such as connection admission, resource allocation and routing.

Efficient network control strategies are beneficial to both the customers/end-users (through satisfying application requirements at a reasonable cost) and network operators/service providers (through maximizing their revenues and boosting economies of scale). Conventional approaches towards carrying out network control operations rely on signalling protocols. Classical examples can be extracted from the ATM technology, where signalling protocols such as ITU-T Q.2931 or ATM Forum UNI 3.1 and PNNI provide the means for control plane operations in most ATM-based networking infrastructures. Similar signalling protocols have been proposed in the scope of frameworks/technologies supporting QoS in the Internet (e.g., RSVP in the IntServ framework). Signalling protocols have contributed considerably to the rapid deployment of networking equipment and have boosted interoperability.

According to the current networking paradigms, signalling incorporates the key intelligence of the network, which is usually based on a few control programs and algorithms. As a result, changing or improving network control strategies in standard signalling-based networks requires the deployment of enriched signalling protocols. Note however, that enhancing signalling protocols is inevitably the outcome of arduous and slow standardisation processes [1]. It is usually hard to adapt the network and service control logic to the different needs of current and emerging services. Network operators and service providers should ideally be provided with a higher degree of flexibility regarding the configuration/establishment of network control strategies and mechanisms. This degree of flexibility becomes even more important given that future networks will have to incorporate more intelligence in order to manage their increased complexity and heterogeneity and to support continuously emerging services and traffic types [6].

Network intelligence and flexibility constitute the main rationale behind proposals for open signalling and programmable interfaces. Most of these proposals aim at replacing fixed signalling protocols and control programs with open programmable and extensible signalling protocols. This is accomplished through a set of software abstractions of network resources, which allow distributed access to low-level control capabilities of network device. It is noteworthy that open signalling and network programmability proposals (e.g., [2], [3], [4], [5]) propose new open broadband control architectures, in most cases oriented towards QoS-aware networks. These proposals constitute another indication that network control is still an open and significant issue. It is also quite remarkable that several organisations (both industrial and academic) started a new IEEE standards development project, namely IEEE P1520 which has already specified a framework for programmability in the network element, network service and application service layers [1].

Network programmability constitutes an important advance towards increasing flexibility in network control. Nevertheless, most of the efforts do not deal with the whole range of network control issues that have to be confronted in the scope of the present composite infrastructures. Rather, they are mostly tailored to specific needs (e.g. [4] considers mostly multimedia communication, [5] speeds up the set-up time for intra-network services and [7] deals with interworking of IP and ATM). Furthermore, network control should nowadays provide additional intelligence, addressing the expectations of all the involved parties (i.e. network operators, service providers, customers, end users). Towards adding extra intelligence, a direct extension to the concept of programmability has been proposed, based on the properties of intelligent agents. Recently there has been interest in applying the agent technology towards tackling network management [20] and network control tasks. Network control and management based on intelligent agents reveals similarities to open programmable control. However, an agent based approach has the additional advantage that the distributed entities possess properties such as pro-activity, re-activity, autonomy, social ability, intelligence and communicate through an Agent Communication Language.

Agent-based approaches have been adopted towards tackling a host of telecommunication issues. These applications include a large number of projects dealing mainly with network management. Agents have not been widely applied to network control yet. Nevertheless, several efforts have addressed issues related to network resource management (NRM). Although network resource management mechanisms reside in the management plane from a computational viewpoint ([8], [9]), they are tightly coupled with network control strategies. This is because they operate in considerably shorter time scales than network planning and provisioning tasks.

Network control based on Intelligent Agents is the main topic of the rest of this paper. In particular, the paper presents work carried out within a project dealing with agent-based control for ATM networks, namely the IMPACT[1] project. IMPACT has designed and developed a prototype software system for open distributed network control of ATM networks, based on Intelligent Agents. The architecture of this system is outlined in section 2, following this introductory section. Implementation

---

[1] The IMPACT project (Implementation of Agents for CAC on an ATM Testbed - AC324), is funded by the European Commission under the framework of the Advanced Communications Technologies and Services (ACTS) programme.

issues and details, along with results from experiments are also discussed in the scope of the same section. In the sequel, section 3 is devoted to describing mechanisms employed towards interfacing the agents with the network. This interfacing is achieved through middleware components that provide a network programmability layer on top of the network elements. Although the paper focuses on the implementation of the system for controlling ATM networks, the architecture is quite generic and can be applied to other QoS-aware network technologies as well. Section 4 gives such an indicative mapping for the case of an IP network that supports Differentiated Services. The paper concludes (section 5) by providing a critical view on the future of open network control systems that are based on programmable networks and their potential dominance over conventional signalling-based approaches.

2   Overview of the IMPACT architecture

2.1   Network Model

The architecture of the IMPACT system for ATM network control and resource management takes into account the need for providing flexibility to network operators and service providers towards establishing their own network control and resource management strategies. The network model envisaged within IMPACT comprises a network provider (NP) and multiple service providers (SPs). The network provider is responsible for operating and managing the entire physical network. Service providers lease bandwidth from the NP, with a view to offering connectivity services to customers. The various service providers are providing alternative options for service provisioning to customers, thus they can be seen as competing parties. Building on the notions of a NP and multiple SPs, IMPACT network control focuses on applying Call Admission Control strategies for connectivity requests through the ATM network. Moreover, IMPACT performs network resource management through employing dynamic bandwidth (re)allocation schemes to virtual paths.

In the scope of the IMPACT project, each SP delegates a special agent, called *Resource Agent* (RA), as responsible for managing the resources allocated for each *source-destination* pair it services. As a result, the flexibility required in an environment where multiple service providers and multiple source-destination pairs exist is attained by having several resource agents. Fig. 1 depicts the notions of a NP, multiple SPs with their RAs managing a source-destination pair. Observe that each SP corresponds to a different plane in the diagram. The NP also provides services and therefore constitutes a plane in the diagram, although it has more flexibility in control options. The NP and the SPs are also represented by one agent each namely; the *Network Provider Agent* (NPA) undertakes network provision tasks, such as establishment of Virtual Paths and allocation of resources to SPs; whereas the *Service Provider Agents* (SPAs) represent the global policy of each SP in managing their resources and negotiate with the RAs for re-allocation of resources from a source-destination pair to another, according to the anticipated demand.

User requests for connectivity from a specific source to a specific destination through the ATM network originate from the *Proxy User Agent* (PUA) and are propagated to the RAs by the *Connection Agent* (CA). The CA is also responsible for collecting the bids of RAs of competing SPs and selecting the most suitable one. A call request is identified by its source-destination pair and, furthermore, is associated with a particular *Class of Service* (CoS), i.e. a class of QoS requirements, which is requested from the network. Along with the CoS indication the request includes also traffic and QoS parameters which are essential for calculating the required resources for Connection Admission Control as well as for charging the connection.

The main vehicle for managing network resources is by exploiting dynamic bandwidth allocation to virtual paths. A Virtual Path (VP) in IMPACT terminology is a path of specified bandwidth from a source node to a destination node in the network using physical links of the network. Another major control function, which is applied before performing Admission Control, relates to the routing of the connection. The system's architecture simplifies the routing problem, through allocating new connections to one of a set of pre-enumerated VPs relevant to the target source-destination pair. Note that the VPs in this set are known, fixed in terms of route though not bandwidth and represent a small manageable subset of the set of possible VPs for that source-destination pair. Whilst this sounds limiting, it is not believed to be so in practice, as the set of pre-enumerated VPs could be changed over time. Apart from easing the routing decisions, the pre-enumeration of routes simplifies the CAC mechanism. This is because the CAC algorithm is not performed in a per-hop basis, but only upon entrance to the selected VP.

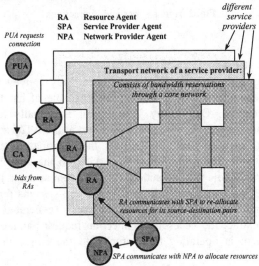

Fig. 1. The Notion of a Network Provider and multiple Service Providers in the IMPACT agent architecture

2.2   Description of Agents Supporting Network Control

Having described the networking environment envisaged within IMPACT, we now review the basic agent types that compose the overall network control system. Fig. 2 depicts the role of each one of the principal agents in the ATM network.

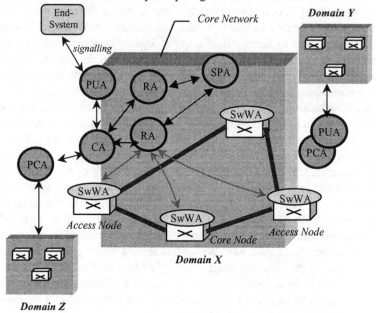

Fig. 2. Agent Types and Interactions

Proxy User Agent (PUA)

Proxy User Agents act as brokers of the end-user. They convey users' connection requests to the agent system, while they are also capable of handling the system's responses and decisions. One Proxy User Agent is actually associated with a particular ATM end-system, i.e. with the (input) port at the ATM access node, which is connected to this end-system. PUAs pass the information associated with a connection request (i.e. destination, class of service) to the appropriate Connection Agent (CA) that will handle the interaction with the agent system. Towards obtaining this information, PUAs can communicate with the end-systems through appropriate wrapping of conventional ATM signalling messages (generated by native ATM applications) and extract the necessary connection request parameters. In this way, the IMPACT system is operating transparently to the conventional native ATM applications.

Connection Agent (CA)

Connection Agents perform the selection of the most appropriate service provider that will offer service for a particular connection request. Specifically, the CA determines the destination and CoS of a call request and instigates requests to relevant RAs (one belonging to each SP) for network resource availability and price. Relevant RAs that will bid are those that can provide the requested connection to the specified destination, taking into account QoS requirements. After receiving the bids, the CA selects a particular bid, based on price and the parameters passed by the PUA, and conveys the results to the PUA. The PUA notifies the terminals as in the conventional case of an accepted connection in a signalling based ATM network. The CA also instructs the selected RA to install the connection, end to end. RAs that were not chosen by the bidding process are also notified, so that they can cancel any reservations made towards offering a bid.

Resource Agent (RA)

Resource Agents encapsulate the core logic of the network control strategies. Each RA implements and applies a particular strategy on behalf of a Service Provider. A Resource Agent is appointed to manage the resources of a particular source-destination pair on behalf of a specific Service Provider. Therefore, in the usual case where a Service Provider can offer service for various source-destination pairs, it "owns" a host of resource agents. The rationale behind this specific degree of distribution is that the system offers flexibility to the service provider in determining specific control strategies at the granularity of a source-destination pair. Nevertheless, other schemes for distributing RAs could have been adopted as well. Resource Agents receive requests (from CAs) for connectivity of a particular class of service from a source node to a destination one. Upon the reception of such a request, a resource agent checks resource availability and then provides a measurable bid (e.g., in terms of price) to the CA.

The particular criteria that drive the participation of the RA in the bidding process are internal to the network control strategy specified by the service provider. These strategies take strongly into account the CoS of every call request. In this way, a variety of network control schemes can be coded within RAs. As a simple example, strategies based on effective bandwidths (e.g., [10], [11]) can easily be applied, along with the charging schemes that stem directly form the notion of the effective bandwidth [1]. Apart from applying a specific admission control algorithm, RAs allow an increased degree of flexibility through allowing negotiation with SPs and the NP towards combating problems such as resource starvation and providing the intelligence in managing network resources. The ways in which such intelligence is applied are outlined in section 2.4 dealing with experimental results.

Note also that RAs invoke operations on the ATM switching nodes (e.g. for installing connections on the switches), through appropriate interaction with Switch Wrapper agents (SwWA).

Service Provider Agent (SPA)

As already outlined, a Service Provider Agent represents a Service Provider. It manages and negotiates with a set of RAs. Negotiation deals with re-distributing the resources (leased from the NP), among the various RAs, so that the Service Provider maximises its profits.

Network Provider Agent (NPA)

The Network Provider Agent constitutes the entity that represents the Network Provider. It incorporates the strategies of the NP, with respect to negotiation with SPs, as well as with RAs that may be "controlled" directly by the NP. The NP is the entity that has control over the physical network. As a result, the NPA interacts with Switch Wrappers towards establishing the logical network topology, i.e. the (pre-enumerated) Virtual Paths for each source-destination pair and for each SP.

Switch Wrapper Agent (SwWA)

SwWAs constitute a "virtual" software abstraction of an ATM switch and its resources and provide a generic, vendor-independent interface for network control and management applications. SwWAs receive requests for operations to be executed on the switch by the RA in the form of agent communication language. These operations are then translated to a sequence of low-level management/control commands, to be executed on the switch. Individual management/control commands are executed by accessing the switch control software either through a management protocol (e.g., SNMP, GSMP [12]) or through some sort of application programming interface (API) in the case of programmable switches. The next section describes a methodology for implementing such Switch Wrappers, elaborating also on the particular implementation carried out within IMPACT.

Proxy Connection Agent (PCA)

PCAs have been devised in order to enable the architecture to scale to wider networking infrastructures (e.g. the Internet). PCAs reside at the edges of the network and propagate connectivity requests to other networks. As a result they act as gateways of requests that have to be conveyed to other (upstream/downstream) networks.

For a detailed specification of the agents and their role, see [17] and [22].

2.3   Inter-Agent Communication

IMPACT agent entities take advantage of the communication facilities provided by an agent platform built within the project, namely the Basic Agent Template (BAT). Several agent platforms were considered, examined and evaluated before proceeding with the development of BAT. Although some of these platforms were adequate for providing the agent communication functionality, it was decided that they imposed too much processing overhead. This overhead was certainly unacceptable, since the project deals with network control tasks which are very demanding in terms of speed.

BAT constitutes a streamlined, purpose-designed system that takes advantage of inherent features in the Java language to implement the distributed communication features required by agents. In particular, BAT is written entirely in Java and relies on Java's Remote Method Invocation (RMI) for the agent communication. Moreover, in order to implement the Agent Communication Language (ACL), BAT makes use of a *signature matching* mechanism. According to signature matching, communication is achieved by the exchange of Java objects using a subset of FIPA [13] communication acts and, thanks to Java's "reflection" mechanism, agents determine at runtime which method is to be used to handle a received object. Note that the implementation of the model follows partly the FIPA specifications, while being deliberately simple in order to support a real-time system.

Using BAT and the Java language, one can easily implement new agent entities featuring the FIPA-compliant communication capabilities. This process was followed towards implementing the agent entities outlined in the previous section.

2.4   Results from Experiments

A proof-of-concept implementation of the architecture has taken place resulting in a software system for control and resource management of ATM networks. Initially, it focused on implementing a robust version of the basic functionality of the agent components, as well as the interactions between them. In the sequel, more capable versions of the agent entities were deployed. It is noteworthy that the IMPACT system has been successfully installed and operates on three different experimental sites, namely at Tele-Denmark's ATM testbed (Aarhus, Denmark) which constitutes the main experimental site of the project, at the Telecommunications Laboratory of the National Technical University of Athens (Athens, Greece), as well as at the EXPERT testbed [14] (Basel, Switzerland). In the scope of this section, we describe scenarios and results from early experiments with the system.

One of the goals of the experiments was to demonstrate that agents can provide more efficient resource utilisation by re-allocating resources either re-actively or pro-actively. These competencies result in better utilisation of the network resources and consequently in more revenue for service providers and network operators and in better service for the customer. A mechanism towards realising these competencies is to re-allocate the unused capacity of VPs. Another approach is to transfer already connected calls to other routes, if possible, thus freeing bandwidth where is needed. Therefore, during the experiments two straightforward cases were targeted:
- transfer of unused capacity of VPs between RAs of the same SP and
- connection re-routing in order to re-allocate VP capacity.

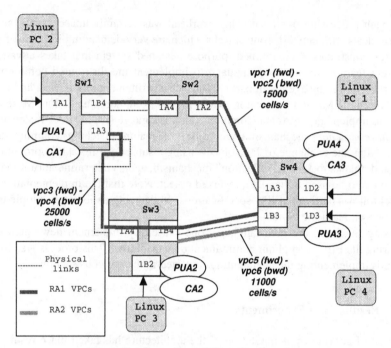

Fig. 3. Configuration of the experiment

We now describe one of the experiments in more detail. The scenario assumed that there was only one SP using some capacity of the network (a fixed capacity on each physical link). The rest of the network capacity was assumed to be used by the NP. Two source-destination pairs and one CoS were established. A SPA (SPA1) represented the sole SP, having two RAs (RA1 and RA2) under its control, each one controlling one of the source-destination pairs. The setting described above is illustrated in Fig. 3.

RA1 controlled the source-destination pair between ATM switching nodes Sw1 and Sw2. Two routes were established, *vpc1(fwd)-vpc2(bwd)* with initial capacity 15000 cells/sec each and *vpc3(fwd)-vpc4(bwd)* with initial capacity 25000 cells/sec each. RA2 controlled the source-destination pair between Sw3 and Sw4. One single route was established, *vpc5(fwd)-vpc6(bwd)* with initial capacity 11000 c/sec each. Simple native ATM applications (an AAL5 "ping" application and an application capturing and transmitting video frames over AAL5) were used for initiating connection requests.

According to the experiment's scenario, the following connection requests were subsequently performed:
(1) From Linux PC 2 (Sw1, port 1A1) to Linux PC 1 (Sw4, port 1D2), with PCR equal to 10000 cells/sec. This connection was routed by RA1 through *vpc3,4* (the rationale being that this VP had larger free capacity than *vpc1,2*).
(2) From Linux PC 3 (Sw3, port 1B2) to Linux PC 4 (Sw4, port 1D3), with PCR equal to 10000 cells/sec. This connection was routed by RA2 through *vpc5,6*.

(3) Again, from Linux PC 3 (Sw3, port 1B2) to Linux PC 4 (Sw4, port 1D3), with PCR equal to 10000 cells/sec. This call was routed by RA2 through *vpc5,6*. At that stage, *vpc5,6* did not have enough capacity to accommodate this connection (because of connection (2)), so after negotiating with the SPA, RA1 allocated 10000 cells/sec from the spare capacity of *vpc3,4* to RA2. Therefore, the spare capacity of vpc5,6 increased to 11000 cells/sec and thus could now handle connection (3). After handling this connection its spare capacity fell to 1000 cells/sec again.

These three connection requests enabled the comprehensive demonstration of a capacity transfer case. Capacity transfer led to accepting a connection that would have been rejected otherwise (i.e. without the IMPACT control system). In the sequel, another connection request was generated, in order to provoke a connection re-routing:

(4) From Linux PC 3 (Sw3, port 1B2) to Linux PC 4 (Sw4, port 1D3), with PCR equal to 10000 cells/sec. This connection was routed by RA2 through *vpc5,6*; *vpc5,6* did not have enough capacity to accommodate this connection (because of connection (2)), but also RA1 did not have sufficient spare capacity on its VPs that share physical links with *vpc5,6* (recall that the bandwidth allocation of the SP on each physical link is fixed). Therefore, after negotiating with the SPA again, RA1 attempted to re-route connections from *vpc3,4*. Indeed, connection (1) was re-routed from *vpc3,4* to *vpc1,2* and 10000 cells/sec spare capacity from *vpc3,4* were allocated to RA2 *vpc5,6*, in order to accommodate this last connection request.

3   Interfacing Agents to Networking Devices

A very important issue towards implementing an agent-based network control system is to devise and implement mechanisms for interfacing the agent software with the networking devices. To this end, it is highly desirable that the target networking devices offer some degree of programmability in the form of some sort of programmable software interfaces (Application Programming Interface – API). In most devices however, such interfaces do not exist. In particular, each device (e.g., ATM switch, IP Router) comes as a "black box" with a full suite of network control schemes, specified and implemented by the particular manufacturer. Internal details regarding these network control schemes are not released and access to the control software is not provided. Even in cases where some commercial devices provide access to their control software through an API, this is limited to configuration tasks and does not allow modification of the core components of network control and resource management mechanisms. In the sequel, we present a quite general architecture, that allows the implementation of network control and resource management functionality other than the "default" behavior supported by the devices. Building on the concepts of this architecture, we have implemented the interface between the software agents and the ATM switches used in the scope of IMPACT experiments.

Having agents controlling network nodes results in altered control functionality and demands that the "default" control behavior of the nodes is overridden. In this context, Fig. 4 presents a framework for embedding custom network control functionality in ATM networks, which overcomes the lack of access to the switch control software of most commercial ATM switches. According to this architecture, the network nodes export appropriate generically defined, switch-independent programmable interfaces to distributed software entities that implement various network control schemes on top of the nodes.

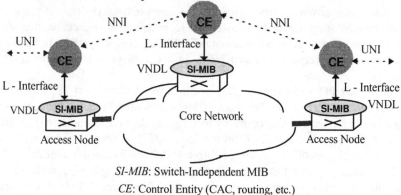

SI-MIB: Switch-Independent MIB
CE: Control Entity (CAC, routing, etc.)

Fig. 4. Architecture for embedding network control and resource management functionality in ATM networks

The idea of implementing programmable interfaces stems from the recent strong interest in network programmability. Since several companies and laboratories have started a new IEEE standards development project, namely IEEE P1520 [1], we strongly believe that the implemented programmability should comply with the proposals and developments of this group. In order to align the architecture depicted in Fig. 4, to the reference model produced by the P1520 initiative (Fig. 5), each node should export an L-interface, allowing network control components to make use of the so-called Virtual Network Device Layer (VNDL). A VNDL constitutes a virtual (software) representation of a network device, comprising any essential piece of information for the deployment of network control and resource management algorithms. In the proposed architecture, we call the VNDL representation (i.e., the collection of appropriate controlled objects) a *Switch-Independent Management Information Base* (SI-MIB).

According to our architecture, distributed *Control Entities* (CEs) deal only with the information contained in the SI-MIBs, which is consistent with the actual state of the switches. When performing the desired network control tasks, the Control Entities exchange network control information by consulting their SI-MIBs. The interaction between a Control Entity and its corresponding SI-MIB takes place through this generic L-interface. In order to have a functional L-interface, a mechanism for accessing the low-level capabilities of the actual switch is needed. Such a mechanism should enable a Control Entity to query or alter the status of a network node. This can be well performed through a general purpose management protocol (e.g., SNMP, GSMP, CMIP), depending on the management capabilities of the target ATM switch. Thus, a control operation on commercial ATM switching nodes can be performed through the SNMP protocol, since most switches incorporate appropriate SNMP daemons and MIBs. As a result, according to our architecture a network control algorithm invokes some generic operations of the L-interface. The execution of these operations on the actual switch takes place through appropriate drivers that interact with the switch using a management protocol. However, it must be noted that the use of a management protocol (such as SNMP) for controlling the switch by a Control Entity residing outside the device can introduce a performance penalty.

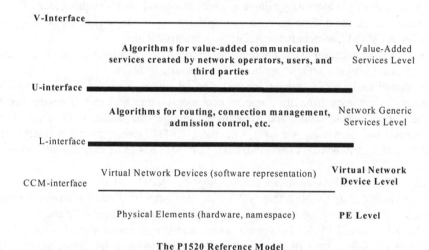

The P1520 Reference Model

Fig. 5. Reference Model of the P1520 standard for Programmable Networks

It is important to note that this architecture allows the network control functionality to be applied on a network of switches from several manufacturers. This is due to the fact that the algorithms interact actually with the SI-MIB, which is independent of the switch model. Therefore, the L-interface can also be generic and uncoupled from manufacturer-specific details. However, for each different switch model a new software driver for accessing the low-level capabilities of the switch has to be implemented. This inevitably implies some additional effort for installing/adapting the described software system to a multi-vendor environment.

This fairly general architecture for interfacing software control entities to the network, has been used by the authors (apart from IMPACT), in the scope of other applications ([15], [16]).

4   Application to other Network Technologies

Although the framework presented in the previous sections has been illustrated here by its application to an ATM network environment, the underlying concepts are fairly generic and can be applied to any network technology that supports QoS or some sort of service differentiation. In support of this argument, we now give a brief example of how the IMPACT architecture can be mapped and applied on an Internet network domain. We assume that the domain supports *Differentiated Services* (DiffServ), enhanced with some constraint-based routing mechanism, which takes into account not only the destination address, but the source and the CoS as well. Such a mechanism can be achieved by using e.g. Multi-Protocol Label Switching (MPLS).

In the sequel, we outline how the principal IMPACT concepts could be mapped from the ATM environment to the IP/DiffServ environment:

- An ATM connection is mapped to an *IP flow*, i.e. a sequence of IP packets exchanged between specific applications of specific IP hosts.
- A Virtual Path (VP) is now a *constraint-based routing path* for IP flows, which enter the network from the same source access node and exit from the same destination access node and, moreover, require the same QoS class. For abbreviation purposes, we still call them "VPs" (with quotation marks to distinguish from actual ATM VPs). Each "VP" still has some capacity allocated to it.
- The NPA and SPAs establish the topology of "VPs" and allocate resources to them. In order to establish the "VPs", they interact with *Router Wrapper Agents* (RWAs) for the purpose of configuring packet forwarding tables in IP routers.
- The RAs control "VPs" between a source and a destination access node. They accept (or reject) new IP flows and they allocate them to the most appropriate "VP", depending on resource availability. For this purpose, negotiation with the SPA for transfer of "VP" capacity, or re-routing of IP flows might be needed. In order to allocate an IP flow to a "VP", the RA has to interact with RWAs so as to properly configure packet forwarding tables.
- PUAs and CAs still interact as intermediaries between the end-systems/users and the network. The end-systems may use RSVP (instead of ATM signalling) for conveying connectivity requests to the PUAs, which would act, in this case, as RSVP "wrappers".

5   Conclusions

This paper has illustrated the need for flexible network control. It has been pointed out that, although current signalling protocols have enabled rapid deployment of interoperable ATM networks, they present some drawbacks when it comes to customising control of networks and services. The need for flexible and efficient network resource management schemes appears more essential than ever before, since networks are becoming more complex and heterogeneous and applications are becoming more resource-demanding and require QoS guarantees. A lot of effort has been allocated towards this direction. Several technologies and frameworks have been developed in order to boost flexibility and efficiency in network control. Among the characteristic examples are initiatives towards network programmability [1] [2] [3] [4] [5], as well as technologies such as active networks [19].

This paper presented an architecture for distributed network control based on Intelligent agents devised and implemented within the IMPACT project for an ATM networking environment. This architecture takes strongly into account the requirements for flexible network control. It also provides an open framework for the deployment of network control schemes. Different service providers are even free to use different control algorithms on the same (logically partitioned) physical infrastructure. We really expect that the system's major contribution is the ability to specify different control policies, operating together on the same physical infrastructure, with the inherent ability for negotiation, resource re-allocation and adaptation to changes in traffic demand.

An experimental scenario has been also described with a view to demonstrating that the architecture is capable of delivering the pledged benefits. The implementation and experiments have been conducted on ATM testbeds. However, the concepts introduced by the project are generic and can be applied to IP networks with QoS support, as well. This applicability is almost direct, when considering enhancements to the IP service model that offer QoS (such as those described within the IntServ and DiffServ frameworks).

Given the research effort on open distributed network control approaches, one may argue that it is time that these developments become deployed in practice. A major concern towards deploying distributed approaches, relates to the performance of the control/management systems. In the management plane distributed systems have already been deployed [21]. Nevertheless, it is still under question whether the performance required in the control plane can be reached. We do believe that network resource management operations could be carried out in such a distributed fashion, since the potential benefits can compensate for some performance degradation.

Finally, another critical point concerns network programmability. The network control mechanisms currently built into networking equipment are not open. Architectures such as the one presented in this paper demand programmability at the network nodes. Although a shift is already taking place, with manufacturers implementing limited programmability in their devices, there is still some way to go. Until fully programmable network devices are available, researchers, network administrators, as well as network architects need in several situations means of customising the control functionality of their nodes. Section 3 constitutes a short contribution to this direction.

Acknowledgements

The authors gratefully acknowledge support from the European Commission under the ACTS Project AC324 "Implementation of Agents for CAC on an ATM Testbed".

The authors also acknowledge valuable help and contributions from the other project partners: Queen Mary and Westfield College, Swisscom AG, Tele-Danmark A/S, Flextel SpA, Teltec, and ASPA.

References

1. Biswas, J., Lazar, A. et al "The IEEE P1520 Standards Initiative for Programmable Network Interfaces", IEEE Communications Magazine, pp. 64-71, October 1998
2. Rooney, S., Van der Merwe, J., Crosby, S., Leslie, I., "The Tempest: A Framework for Safe, Resource-Assured, Programmable Networks", IEEE Communications Magazine, pp. 42-53, October 1998
3. Rooney, S., "The Hollowman: an innovative ATM control architecture", *Proceedings of IM'97* (1997), San Diego
4. Lazar, A., Lim, K. and Marconcini, F., "Realizing a Foundation for Programmability of ATM Networks with the Binding Architecture", *IEEE J. on Sel. Areas Comm.*, vol. 14, no. 7, pp. 1214-1227, September 1996
5. Cidon, I., Hsiao, T., Khamisy, A., Parekh, A., Rom, R., and Sidi, M., "An Open and Efficient Control Platform for ATM Networks", Sun Microsystems laboratory, (1996), available from: http://www.sunlabs.com/research/hsn/
6. QoS Forum 1999, "The Need for QoS", White Paper, electronically available at: http://www.qosforum.com/tech_resources.htm
7. Newman, P., Minshall, G., Lyon, T., "IP Switching: ATM under IP", IEEE/ACM Trans. on Networking, vol. 6, no. 2, pp. 117-129, April 1998
8. Griffin, D. and Georgatsos, P., "A TMN System for VPC and Routing Management in ATM Networks", Integrated Network Management IV, Proc. 4th Int'l. Symp. on Integrated Network Management, A. Sethi, Y. Raynaud and F. Faure-Vincent, eds., (1995)
9. Aneroussis, N. and Lazar, A., "Virtual Path Control for ATM Networks with Call Level Quality of Service Guarantees", IEEE/ACM Trans. on Networking, vol. 6 no. 2, pp. 222-236, April 1998
10. Guerin, R., Ahmadi, H., and Naghshineh, M., "Equivalent Capacity and its Application in High Speed Network", IEEE J. on Sel. Areas Comm., vol. 9, no. 7, pp. 968-981, September 1991
11. Kelly, F., "Notes on Effective Bandwidths", in Stochastic Networks: Theory and Applications (1996) (Editors: F.P. Kelly, S. Zachary and I.B. Ziedins), Oxford University Press, pp. 141-168
12. Internet Engineering Task Force: Request For Comments 1987, P. Newman, W. Edwards, R. Hinden, E. Hoffman, F. Ching Liaw, T. Lyon, G. Minshall, "Ipsilon's General Switch Management Protocol Specification, Version 1.1", August 1997
13. Foundation for Intelligent Physical Agents, FIPA 97 Specification Part 2, "Agent Communication Language", http://www.fipa.org
14. Electronic information on the EXPERT platform, available on the WWW through the URLs: http://www.telecom.ntua.gr/watt/wdb/Expert and http://www.snh.ch/projects/watt/wdb/expert/
15. Vayias, E., Soldatos, J., Kormentzas, G., Mitrou, N., Kontovassilis, K., "Monitoring Networks over the Web: Classification of approaches and an implementation", in Proc. of International Conference on Telecommunications (ICT'98), Vol. IV, pp. 451-456, Chalkidiki, Greece, (1998)

16. Soldatos, J., Kormentzas, G., Vayias, E., Kontovasilis, K., Mitrou, N., "An Intelligent Agents-Based Prototype Implementation of an Open Platform Supporting Portable Deployment of Traffic Control Algorithms in ATM Networks", Proc. of the 7th COMCON Conference, Athens, Greece (1999)

17. IMPACT AC324 Project Deliverable 03 "Specification of Agents", July 1998, available at http://www.acts-impact.org

18. IMPACT AC324 Project Deliverable 04 "Specification of Signalling Implementation", November 1998, available at http://www.acts-impact.org

19. Calvert, K. L., Bhattacharjee, S., Zegura, E., Sternebz, J., "Directions in Active Networks", IEEE Communications Magazine, October 1998, pp. 72-78

20. Hayzelden, A., Bigham, J. "Agent Technology in Communications Systems: An Overview", Knowledge Engineering Review Journal, Vol.14:3, (1999) pp. 1-35

21. Millikin, M. "Distributed Objects: A New Model for the Enterprise", Data Communications Magazine, February 1997

22. Bigham, J., Cuthbert, L.G. , Hayzelden, A., Luo, Z. and Almiladi, H., "Agent Interaction for Network Resource Management", in the Proc. of the Intelligence in Services and Networks '99 (IS&N99) Conference (1999)

23. Mandeville, R., and Newman, D., "Empowering Policy", Data Communications Magazine, May 1999

Providing Customisable Network Management Services Through Mobile Agents

David Griffin[1], George Pavlou[2], Panos Georgatsos[3]

[1] Department of Electronic and Electrical Engineering, University College London, Torrington Place, London WC1E 7JE, United Kingdom
D.Griffin@ee.ucl.ac.uk
[2] Centre for Communication Systems Research, School of Electronic Engineering, IT and Mathematics, University of Surrey, Guildford, Surrey GU2 5XH, United Kingdom
G.Pavlou@ee.surrey.ac.uk
[3] A.T.T.S. Dept., Algosystems S.A., 4, Sardeon St., 171 21 N. Smyrni, Athens, Greece
pgeorgat@algo.com.gr

Abstract. Telecommunications network management has attracted a lot of attention in terms of research and standardisation in the last decade. TMN and TINA architectural frameworks try to address the management needs of broadband networks and services. They both cater for multi-domain, multi-operator environments while they also allow for electronic customer access to management services. These services, though, are fixed in the sense that new features can only be added after a full research-standardisation-deployment cycle and are static as far as their use: they execute according to their in-built logic and clients may customise their operation only through tuning standardised operational parameters before service execution. The advent of mobile agent technologies opens new possibilities, allowing dynamic and customisable services to be offered to clients. In this paper we explain how mobile agents can enhance traditional connectivity management services, presenting the relevant architecture and design of a customisable network management system based on mobile agents.

1 Introduction and background

A significant amount of research and standardisation activity took place in the area of telecommunications network management over the last decade. The culmination of this work is the Telecommunications Management Network (TMN) [1], with systems based on the relevant standards currently being deployed. More recently, the Telecommunications Information Networking Architecture (TINA) [2], has tried to integrate management with service control for future multimedia teleservices, integrating and evolving the concepts of TMN and the Intelligent Network (IN). While the TMN currently uses the manager-agent model and OSI Systems Management (OSI-SM) [3] protocols as the basis for its interoperable interfaces,

TINA advocates the use of more general purpose distributed processing technologies such as the OMG Common Object Request Broker Architecture (CORBA) [4].

Both the TMN and TINA architectures cater for multi-domain multi-operator deregulated telecommunications environments, which allow also for customer access to management services. The TMN defines the inter-domain X interface, specific classes of which can be used between operators in a peer-to-peer fashion; they may also be offered to other Value Added Service Providers (VASPs) and customers e.g. the EURESCOM Xcoop and Xuser interfaces respectively. TINA defines the LNFed and Ret2Ret reference points for peer-to-peer network and service management co-operation respectively; it also defines the ConS reference point for providing connectivity services to Retailers and the Ret and TCon reference points for providing services to end-users [5]. Both TMN and TINA based management systems have been designed, developed and demonstrated in a number of research projects in Europe and elsewhere while management systems offering Customer Network Management (CNM) services have started to be deployed.

In the emerging multi-service telecommunications environment, customers will be able to search for the services they require through electronic brokers. They will subsequently be able to subscribe and customise their services electronically, while the whole set of interactions required for service provisioning will take place almost instantly, in an automated fashion supported by integrated management systems. Customers will be able to monitor the usage of their services, peep/pay their bills, modify service features and unsubscribe electronically. An example of such a service that may be offered to corporate users is a Virtual Private Network (VPN) service.

In such environments, the features of management services offered to customers are first researched, standardised and eventually implemented. This process takes a long time e.g. research on TMN-based VPN services over SDH/ATM has taken place over the last five years but such services are not yet offered to customers. In addition, any modification to the interfaces that support those services, e.g. for providing more sophisticated features that were not thought out in advance, needs to go through the full research, standardisation and deployment cycle.

The recent emergence of "execute-anywhere" languages like Java has made code mobility possible and has given rise to research and development into frameworks for mobile software agents. TMN and TINA use static manager-agent and client-server approaches respectively, static in the sense that the capabilities of the agent or server part are statically defined and cannot be changed. The use of mobile agents offers new possibilities in the sense that logic may be sent to execute in locations which are largely "unaware" of the functions and capabilities of those agents. Such a facility may enable the provision of management services in a different, more flexible and dynamic fashion and this is the subject of intense current research.

Mobile agent frameworks are currently addressed by two standards bodies. The Federation of Intelligent Physical Agents (FIPA) [6] looks at high-level semantically rich interactions between software agents that deploy some form of intelligence and adaptability, having its roots in Distributed Artificial Intelligence (DAI). OMG looks mostly at the issue of mobility according to a standard interoperable framework through its Mobile Agent System Interoperability Facility (MASIF) [7]. In the latter, *agent systems[1]* model the execution environment able to host mobile agents and correspond roughly to OSI-SM *agents* in the TMN manager-agent framework or to the TINA concept of the Distributed Processing Environment (DPE) *node*. Within an agency, *fixed* or *static* agents provide the bare-bones functionality which is statically defined but this can be augmented dynamically through *mobile* agents which are sent to execute there.

The flexibility potentially offered by mobile agents has led a number of researchers to consider their applicability to network and service control and management environments. Breugst [8] considers mobile agents in the context of the Intelligent Network and proposes an agent-based architecture for "active" IN service control. Biesz [9] discusses the general issues of using agents for network management while other researchers have presented specific case studies using mobile [10] and intelligent agents [11]. In this paper we examine how mobile agents can enhance traditional connectivity provisioning management services, allowing flexible customer network management with respect to performance, fault and dynamic reconfiguration which are driven by customer-defined policies. This work examines also the applicability of mobile agents to TMN and TINA-based customer network management services.

The proposed enhancements to connectivity provisioning management services make these services usable in a 'dynamic' rather than 'static way', where the terms 'static' and 'dynamic' are used as follows. In traditional service provisioning systems, customers subscribe to certain network management services according to their needs. During service subscription, customisation of certain management aspects to a particular set of customer requirements may take place - by selecting a subset of offered service features. However, this presents a 'static' approach with respect to the use of the services by the customers, who use the services as they have been subscribed, and services execute according to their 'built-in' logic. This approach is valid as long as the management needs of the customer environment are static, which however is not generally the case. Alternatively, a more flexible approach should be followed. Network management services should be built and offered to customers so that they leave several 'degrees of freedom' for use. Customers will modify the use of the offered services not only during subscription time but also during service operation, according to the dynamics of their environment.

In this paper, we demonstrate how customers may be offered an initial set of management services which could then subsequently be customised by sending customer-owned mobile agents to execute in the provider's environment. With mobile agents it is possible for a client to deploy specific functionality *at run-time* in a server to add value above and beyond the server's basic facilities. Following deployment, the mobile agent may act autonomously to interact with the local environment and make *local* decisions which may then be implemented as *local* management actions without needing to interact with the remote client every time a

---

[1] While MASIF uses the term *agent system*, some implementations - notably Grasshopper - use the term *agency*. Our implementation is based on Grasshopper and we use the term *agency* in our system and in the remainder of this paper.

decision is required. The provided facilities correspond roughly to the TMN Xuser interface and the TINA ConS reference point but they support extensible and customisable functionality. The proposed architecture and associated realisation are being pursued in the ACTS MIAMI project (Mobile Intelligent Agents in the Management of the Information infrastructure) [12], which examines the impact and possibilities of mobile agent technology to network and service management in general. MIAMI has defined a case study and associated environment which will allow co-operating customers dynamically to form Virtual Enterprises (VEs) for providing a particular service to end-users [13]. The VE makes use of services offered by a Active Virtual Pipe Provider (AVPP), a business role similar to the TMN VASP or the TINA retailer. The AVPP provides a programmable, dynamic virtual private network, as needed by the virtual enterprise. The AVPP makes use of connectivity services offered by a Connectivity Provider (CP), which is a business role similar to the TMN Public Network Operator (PNO) or the TINA Connectivity Provider.

The architecture presented in this paper concentrates in the CP domain of the MIAMI system. Mobile agents are used both in reference points offered to customers and also internally within the CP domain, trying to implement better performance and fault management functionality. The work presented here is being verified and validated through an implementation which will be followed by a field trial. The status of the implementation work, which aims to provide a proof of concept for agent-based customisable network management services, is discussed in more detail in the summary and conclusions section. It should be finally noted that the types of agent-based services discussed in this paper require strong security guarantees, especially as they target telecommunications environments which allow customers some degrees of freedom. Security is orthogonal to the architecture presented here and is not considered in this paper; the reader may consult [14] for a general consideration of security issues and associated mechanisms in agent-based systems.

The rest of this paper is structured as follows: section 2 introduces an example network environment to be managed; section 3 introduces our architecture for agent-based network management systems; section 4 discusses the way in which the management services offered by a Connectivity Provider may be customised by their users; section 5 illustrates the flexibility of customisable management services with an example; section 6 discusses our implementation approach; and section 7 presents the summary and conclusions, examining the benefits and drawbacks of such an approach.

2   A typical networking scenario

In the MIAMI project, and in this paper, we assume that the network scenario consists of an end-to-end IP network which interconnects the participants of a Virtual Enterprise (VE). In addition to standard best-effort, Internet quality connectivity for general email and web browsing the VE users also require access to higher quality connectivity facilities for real-time services such as high bandwidth video conferencing or for high speed access to large files. The users expect to pay a premium rate for guaranteed quality services, but they also wish to use lower cost, and correspondingly lower quality services for more general purpose communications.

In addition to providing access to the Internet, the IP CP also makes use of the services of an underlying CP - offering semi-permanent ATM connections in this case - who is able to provide guaranteed quality leased lines between the IP CP's routers. Figure 1 shows the network scenario we assume in the rest of this paper.

The service provided by the ATM CP is an end-to-end Virtual Path (VP) service offering PVPs (permanent VPs) between specified termination points. Associated with each VP is a number of parameters which defines the capacity of the connection and the level of performance to be provided in terms of end-to-end delay, delay variation and tolerable cell loss ratio. VPs may be created, deleted and modified through client management actions. The clients may monitor VP usage and performance statistics, and initiate fault monitoring activities on their resources.

Issues associated with inter-administration connectivity and federation of management systems are outside the scope of this paper although the dynamic customisable approach presented for a single CP domain could also apply to a multiple CP, inter-domain environment. The remainder of this paper will concentrate on the ATM CP, although the issues discussed are also generally relevant to CPs offering managed services for any network technology.

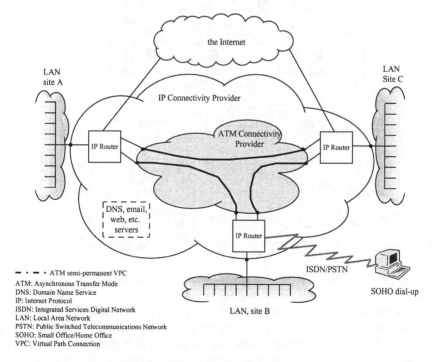

Fig. 1. A possible networking scenario

3   An architecture for agent-based customisable management systems

When treated as a black box, the CP's management system can be seen as offering two main classes of interface (see Figure 2). Firstly the interfaces to its clients (Dynamic Connectivity Management (DCM) and Contract Negotiation (CN) interfaces in the figure which are described later in section 3.2) over which it offers a set of services for the management of ATM connectivity. Secondly the interfaces to the underlying Network Elements (NEs) which are ATM VP cross-connects in our scenario. The interfaces to the NEs will be based on the technology offered by the vendors of the switches - Simple Network Management Protocol (SNMP) or Common Management Information Protocol (CMIP); if CORBA-based interfaces are available then this is also an option. It is unlikely that commercial switches will have embedded agent enabled interfaces in the immediate future.

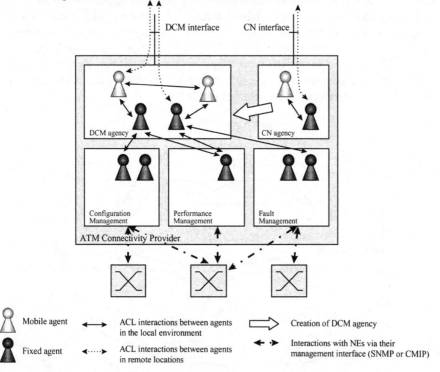

Fig. 2. High-level architecture of an agent-based CP

The architecture for our system draws heavily on traditional network management systems, TMN [1] and TINA [2] in particular. There is a network management level which contains functionality for Fault, Configuration and Performance management activities; this is above the network element management level which performs element specific management activities on the set of network elements below it (with possibly one or more sub-network layers between (see Figure 3). The management software at the network and element management levels resides in general-purpose computing workstations which are interconnected and also connected to the underlying network equipment by a communications network. The latter network, termed the Data Communications Network (DCN) in TMN and the Kernel Transport Network (kTN) in TINA, is logically, if not physically, separate from the underlying *managed* telecommunications network.

The main difference of our architecture compared to traditional network management systems is that we have replaced a *static* service management level with a programmable layer - the DCM agency - above the network management level. This is where client mobile agents may execute and customise the offered management services according to the requirements of the clients. In summary, compared to a TMN system, the architecture shown in Figure 2 may be mapped as follows: the Configuration, Fault and Performance Management agencies each span the Network- and Network Element Management layers while the DCM and CN agencies map to the Service Management layer.

3.1   Agent-based communications

We use the term Agent-Based Communication (ABC) to refer to the mechanism by which agents communicate with one another. The term implies that specific protocols and interface definitions are used which could either be based on general distributed systems techniques for remote method invocation (CORBA or Java RMI, for example) or on higher level semantic/AI languages (such as KQML [15] or FIPA's Agent Communication Language (ACL) [16]) which support interactions with "semantic heterogeneity". The specific languages and communications protocols are dependent on the chosen agent platform and are not of direct concern in this paper. Mobile agents may communicate with fixed agents and mobile agents may communicate with other mobile agents using ABC.

The communication between agents may take several forms: to raise asynchronous notifications, to query specific agents to retrieve information and to invoke operations. It is assumed that there is a mechanism for publishing the facilities offered by an agent operating in the server role - i.e. a way of formally specifying an agent's interface. In the design of our system, the Unified Modelling Language (UML) [17] is used to specify formally an agent's interface, being subsequently mapped to the Java language. It is further assumed that an event/notification service is offered by the host environment for disseminating the events raised by agents based on filtering criteria. Today's agent platforms, including those based on OMG's MASIF specifications, do not currently offer event/notification services. In our environment it has been necessary to implement notifications in a non-generic and fairly inelegant way, on a case-by-case basis. It should be noted that the MIAMI project is currently in the process of extending the Grasshopper agent platform [18] to allow communication between agents using FIPA's ACL.

ABC extends beyond the local agency to allow communication with agents in remote execution environments. This implies two methods for communications in agent systems: either remote operations may be invoked through ABC (in a similar way to traditional distributed systems based on statically located objects); or mobile agents may physically travel to the remote agency where they may run in the local environment and invoke the *same* operations through *local* (i.e. intra-node) rather than remote (i.e. inter-node) ABC mechanisms.

By relying on mobile agents, an active and dynamically adaptive management system can be built which is not fixed and limited by initial deployment decisions at system design or build time. The choice of which communications method to use - remote or through mobile agents - is an issue which may even be decided dynamically, even at system *runtime*. It is possible to create and deploy a mobile agent when the communications overhead between remote systems rises above a certain threshold, for example. This, however, would be at the cost of physically transferring the agent to the remote execution environment.

3.2   High-level architecture

Figure 2 shows the overall architecture of the CP's management system. The management system consists of three separate agencies for the main management activities of the CP: one each for configuration management, performance management and fault management[2]. In addition there are two agencies which represent the two classes of interface to the clients of the CP: the CN (Contract Negotiation) agency supports the CN interface, and the DCM (Dynamic Connectivity Management) agency supports the DCM interface. Within these latter two agencies a number of fixed and mobile agents may execute. The fixed agents are provided by the CP, at initialisation time, and form the agent-based interfaces to the basic management services of the CP. The mobile agents belong to the clients of the CP and are dynamically created by remote clients.

The role of the CN interface is to provide access to the CN agency for *negotiating* the contract between the CP and its clients. The contract defines the set of agreed basic management services to be provided through the DCM interface.

Following the completion of the contract negotiation phase, a DCM agency and a DCM interface will be instantiated to provide the client with access to the agreed management services. It is through the DCM interface that the client may dynamically invoke management operations and that active network management services are realised through client programming and customisation via mobile agents.

---

[2]   Other management functional areas and management services such as accounting and security are also envisioned, but we currently limit ourselves to these three.

3.3   Operational scenario

Initially, a potential client is unable to invoke CP management services for two reasons: physically it does not have access to a management interface, and legally it does not have a contract with the CP. The first step is to negotiate a contract. A fixed agent in the CN agency offers an interface to allow contract negotiation. This negotiation can be achieved in two ways: either the client creates a mobile agent to move to the CN agency and negotiate locally with the fixed agent; or the client may communicate remotely with the fixed agent.

Following successful contract negotiation, the CP creates an agency and DCM interface for the client. This involves the creation of one or more fixed agents in the DCM agency to offer specific interfaces to the management services which feature in the contract. The fixed agents tailor (in a *static* sense) the management services of the CP to the requirements of the client and to limit access to the services according to the terms of the contract. For example, not all management services may be made available to all clients, or the geographical coverage of, say, configuration management may be limited to specific locations. In other words, the fixed agents operate as *proxies* to the configuration, performance and fault management services of the CP.

When a management service - to create a new VP, for example - has been invoked, either by a locally running mobile agent or by a remote operation from the client, the fixed (proxy) agent invokes the corresponding operations on the agents in one of the configuration, performance or fault management agencies within the CP. It is within the latter agencies that the real management work - such as the creation of a VP - is achieved. Through the activity of the CP's configuration, performance or fault management systems, modifications are made to the network elements through their management interfaces (SNMP, CMIP) to reflect the original requests made by the clients at the DCM interface.

3.4   Implementation approaches to agent-based management systems

In the scenario above the interactions between the client and the CP were discussed for contract negotiation, tailoring of offered management services and dynamically invoking specific management services. This section discusses the way in which the configuration, performance and fault management systems *within* the CP are organised.

In general, network management systems are hierarchical with a network-wide view at the top of the hierarchy and an element-specific view at the bottom with zero, one or more intermediate levels according to the needs of the system. This hierarchical approach can be seen in both TMN and TINA architectures. It is assumed that the configuration, fault and performance management systems in the agent-based CP will also follow a hierarchical architecture for many reasons including scalability and compatibility with existing management architectures and information models. There are two ways in which the CP's management systems could be deployed: either through building agent wrappers on existing management software - to represent the highest level of the TMN or TINA compliant system in the agent environment; or through building the entire management system from scratch in an agent based way and through building agent wrappers to represent the SNMP or CMIP interfaces of the network elements in the agent environment.

These two approaches are shown in Figure 3. Although the figure shows options for configuration management, the same holds for fault and performance management. Option A shows the approach of wrapping legacy systems with agents at the highest level; option B shows agent wrappers at the NE level and the entire system being built using agent technology. Option A retains existing management systems and therefore builds upon the operator's previous investment while option B allows a more agent-centric management system to be designed and built without being constrained by interworking with existing systems.

It is also possible that a hybrid approach (not shown in Figure 3) may be taken. In this case agent wrappers would be provided *at each level* of the hierarchy of a legacy management system. Mobile agents would then be able to visit the hierarchical level that is relevant to their operation and interact with fixed agents representing the legacy software at that level. Mobile agents in such a hybrid environment could relocate by traversing the hierarchy horizontally between subnetworks, or management functional areas, or they could migrate vertically to "zoom-in" or "zoom-out" the level of detail with which they are concerned.

The hierarchical nature of management systems in TMN and TINA (option A, Figure 3) is fixed at system design time and to a certain extent at standardisation time. For each management service to be deployed, the system designers make decisions on the placement of functionality at each hierarchical level and whether to distribute or centralise functionality within a particular hierarchical layer. These decisions are based on many factors, including: the degree of parallelism required; the quantity and complexity of information to be passed between components and whether existing information models need to be modified to support the required information flows; the scalability of the solution; balancing of processing load between management workstations; the complexity of each component.

A promising application of mobile agents for network management is in deploying each management component as a set of co-operating agents (Option B in Figure 3). The way in which these agents are grouped and placed is initially determined by the system designers according to similar criteria as those for the design of static hierarchical TMN or TINA systems. However, now it is possible to revise the grouping and placement decisions during the operation of the management system through the mobility of agents. There could be a number reasons for migrating agents on-the-fly or for spawning new agent instances: e.g. to reduce the processing load on an overloaded management workstation, to cater for an expanding set of managed resources or to reduce the quantity of management traffic, or information lag, between remote systems when it crosses a certain unacceptable threshold. This aspect of agent mobility for network management deserves to be studied further but is not the subject of this paper.

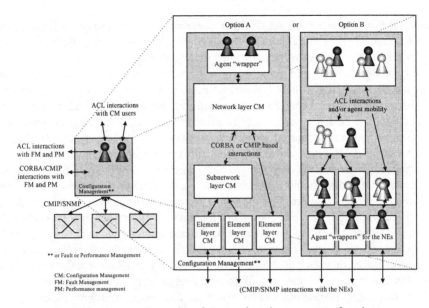

Fig. 3. Implementation options for agent-based management functions

4   Customisation of management services

So far the basic operation of the CP's configuration, performance and fault management systems have been presented together with the means by which clients in the IP CP or AVPP may access them. This section presents the mechanism by which the management services may be customised and the following section presents three integrated scenarios to demonstrate the power of the mobile agent based approach to network management.

With the basic operation described in the previous section there is very little apparent advantage in adopting an agent based management system - very similar facilities are available in a traditional system based on distributed systems: TMN or TINA for example. However, there was one distinguishing advantage in the system as presented above, that was the way in which the DCM interface could be *customised* - with client-owned agents - to *tailor* the services offered to specific clients - this is one area where agents gain an advantage over traditional software systems.

In the scenarios below mobile agents are able to autonomously interact with one or more management functional area in the server to add value to the original management services. In traditional client-server[3] distributed systems for network management (e.g. SNMP, TMN, TINA) the client is limited to working with the in-

---

[3] The terms "client" and "server" are used rather than "manager" and "agent" to avoid confusing mobile or intelligent *agents* with OSI management or SNMP *agents*.

built facilities of the server. Any further manipulation of management information, beyond that which was generically provided by the relevant standards or by the developer of the server, must be performed in the client application code.

If customers wish to add value to the basic management services offered by the provider they must have the capability of running a management platform in their premises which supports the protocols and information models offered by the server. In addition, the customers must deploy suitable applications running on their local platform to house the required logic. The client applications must interact with the remote server to receive notifications and to initiate management operations. The quantity of information to be exchanged between client and server is a function of the management activities being undertaken and on the efficiency of the protocols and information models supported by the server for the task in hand. The delay and cost associated with each remote operation is a function of the network interconnecting the client and server. Given the particular protocols and information models supported by the server and the characteristics of the network interconnecting them, it may be not be cost effective or even possible to perform certain management tasks remotely in the client. For example, the cost of communication may outweigh the benefits of performing some management tasks (such as fine grain, real-time monitoring of performance parameters) or the information may be out of date by the time an appropriate course of action has been determined by the client application, if the network delays are too large.

With mobile agents it is possible for a client to program a mobile agent with specific functionality which may then be deployed *at run-time* in the server to add value above and beyond the server's basic facilities. Following deployment, the mobile agent may act autonomously to interact with the local environment and make *local* decisions which may then be implemented as *local* management actions without needing to interact with the remote client every time a decision is required. Through this approach it is possible for a remote client to deploy management behaviour and algorithms *inside* the remote server. Note that there are issues of server scalability and/or distribution to be considered further but are not tackled directly in this paper.

5   Examples of customisable network management services

To illustrate the use of dynamically deployed agents we have identified the following three integrated scenarios which are being studied further in the MIAMI project:
- *Intelligent reporting*: Mobile agents may respond to reports from both the performance and fault management systems. According to their programmed policies, rather than relaying *all* fault and performance reports back to the remote client, they will only report when certain conditions have been fulfilled. An example might be a performance degradation on one connection following the failure of a connection in a remote part of the network which forms an alternate route. Only the correlation of these two events might be relevant to the client. Alternatively, observed performance degradations might cause the agent to initiate

tests to verify that unreported failures have not taken place. This scenario integrates the facilities of fault and performance management.

- *Fault repair*: A mobile agent is programmed to listen to fault reports from the fault management system when connections have been interrupted by network failures. According to a pre-programmed policy it can initiate new connection requests between the same end points as the failed connection to restore connectivity. This scenario integrates fault and configuration management.
- *Bandwidth management:* Assuming that performance monitoring agents have been deployed (either by the client or the bandwidth management agent itself) to monitor and report on the utilisation of connections, a bandwidth management agent will be programmed to listen to utilisation reports for certain connections. Depending on the policy for a specific connection the bandwidth management agent may decide to request increased bandwidth on highly utilised connections or to reduce the capacity of a lightly utilised connection. The decision may depend on the cost of changing the bandwidth and so a negotiation between the configuration management agents and the bandwidth management agent may take place. This scenario integrates performance and configuration management.

The integrated scenarios introduced above combine the basic facilities of the configuration, fault and performance management services offered by the server with additional customer specific logic. In other words the customer is able to program the offered management service to a certain degree. This concept has its parallel in traditional management systems through the use of the OSI management Systems Management Functions (SMFs) [19] for event forwarding and logging, resource monitoring [20] [21], and testing [22], albeit in a more limited way. Previous research work [23] [24] has demonstrated how clients can take advantage of these generic facilities to simplify the construction of intelligent clients.

We now consider how these basic facilities could be implemented through the use of mobile agents. Rather than being restricted to standardised capabilities such as the SMFs it is now possible to build entirely arbitrary and powerful behaviour into mobile agents which will be physically located in the managed system's environment. This embedded intelligence not only allows event reports tailored to the client's requirements to be emitted but it enables the migration of the client's logic and decision making algorithms to the server. This has an obvious impact on reducing the quantity of management traffic between remote systems and achieves more timely access to information generated by the remote server.

Considering the network example in Figure 1 a number of ATM connections (VPs) are required between a limited set of endpoints to interconnect the IP routers in the IP CP domain to support higher quality VE traffic. The degree of interconnection and the characteristics of each connection are subject to change throughout the lifetime of the VE. To achieve this a fixed contract for specific connections is not appropriate. The contract between the ATM CP and its customer should be flexible to allow the resources to be managed in a dynamic way: e.g. to allow the creation, deletion and modification of connections and to allow them to be monitored and tested.

The initial contract negotiation over the CN interface results in the creation of a DCM interface and agency where semi-permanent VPs may be established and subsequently managed. The contract should specify the allowed connection termination points and the set of basic management services to be offered across the DCM interface. The contract may limit the number of simultaneous VPs that may exist at any one time and determine the costs associated with resources and management operations.

In our MIAMI work we have used UML for object specification with mappings to the Java language in the context of the Grasshopper platform for implementation. However, in the following example we are using a pseudo-procedural style to represent interfaces and logic. Example service contract:

```
List of termination points: a, b, c ... n

Management services:

    request_trail(source, destination, capacity, qos parameters)
    delete_trail(trail_id)
    modify_trail_capacity(trail_id, new_capacity)
    monitor_trail(trail_id, performance_parameter, polling_interval,
    thresholds)
    modify_monitoring_characteristics(trail_id, performance_parameter,
    polling_interval, thresholds)
    test_trail(trail_id)

Costs: ...
```

Given that these basic management services are available, the customer could create a mobile agent to capture his logic and execute in the remote domain. The following example uses the configuration and performance management services to dynamically manage the capacity of a trail:

```
monitor_trail(trail_1, utilisation, 10 seconds, {upper threshold: 85%,
lower_threshold: 45%})
during_the_hours_9am_to_5pm:
{
        if (upper threshold crossing on trail_1)
        if (old_capacity+20% > some_limit) then raise_notification_to_user()
        else modify_trail_capacity(trail_1, old_capacity+20%)
        if (lower threshold crossing on trail_1) then
        modify_trail_capacity(trail_1, old_capacity-10%)
}
during_the_hours_5pm_to_9am:
{
        if (upper threshold crossing on trail_1)
        if (old_capacity+10% > some_limit) then raise_notification_to_user()
        else modify_trail_capacity(trail_1, old_capacity+10%)
        if (lower threshold crossing on trail_1) then
        modify_trail_capacity(trail_1, old_capacity-50%)
}
```

During working hours the customer wishes the capacity of his trail to increase if his users consume more than 85% of the trail's capacity unless this would push the capacity above a pre-defined limit which might cause the cost of the trail to be greater than he is prepared to pay, even if this causes his users to experience reduced quality. In this case a notification is raised to either the customer or to other agents in the DCM.

In addition to the logic in the above example the agent could be programmed to offer an interface to other agents or the remote customer to set and modify some of the parameters. For example the customer could dynamically change the value of the utilisation thresholds and the percentage increase/decrease to be applied in the case of threshold crossings.

In OSI management, SMFs were standardised by international organisations and encapsulated in the compiled functions of OSI agents; in CORBA-based management systems the SMF-like facilities could be determined by the designers of the systems and embedded at design and system-build time; with mobile agents and intelligent reporting in agent-based management systems the SMF-like facilities can be enhanced and extended almost infinitely and deployed *at run time*!

As seen in the example above, it would be very difficult to capture such behaviour in traditional TMN or TINA systems without standardising such a bandwidth management service at the Xuser or ConS interface. If such a service was to be standardised it would be difficult to capture all possible potential behaviours that clients may request without making a comprehensive and therefore complex specification of the service in GDMO or IDL. However, through the use of programmable, intelligent agents based on mobile code for dynamic and customisable network management this is achievable and deployable on the fly and at the whim of the client. This is clearly a very powerful application of mobile agents for telecommunications management.

6   Implementation issues

The initial approach for the configuration management domain was to base it on an existing TMN system for ATM PVP set-up originating from the ACTS MISA project [25], to which an agent interface would be added (option A in Figure 3). For logistical reasons, the final implementation is based on *static* agents internally (Figure 3, option B), which communicate using remote method calls. This implementation could be also based on distributed object technology e.g. CORBA. Agents were chosen for two reasons: first for uniformity, since there is no need for an adaptation agent-based interface; and second for evaluating mobile agent platforms in the same role as distributed object frameworks. It should be finally noted that immediate benefits from applying mobile agent technology to configuration management are not evident.

On the other hand, the performance and fault management systems use agents internally (Figure 3, option B) in a way that mobility is exercised and exploited. In the performance management domain, customised agents replace, augment and allow customisation of the functionality of the TMN/OSI-SM metric monitoring and summarisation objects [20][21], while in the fault management domain customised agents replace, augment and allow customisation of the TMN/OSI-SM testing objects [22]. In both domains, mobile agents are instantiated at the "network management level" of a management hierarchy according to requests originating from the DCM domain, migrate to network elements and perform relevant tasks locally, enjoying minimal latency and reducing network traffic. Details of the performance and fault management approaches are described in [26] and [27] respectively.

The DCM domain uses agent mobility in a similar fashion to the performance and fault management domains, supporting the programmability of connectivity services by clients across the DCM interface as described in section 5.

At the time of writing, this system is in the final integration and testing stage of the various components with a field trial planned for December 1999. On the other hand, some experimentation with individual domains has already taken place, with results reported in [27] for the fault and in [26] for the performance management domain. These results point to a performance and resource overhead for mobile agent platforms in comparison to distributed object platforms, which partly outweighs the programmability advantages. On the other hand, there is ongoing research work to integrate distributed object and mobile agent platforms. Such an integration would allow the performance benefits of distributed object platforms for static objects/agents, with the additional flexibility of mobility for mobile agents/objects.

7   Summary and Conclusions

Traditional and emerging frameworks for network management such as TMN and TINA allow customers electronic access to management services. These services, however, are fixed in the sense that new features can only be added after a lengthy research-standardisation-deployment cycle. In this paper we have discussed the advent of mobile agent technologies and how they may enhance traditional connectivity management services making them dynamically customisable by clients. We presented three examples which add value to the offered management services as perceived by and required by the *client* rather than by the researchers, standardisation bodies, equipment vendors or service providers. Clients may introduce their own value-added logic during service operation to cater for the dynamics of their environment and to enforce their own policies.

This prompts for a new paradigm for building network management services. Instead of providers building services that attempt to encapsulate the requirements of all clients, they build the necessary hooks and let the clients apply their logic. Customisation and programmability of management services was always possible in traditional systems based on client-server paradigms through the development of client applications in the customers' premises management platform to capture the required logic and intelligence. However, the cost and efficiency associated with such remote operations compared to the proposed agent-based approach should be considered.

The advantages and disadvantages of agent based methods for tailoring, customising and programming management services compared to more traditional client-server management approaches are currently being evaluated through prototype development and subsequent experimentation. This work is being undertaken in the context of the ACTS MIAMI [12] project.

In summary, agent mobility in the presented network management architecture is used in a fashion which we would term "constrained mobility": mobile agents are instantiated at a control point by a master static agent and then move to another point (i.e. network node) where they stay until their task is accomplished, this can be considered as an intelligent software deployment activity. The key benefit of this approach is *programmability*, allowing clients to "push" functionality to a point offering elementary hooks which can be accessed to provide derived, higher-level services. In a similar fashion, we could term "full mobility" as a situation in which a mobile agent moves from point to point using its built-in logic, adapting to changing situations in the problem domain where it is involved. We have not yet found convincing cases in our network management research where full mobility could offer tangible benefits.

In summary, we believe that agent-based, customisable, programmable services provide a new range of opportunities to service providers. The three examples presented are just the tip of the iceberg, an enormous number of potential services are possible. By combining the concepts of intelligent reporting agents with fault repair agents and bandwidth management agents with other possibilities, it is certainly possible to deploy significant, complex, "active" management applications in a dynamic and flexible manner. This would move away from the client-server, manager-agent paradigms of today's systems to a fully dynamic system of interacting agents which can be built layer-by-layer with increasingly complex behaviour to fulfil the demands of sophisticated clients.

Acknowledgements

This paper describes work undertaken in the context of the ACTS MIAMI (AC338) project. The ACTS programme is partially funded by the Commission of the EU.

References

1. ITU-T Rec. M.3010, Principles for a Telecommunications Management Network (TMN), Study Group IV,1996.
2. TINA consortium, Overall Concepts and Principles of TINA, Document label TB_MDC.018_1.0_94, TINA-C, February 1995.
3. ITU-T Rec. X.701, Information Technology - Open Systems Interconnection, Systems Management Overview, 1992.
4. Object Management Group, The Common Object Request Broker: Architecture and Specification (CORBA), Version 2.0, 1995.
5. TINA Business Model and Reference Points, version 4.0, TINA Consortium, May 1997.
6. Foundation for Intelligent Physical Agents, web page: http://www.fipa.org/
7. Mobile Agent System Interoperability Facilities Specification OMG TC Document orbos/97-10-05, November 10, 1997. ftp://ftp.omg.org/pub/docs/orbos/97-10-05.pdf
8. Breugst, M., Magedanz, T., Mobile Agents - Enabling Technology for Active Intelligent Network Implementation, IEEE Network, Vol. 12, No. 3, May/June 1998.
9. Bieszczad, A., Pagurek, B., White, T., Mobile Agents for Network Management, IEEE Communications Surveys, Vol. 1, No. 1, 4th Quarter 1999.

10.M. Zapf, K. Herrmann, K. Geihs, Decentralised SNMP Management with Mobile Agents, in Integrated Network Management VI, Sloman, Mazumdar, Lupu, eds., pp. 623-635, IEEE, 1999.

11.Gurer, D., Lakshminarayan, V., Sastry, A., An Intelligent Agent Based Architecture for the Management of Heterogeneous Networks, in Proc. of the 9th IFIP/IEEE International Workshop on Distributed Systems: Operations and Management, Sethi, ed., Univ. of Delaware, USA, October 1998.

12.Web pages: main project page: http://www.fokus.gmd.de/research/cc/ima/miami/
page at UCL: http://www.ee.ucl.ac.uk/~dgriffin/miami/
page at UniS: http://www.ee.surrey.ac.uk/CCSR/ACTS/Miami/

13.Covaci, S., Zell, M., Broos, R., A Mobile Intelligent Agent Based Virtual Enterprise Support Environment, in Intelligence in Services and Networks - Paving the Way for an Open Service Market, Zuidweg, Campolargo, Delgado, Mullery, eds., pp. 536-549, Springer, 1999.

14.Vigna, G., (ed.), Mobile Agents and Security, LNCS 1419, Springer-Verlag, 1998.

15.Labrou, Y., and Finin, T., A semantics approach for KQML -- a general purpose communication language for software agents , Third International Conference on Information and Knowledge Management (CIKM'94), November 1994.

16.Agent Communication Language, FIPA 97 Specification, Version 2.0, October 1998, http://www.fipa.org/spec/FIPA97.html

17.Booch, G., Rumbaugh, J., Jacobson, I., The Unified Modelling Language User Guide, Addison-Wesley, 1999.

18.http://www.ikv.de/products/grasshopper/index.html

19.ITU-T Recommendations X.730-750, Information Technology - Open Systems Interconnection - Systems Management Functions.

20.ITU-T X.738, Information Technology - Open Systems Interconnection - Systems Management: Metric Objects and Attributes, 1994.

21.ITU-T X.739, Information Technology - Open Systems Interconnection - Systems Management: Summarisation Function, 1994.

22.ITU-T X.745, Information Technology - Open Systems Interconnection - Systems Management: Test management function, 1994.

23.Georgatsos, P., Griffin, D., Management Services for Performance Verification in Broadband Multi-Service Networks, in - Bringing Telecommunication Services to the People - IS&N'95, Clarke, Campolargo and Karatzas, eds., pp. 275-289, Springer, 1995.

24.Pavlou, G., Mykoniatis, G., Sanchez, J., Distributed Intelligent Monitoring and Reporting Facilities, IEE Distributed Systems Engineering Journal, Vol. 3, No. 2, pp. 124-135, IOP Publishing, 1996.

25.Karayannis, F., Berdekas, K., Diaz, R., Serrat, J., A Telecommunication Operators Inter-domain Interface Enabling Multi-domain, Multi-technology Network Management, Interoperable Communication Networks Journal, Vol. 2, No. 1, pp. 1-10, Baltzer Science Publishers, March 1999.

26.Bohoris, C., Pavlou, G., Cruickshank, H., *Using Mobile Agents for Network Performance Management*, to appear in the Proc. of the IFIP/IEEE Network Operations and Management Symposium (NOMS'00), Hawaii, USA, IEEE, April 2000.

27.Sugauchi, K., Miyazaki, S., Covaci, S., Zhang, T., Efficiency Evaluation of a Mobile Agent Based Network Management System, in Intelligence in Services and Networks - Paving the Way for an Open Service Market, Zuidweg, Campolargo, Delgado, Mullery, eds., pp. 527-535, Springer, 1999.

Accounting Architecture Involving Agent Capabilities

Kostas Zygourakis[1], Didoe Prevedourou[1], George Stamoulis[2]

[1]INTRACOM S.A., Greece
[2]Institute of Computer Science Foundation for Research and Technology Hellas

Abstract. This paper proposes an Accounting Management architecture, based on TINA principles. Accounting management is defined and positioned in the service provision chain. A relevant computational architecture is proposed and some of the involved computational entities are modeled as agents. A discussion follows on the merits of introducing agents to cater for certain accounting management activities. The ACTS AC325 MONTAGE project provides the main source of the presented work and results.

1 Accounting in the Service Provision Chain

The information and communication domains gradually converge, in an evolutionary path characterized by Internet-based access networks and high capacity terminals augmented with integrated multimedia service access capabilities, which can take advantage of the increased network capabilities and emerging service features available to the user. Considering this evolution, future communication systems ambitiously promise to offer a wide variety of highly sophisticated and personalized services over the widest possible coverage area. Several providers are involved in such an ambitious yet realistic service provision scenario in which competition will be mostly played at the service level, with multiple providers offering services over a variety of networks and end systems.

TINA[1] has defined and documented in the Service Architecture [3] an adequate business model that involves multiple stakeholders, acting in various business roles in the envisaged service provision chain. In a single instance of service provision to a user, several *administrative domains* may be involved, each controlled by a different stakeholder. Stakeholders may compete and collaborate for attracting the users. The administrative domains may embody several *management domains* to cater for the necessary management activities, which may span to more than one federated administrative domain.

---

[1] The Telecommunications Information Networking Architecture (TINA) is an open architecture for future telecommunications and information services. TINA involves a set of principles, rules, and guidelines for constructing, deploying, and operating services.

A management domain can be modeled by an information object associated with certain FCAPS (fault, configuration, accounting, performance and security) management functionality [1]. Each management domain is governed by one (or more) management policies (such as accounting, security etc.), implemented by manager and managed objects. Additionally, a management domain involves *management contexts* associated to particular service transactions, setting the management requirements for the delivery of an instance of a service.

One of the FCAPS management areas is that of accounting, which is concerned with the collection of resource consumption data for the purpose of capacity and trend analysis, cost allocation, auditing and billing. In capacity and trend analysis, the goal is typically a forecast of future usage. Auditing tasks aim at verifying correctness in billing mechanisms, conformance of resource usage to stakeholder policy established agreements and security guidelines. Cost allocation caters for the appropriate charge assignment to specific resource usage. Billing is responsible for collecting issued cost allocation data, reconstructing missing entries, preparing invoices and distributing them to the corresponding charged domain [4].

TINA introduces the concept of *Accounting Management Context* (*AcctMgmtCtxt*) which specifies quality and quantity details for the accounting management. AcctMgmtCtxt is a specialization of the management context and constitutes an aggregate of a service transaction. Its purpose is to guarantee that accountability is preserved across a set of activities of distributed objects, which constitute a service. Accountability is not defined exactly, but TINA surmises that it is preserving a set of quantities corresponding to the service being received [5].

The AcctMgmtCtxt is part of the set-up and wrap-up phases of a service transaction [1]. In the set-up phase, the user presents the desired accounting schema in the form of AcctMgmtCtxt, which is analyzed and resolved[2] with the schema of the provider. The context is then interpreted referring to the service session specification and the accounting environment of the domain (e.g. tariff, billing). The service resources are set following the results of the interpretation. In the wrap-up phase, the service transaction is considered as successfully concluded if the accounting information conforms to the agreed AcctMgmtCtxt [5].

The paper presents a general accounting management model and proposes a TINA-based computational architecture that caters for accounting management in a federated context. A discussion then unfolds on the merits of considering agent capabilities in support of certain accounting management functionalities. Relevant concerns are discussed, justifications of choices are presented and expected advantages are advocated. Conclusive remarks complete the paper.

---

[2] For a more elaborate description on the accounting schema resolutions, the reader is referred to TINA Accounting Management Architecture [5, p11-12].

2   Accounting Management Model

In the highly evolved and competitive world of communication services provision, accounting management plays a vital role within the operations of each involved stakeholder. Supplementary functionality needs to augment accounting management in order to enable more accurate, efficient and profitable operations for the administrative domain it applies.

Based on the above, we have developed an accounting management model, which has served as a basis for the relevant implementation of MONTAGE [2]. This model comprises the following processes:

- *Policy establishment process.* This process sets the rules governing the whole accounting management function. It must be able to precisely determine the origin, formation, direction and collection of all actions pertaining to the accounting scope of the particular administrative domain.
- *Accounting Management Context process.* During this process, the context for the accounting management function pertaining to a service session instance is communicated among the appropriate parties, negotiated, finalized and bound for the duration of the service transaction.
- *Usage process.* This process meters the usage of the available resources and creates records in an administrative domain [6]. The metering scheme is prescribed by the existing policy and the appropriate accountable resources trigger the metered events.
- *Charging process.* This process collects data generated by the usage process and classifies them into a set of classes relevant to the accounting scheme applicable to the administrative domain. Additionally, the existing tariff function is utilized in order to assign monetary cost to the generated classes.
- *Billing process.* This process is responsible for the collection of charging information, the appropriate indication of the charged entity and the delivery of the issued charge.
- *Supplementary accounting process.* Under this item, add-on capabilities for the accounting management are prescribed. These may concern dynamic behavior of the management domain in order to analyze past and present usage, forecast future demand, dynamically allocate or negotiate applicable tariffs, intelligently perform economic analysis on the service commodities offered, and promote when needed, service offers.

The following figure (fig. 1) depicts these accounting management processes within a multiple stakeholder environment.

The accounting management function is not expected to span over different administrative domains due to the security-sensitive and proprietary data manipulated within its scope. Therefore, all necessary information to be exchanged between domains is done through federated and secure contexts (indicated in fig. 1 by the 'dashed' oval). Examples of such contexts are the notification for session usage (i.e., connectivity session offered by a connectivity provider notifies the service provider) or the on-line-issued bill to a domain for utilization of another domain's resources.

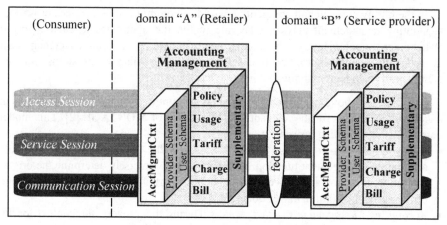

Fig. 1. Accounting management model

3   Proposed Accounting Management Computational Model

Within an administrative domain, the accounting management can be substantiated by a set of computational objects, interacting through well-defined interfaces. A proposed computational model (partially implemented in the MONTAGE project) is shown in figure 2.

This model is based on the axiom that only those resources (objects) that provide the "accountable-object" capability are relevant to the accounting management. An accountable object is a managed object representing a resource or another entity for which usage data are to be maintained, requiring its usage to be attributed to a user [6]. In implementation terms, such object supports a management interface by which the accounting scope is managed and its metered triggering events can be controlled.

The following computational roles are defined in the proposed accounting management system architecture:

- The *Accounting Policy Manager* (APM) informs the object of its expected behavior pertaining to the accounting management scope. It is the component, which negotiates for and prescribes the accounting management context and behavior for the established sessions. It is the primary component responsible for the establishment of the accounting policy process.

- The *Accountable object Factory* (AoF) creates and initializes the necessary usage objects based on the instructions received by the APM. Additionally it implements and sustains the concept of session component[3], particularly useful for the accountability of multi-party service sessions.

- A usage object named as *Accountable-object Agent* (AoA), focuses on collecting and classifying several events pertaining to accounting valued specific classes. It is expected that monitored usage can be passed to another AoA, thus establishing a hierarchy in usage collection activity. For example, the usage of a communication session can be communed to the corresponding service session AoA.

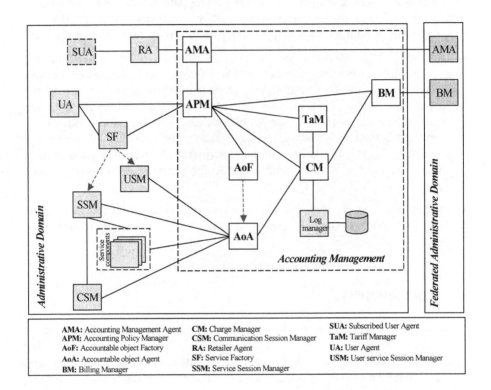

Fig. 2. Computational model for Accounting Management

- The *Charging Manager* (CM) retrieves the information collected by the AoA, at some point in time (also specified by the policy of the accounting management). It

---

[3] A conceptual component entity, which constitutes a session covered by a service transaction, and which maintains the interrelation and state of session events. Its use is evident in multi-party service sessions, where multiple parties may join or leave a specific service transaction and for which accountability must be maintained. For additional information, the reader is referred to [5 pg.31].

then assigns monetary charge to the classified classes, based on predefined or dynamically altered tariffs controlled by the *Tariff Manager* (TaM).

- The *Billing Manager* (BM) is the component within the accounting management domain responsible for issuing and distributing the service utilization invoices. This function can be (and currently is) performed with off-line procedures. However, in a dynamically expanding service provision environment, distributed charge allocation is foreseen, imposing requirements for reliable and fast on-line distribution of service usage invoices to the appropriate stakeholders. The BM can also be exploited by consumer representative components (PA or UA) for providing on-line billing and/or on-line charge information per session or per user.

- *Accounting Management Agent* (AMA) is introduced to materialize the supplementary accounting processes. AMA is a component whose behavior may differentiate the parameterized factors of accounting management among the various administrative domains. Some of AMA potential capabilities are discussed in the subsequent section. It maintains contacts and negotiates with service supporting administrative domains (e.g. connectivity provider) for accounting issues (e.g., connectivity tariffs). It provides Advice of Charge[4] to the user for a particular service session. It formulates the acceptance policies for visiting mobile agents.

The computational model presented above may be engineered as a whole or it can be partially implemented, allowing certain activities to be carried by co-operating systems.

4   Agent Support

An intriguing objective is the investigation and definition of intelligent and/or mobile agent behavior in accounting management, for a TINA-based service enabling architecture. Proceeding through a careful examination of agent capabilities[5] and having in mind the accounting management model, certain thoughts are extrapolated and subsequently presented.

Agent behavior can not be realized in administrative specific, data sensitive and statically operated objects such as the Charging Manager, Tariff Manager, Billing Manager and the Accountable-object Factory. The assumed algorithmic operation of these components does not advocate the usage of agent capabilities. These components are designed to sense their "environment" via their input (materialized by the called methods) and act on it via their output, but they are not considered "agents"

---

[4] Advice of Charge provides the user with an estimate for the service usage. This capability is a service provided by the accounting system and can be activated by the user.

[5] Agent capabilities refer to software-based computer systems comprising one or more of the following properties: a) *autonomy*, b) *social ability*, c) *reactivity*, d) *pro-activity*, e) *mobility*, f) *ethical* and *rational contact*. For additional information, the reader is referred to [8].

because their output would not normally affect what they sense or process later. Additionally these components fail in showing behavior continuity. They run once, carry out their tasks and then stop, waiting to be invoked again.

Agents in Support of AcctMgmtCtxt Establishment

Within the TINA documentation, the concept of management context[6] is addressed in a purely conceptual matter, allowing the implementers to approach it in multiple ways. When a service session is requested, several management contexts applicable to the specific request, grouped in Terms of Management, ToM [5], are resolved (negotiated and/or bound) between the "client" and "provider". One of the management contexts is the AcctMgmtCtxt, which is somehow closely related to accountability as well as financial obligations for the service provision. This particular context has to resolve information of the type:
- "what" (e.g. what exactly the service will provide)
- "who" (e.g. who, which party/domain will provide the service)
- "when" (e.g. the time of the service provision, start/stop/duration/scheduled)
- "how" (e.g. by which means of QoS, content, cost)
- "where" (e.g. the location of the user)

In the TINA Service Architecture, AcctMgmtCtxt is conceived as either statically (context consists of a fixed set of attribute-value pairs) or dynamically (originally statically but the scheme is dynamically expandable) implemented. An additional way, that of being executable, is foreseen involving the use of intelligent mobile agents for negotiating and establishing the precise context scheme [1].

The MONTAGE project has experimented with agent-based behavior complying with the latter method above. The definition of the Subscribed User Agent (SUA) [2] provides the means for negotiating user-desired schemes for service provision. These schemes aim to *utility maximization, charge minimization* or *net-benefit* of services based on certain user-perceived QoS characteristics. This is a major and important part of an AcctMgmtCtxt. It may be complemented by a context set (on a per service basis) from the accounting policy established in a particular administrative domain. For the latter issue, further investigation is needed on the service particular attributes, which could be part of AcctMgmtCtxt, and negotiated on provider's behalf.

As a paradigm, within MONTAGE, SUA has been specified to acquire knowledge on the user's profile (aided by the User Agent) and to formulate a schema on the desired (by the user) service type and characteristics, as these answer to the particular information constituting the AcctMgmtCtxt. Subsequently, SUA follows an itinerary and migrates to the available (within its perceived environment) *service retailer domains* where it triggers conversation with the corresponding agents (Retailer Agents) that represent the visited domains. The negotiation that takes place among

---

[6] Management contexts are scoped by the Fault, Capability, Accounting, Performance and Security areas.

the SUA and the RA concludes with a specific offer that needs to be weighted in the prospect of all received offers (from the visited retailer domains). This final decision is made by the SUA, thus triggering the start of the service session based on the negotiated parameters. This is also what the AcctMgmtCtxt notion addresses. Thus we may presume that SUA assuming agent capabilities contributes to the set-up phase of a service transaction and substantiates the notion of accounting management context, although it is not part of the accounting management model. The APM can play a vital role in complementing the negotiation and binding of AcctMgmtCtxt by enhancing and updating parameters that pertain to the service provision, from the provider's point of view (i.e. overhead charges, restricted content delivery, service delivery time boundaries, etc).

Agents in Support of Service Offer Making

The Accounting Management Agent is another component that can be enhanced with agent capabilities. During the service-offer negotiation (mentioned in previous paragraph) between RA and SUA, some interaction is envisaged between the RA and the AMA. In reality, the RA can be a subset of the AMA or vice-versa. The reasoning behind this is that charging data pertaining to a service offered by an administrative domain must be maintained and controlled by the established accounting management. During this interaction and based on time, resource availability, economy and marketing aspects AMA can alter the applicable charge for a service offer. This is considered as an "autonomous" and "pro-active" property of the accounting management domain, in order to win the user choice for providing the service. This charge alteration can be on a per user or on a per usage basis.

In this type of operation, AMA must be able to differentiate the service requests among the various users. This can be done on either a per user or on a per home-domain basis (if the user being served is roaming to the current domain). This information is known and can be passed from the SUA to AMA, which based on its knowledge can issue certain privileges for the service provision. Certain type of privileges can also be explicitly "demanded" by the user's home domain (in case of roaming) via the profile conveyed by the User Agent.

AMA can additionally realize various investigations pertaining to the demand for the offered services. This analysis can be based on factors applicable to the service nature (e.g. monopolistic versus competitive, necessity versus luxury).

An example of such analysis is the determination of how _elastic_ a service is considered. With the *(price) elasticity* concept, measurements can be taken on how responsive or sensitive consumers are, to a change in the price of a product [7]. The degree of elasticity or inelasticity is defined by the *elasticity coefficient* E_d, shown in the formula:

$$E_d = \frac{percentage\_change\_in\_quantity\_demanded}{percentage\_change\_in\_price} \qquad (1)$$

If $E_d > 1$, then demand is referred to as elastic, while if $E_d < 1$, then it is referred to as inelastic. This can be implemented having AMA slightly adjusting the applicable charges for a service or a set of services and observing the E_d. This may lead to a determination of permanently changing the applicable tariff of a service, if for lower charges more users respond to the service offer.

AMA based on the observed service demand can initiate negotiations with the supporting administrative domains (assuming federation is in place), for variation of the applicable fees. This will result to changes in the tariffs maintained by the TaM. This type of activity may originate either from the user's or the provider's part.

An example of such behavior is provided in relevant MONTAGE implementations [2]. A variation of *Derivative Following* strategy[7] is developed. Prices are adapted at the end of each day according to a profit factor estimated over a period. This factor should reflect both profit and market share in order to make sure that the retailer serves a satisfactory fraction of the consumer population and obtains sufficient profits. The period this factor is estimated over is selected to be the previous seven days, thus ensuring comparison fairness. For example prices of Thursday 22/7 are determined by the observation of the period Thursday 15/7 to Wednesday 21/7. In every such period, five working days and a weekend are included.

AMA can be further enhanced with marketing capabilities in promoting service offers. Such dimension can be deployed by the usage of "rebate-coupons" and "loyalty-awards". By offering rebate-coupons or loyalty-awards, AMA will "push" (via a secure interface) into the user's profile options enabling him/her to access the same or some other service with certain privileges (in terms of cost, service characteristics etc). This is advantageous during the service-offer negotiation given that similar offers are examined by an SUA. Actually, such an implementation may provide a competitive advantage, especially in today's businesses where sales promotion techniques occupy a greater portion of the marketing budget than that of advertisement [10].

Agents in Support of Efficient Event Collection

The instantiated Accountable-object Agents can be enhanced with agent capability when tackled from an engineering viewpoint. Mobility can be an inherent behavior of such components in order to move to the node where the monitored resource is

---

[7] *Derivative Following* states that, retailers adjust prices according to profit-derivative. They increase, by a number selected randomly from a uniform distribution, until the observed profitability is reduced, at which point the direction of change is reversed [11].

located within a distributed environment. Based on the assumption that the AoAs are mainly used for event collection, in distributed component engineering, the migration of these components close to the event-generating components would ideally reduce the network traffic[8]. However it does not seem useful or mandatory for every AoA to migrate in every situation. Thus logic dependent on the "environment specific parameters" may be implemented within such objects, determining whether migration will be assumed and if so the exact itinerary along with the action path to be followed.

Agents in Support of Auditing and Trend Analysis

Accountability[9], usage and capacity trends are issues that need further examination as candidate functions to be assumed by software agents. Based on presumptions, agent properties can be applied in serving such issues, and major contributors can be the AoA and APM. The idea behind this concept is that resource utilization monitored by AoA and predetermined contact defined by APM can be cross-checked for accountability reasons and appropriate actions can then be taken by the appropriate management domain (security or accounting)[10]. Furthermore, usage load statistics can be regularly taken and maintained within accounting management, which shall enable the administration domain to configure and size its resources in the most appropriate and profitable manner.

5   Conclusive Remarks

The paper has presented an overall architecture and a detailed computational model for accounting management. The role of intelligent and mobile agents in support of certain aspects of accounting management has been investigated and elaborated. It has been advocated that agent behavior can not be realized in administrative specific, data sensitive and statically operated objects such as the Charging Manager, Tariff Manager, Billing Manager and the Accountable-object Factory. It cannot either be applied to components whose assumed algorithmic operation neither necessitates nor justifies the usage of agent capabilities. On the contrary, exploitation of agent capabilities in a number of accounting management-related aspects is presumed to be or has been found advantageous. Such aspects include:

---

[8] This is based on empirical hypothesis and has not been validated by the authors within the MONTAGE project framework.

[9] An exact definition of accountability has not been given. However it is understood as a set of quantities corresponding to the service being received or served [5].

[10] Relationship between accounting and security management is two-fold. Part of accounting management must be properly protected by security mechanisms. Security management can benefit from accounting information.

- User and Provider negotiations towards the establishment of the Accounting Management Context.

- Charge-based service marketing that is attractive to users.

- Efficient event collection.

- Accountability, usage and capacity trends analysis.

Implementation of the proposed computational architecture for accounting management, that integrates the prescribed herein agent support, has been partly carried out in the MONTAGE project. Assessment results have not been collected yet. Implementation experiences and assessment results will be the subject of a subsequent publication by the authors on the same topic.

Acknowledgements

This work was supported by the Commission of the European Communities within the ACTS Programme MONTAGE (AC325). The authors would like to warmly thank their MONTAGE colleagues.

References

1. TINA-C, "Service Architecture 5.0" Baseline document, June 1997
2. AC325 MONTAGE, "Implementation of agent-based accounting and charging", Deliverable 26, Report, September 1999
3. TINA-C, "Business Model and Reference Points", Ver. 4.0, May 97
4. Aboba & Arkko, "Introduction to Accounting Management", Internet-draft (work in progress), August 98
5. TINA-C, "Accounting Management Architecture", Ver. 1.2, December 96
6. ITU-T, "IT- OSI- Systems Management: Usage Metering Function for Accounting Purposes", X.742, April 95
7. C.R.McConnell, "Economics: principles, problems and policies", 8th edition, McG-raw-Hill Inc.
8. S.Franklin & A.Graesser, "Is it an Agent, or just a Program?: A Taxonomy for Autonomous Agents", Proceedings of the Third International Workshop on Agent Theories, Architectures, and Languages, Springer-Verlag, 1996
9. M.Kumar, A.Rangarchari, A.Jhingran & R.Mohan "Sales Promotions on the Inter-net"
10. P.Kotler, "Marketing Management - Analysis, Planning, Implementation & Control", 7th edition, Prentice Hall
11. J.Sairamesh & J.Kephart, "Price Dynamics of Vertically Differentiated Information Markets", 1998, (http://www.research.ibm.com/infoecon/researchp-apers.html).

IN Load Control Using a Competitive Market-Based Multi-agent System

A. Patel, K. Prouskas, J. Barria and J. Pitt

Department of Electrical and Electronic Engineering,
Imperial College of Science, Technology and Medicine,
London SW7 2BT, UK
{a.patel1,k.prouskas,j.barria,j.pitt}@ic.ac.uk

Abstract. *Intelligent Networks* (IN) are used in telecommunication networks to provide services that require a decision-making network element. This element is the *Service Control Point* (SCP). An *overload* of an SCP can result in a great reduction in the *Quality of Service* (QoS) provided by the IN. While traditional IN load control algorithms assume a single service network model or make use of a centralized controller, in this paper we propose and investigate a market-based model for solving the distributed IN load control problem for a multi-service network, where any service can be provided by any SCP. Furthermore, we study a realization of this model based on a *multi-agent system* (MAS) and finally draw conclusions as to both its efficiency and effectiveness.

1. Introduction

Technological advances have lead to the increased usage of telecommunication networks which has been driven in part by the use of *Intelligent Networks* (IN). An IN is an overlay network that is designed to provide services for telecommunication networks. An example of an IN architecture is shown in Fig. 1 This structure consists of several distributed components; these include the *Service Switching Points* (SSP), the *Service Control Points* (SCP), the *Service Data Points* (SDP), *Intelligent Peripherals* (IP) and the *Service Management System* (SMS). An IN service is defined using Service Logic (SL) code. Users access IN services through the SSP, which must decide which SCP is capable of providing the service requests and whether it has the right to access that SCP. The SSP can then contact this SCP, which will execute the SL code for the service requested. The SL code may make use of additional IN resources, such as an SDP for database access and IPs for stream data access, in order to complete the service requested. An SMS is introduced to manage the IN services.

The increasing number and variety of services being offered has fuelled the demand for greater network capacity. During peak periods of resource usage, an IN can become overloaded with service requests which leads to a degradation in the *Quality of Service* (QoS) provided by the IN. For example, user calls may be dropped due to buffer overruns, thus significantly reducing network throughput, and consequently increasing the total call set-up time. In order to prevent this, an IN must use a load control algorithm to provide a high level of QoS and protect the network from instability.

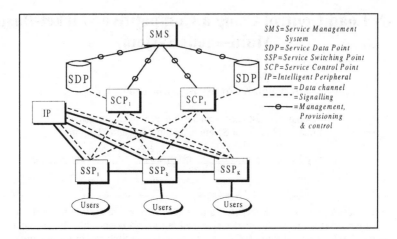

Fig. 1. An example IN architecture.

Current IN load control algorithms can be categorized in a number of ways. For example, algorithms can be categorized according to the throttling mechanism that they use. A throttling mechanism determines which calls the SSP must reject and which ones to accept. Potential forms of throttling mechanisms include the *window-based* and *rate-based* mechanisms. For a window-based mechanism, a limit is set on the maximum number of calls that can be simultaneously accepted. Pham and Betts [1] define such an algorithm where the SSPs monitor time-outs from the SCP to increase or decrease the window size. For a rate-based mechanism, a limit is set on the number of accepted calls per second or some other time period. Smith [2] investigates a rate-based algorithm known as *Automatic Call Gapping* (ACG). This algorithm forces a minimum time separation between call requests.

Load control algorithms may also be categorized by their triggering mechanism. Bedoy et al. [3] suggest that the SSP can monitor the time interval between sending a service request and an acknowledgement from the SCP. Hac and Gao [4] use a mechanism that is always-on. The always-on mechanism requires periodic updates from the SCP specifying which throttling parameters to use. The *always-on* mechanism may be able to respond to sudden overloads quicker. The Automatic Call Gapping mechanism monitors the SCP processor utilization or dropped messages before activating the throttling mechanism. This mechanism requires less processing time than the others but may be slower to respond to a sudden overload condition.

Algorithms may also be categorized by the way the parameters required by the throttling mechanism are obtained. Pham and Betts [1] and Smith [2] suggest using a dynamic search in which the parameters are adjusted incrementally until an appropriate value is found. Any information that may be transferred between SSP and SCP is very simple. Alternatively, more complex information may be passed between network elements, for example, Kihl and Nyberg [6] and Lodge et al [5] use optimization techniques to find the solution to the problems. These algorithms are shown to perform better but require more complex information to be transferred between SSPs and SCPs. Further comparison of these algorithms may be found in [1] and [6].

Many of these algorithms are limited to an IN architecture where a single service can only be provided by a single SCP. During periods of focused overloads, when a single service may cause a sharp increase in resource usage and overload the network, this architecture limits the availability of the overloaded IN service. Kawamura and Sano [7] and Arvidsson et al [8] investigate an advanced IN architecture where each service can be provided by multiple SCPs. However, these algorithms rely on the use of a centralized controller which reduces the reliability and limits the scope for distributed decision making.

Contemporary increases in IN network complexity and information volumes suggest that approaches to the management of telecommunications systems, services and networks will need to be more responsive, adaptive, proactive, and less centralized [12]. These are properties of agents and multi-agent systems; many in the telecommunications research community have recognized that agent-based technology appears to offer a timely solution to the growing problem of designing efficient and flexible network management strategies. Agents have been shown to be an effective tool in the field of distributed resource allocation in general and network traffic control in particular in numerous instances [12][14][15].

In this paper we present a distributed market-based solution to a resource allocation problem which represents the IN load control problem. This model makes use of agents to form a multi-agent system. We investigate issues related to the performance of this algorithm, its overhead (communication and otherwise) and additional issues related to its practical realization. Chatzaki et al [16] investigate the use of *Implied Costs*, which is similar to the concept of a *price* in this paper, to allocate QoS for IP classes. In [16], a two level scheme for resource partitioning is introduced, firstly to allocate routes to services and secondly partitioning of available bandwidth amongst various service classes by using implicit cost feedback from the links in the network to the originating nodes. [16] investigates how to minimize the number of implicit cost messages in the network. In this paper we investigate a general framework for partitioning SCP processor capacity using a distributed auction protocol which is general enough to consider e.g. a joint routing and partitioning scheme. Moreover this algorithm may be extended to deal with scalability and multi-operators issues when a multi-domain environment is taken into account. Price values are generated using an explicit exchange of demand values, while the price messages are fed back to the agents that have participated in a particular auction.

In Section 2 we model the problem of load control in INs. In Section 3 we discuss agents and their applicability to real-time IN load control. In Section 4 we present the market-based solution to the problem, followed by the agent architecture in Section 5. In Section 6 we discuss issues related to the realization of the system. In Section 7 we present results and evaluation of this algorithm. Finally, we conclude in Section 8.

2. Load Control of an Intelligent Network

The QoS provided to users of an IN is affected by the utilization of the SCP processors. It is known from queuing theory that, as the SCP processor utilization approaches its capacity, processing delays increase significantly and buffer overruns are inevitable. The objective of a load control algorithm is to control this processor utilization and to limit it to a desirable maximum, ρ_{limit}. The value of ρ_{limit} is expressed as a fraction of the total SCP capacity.

In our problem formulation, each service type is defined by a single parameter s_j which represents the average processing requirement for the service. The actual processing for a service may be done in several stages and over a time period of several seconds. This simplification may lead any load control algorithm to believe that processor utilization will change faster than would actually be the case. However, this simplification allows us to minimize the information required to define a service.

We can control the processor utilization by using an *input rate selection* throttling mechanism. As the name suggests, this is a rate-based mechanism that defines a parameter x_{ijk}, which specifies the maximum number of call requests that SSP k is allowed to send to SCP i for SL type j per unit of time. This mechanism does not explicitly define which calls to throttle. Using the simplification introduced in the previous paragraph, we can estimate the average processor utilization as the total expected use of the SCP divided by the capacity of the SCP, C_j:

$$\overline{\rho}_i = \frac{\left(\sum_{j=1}^{J} \sum_{k=1}^{K} s_j x_{ijk} \right)}{C_i} \tag{1}$$

We can then formalize the problem as the maximization of the expected revenue generated by the network per unit of time:

$$\max_{x_{ijk}} E\left[\sum_{j=1}^{J} \sum_{k=1}^{K} R_j A_{jk} \right] \tag{2}$$

$$\text{Such that } \sum_{j=1}^{J} \sum_{k=1}^{K} s_j x_{ijk} \leq \rho_{\lim it} C_i \quad \forall i, j, k$$

$$x_{ijk} \geq 0 \quad \forall i, j, k$$

Where I, J, K = The number of SCPs, SL types and SSPs respectively.

R_j = This is the revenue generated by a call of SL type j.

A_{jk} = The total number of accepted calls of SL type j at SSP k.

In order to simplify the solution we assume that the values of x_{ijk} are divisible such that an allocation of 1.5 means that the SSP is allowed to send 3 call requests every 2 seconds.

3. Agents in Intelligent Network Load Control

Agents have been successfully applied to the field of distributed resource allocation in general and network traffic control in particular in numerous instances [12][14][15]. Agents provide a powerful engineering abstraction by operating at a higher level than traditional load control techniques. Agents can make use of network-level information in making their decisions, they allow for different algorithms to be quickly applied as different architectures are introduced and, since they reflect the real-life goals of their owners, can provide a network solution which is more acceptable for all stakeholders.

Furthermore, the solution of a *multi-agent system* is preferred over a single agent. There is always a danger of using multi-agent systems as a fashionable substitute for traditional modularization techniques; that is, to enhance encapsulation or reduce overall system complexity or to simply to take advantages of decentralization (flexibility, redundancy, parallelism etc.) [11]. However, the IN load control problem domain presents the following characteristics [10], which make a multi-agent solution to be an attractive and, indeed, a desirable alternative to traditional approaches:

- The problem is inherently geographically distributed.
- The existing infrastructure contains legacy components (SSPs, SCPs etc.), which must be made to behave or interact in new ways.
- The system can be easily thought of or can be naturally mapped into a society of autonomous, co-operating components. In our case, this mapping comes directly from the real-world market analogy (described in detail in the next section) the system is modeling.

Although it is a well-known fact within the agent research community that there is no single-valued, clear-cut definition of an agent, the intersection of every mainstream definition is generally accepted to contain the following characteristics of an agent: (i) autonomy, (ii) communication (with other software, agents, people), and (iii) social behavior in a community setting. In our context, these apply as:

- *Autonomy*: Each agent acts on behalf of its associated owner in a self-interested manner. The agent effectively represents and protects the interests of its owner, albeit without continuous monitoring and direction.
- *Communication*: Communication is essential in a multi-agent system such as the one we are considering. Each agent has partial local information at its disposal. To solve the problem, it has to communicate with other agents and combine information to build a (whole or partial) view of the overall system.
- *Social behavior*: This can be derived directly from the market analogy, which the system is mirroring. Each agent has a certain role in the society and interacts with other agents with the aim of maximizing its own benefit.

In summary, agent technology presents an attractive vehicle with which to realize IN load control techniques.

4. Market-based Solution for Intelligent Networks

The problem defined by Equation 2 can be converted into a market problem using the *market-oriented programming* paradigm presented by Wellman [9]. In this type of model, a computational economy is created that represents the problem to be solved. This economy consists of commodities that represent real-life resources and *economic agents* that represent the individuals that will be trading these commodities. Within the economy each commodity is associated with a price, which is known by all the economic agents. Using this information, the individual economic agents trade commodities until their preferences are satisfied. The choice of a price mechanism is motivated by the need to find an information ally efficient method of determining demand levels across the network. Prices of commodities are continually adjusted to try and clear the balance of supply and demand for each commodity. No commodities are actually exchanged until all of the commodities have reached this balanced simultaneously.

The computational economy that can be used to solve the problem given by Equation 2 is shown in Fig. 2.

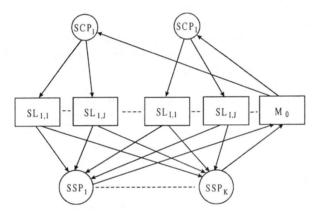

Fig. 2. Market model for IN load control algorithm

In Fig. 2, the circles represent the economic agents and the rectangles represent *markets* where commodities may be traded. Each market, or *auction*, is controlled by an *auctioneer* that performs the exchange of commodities between the economic agents. The directed edges show which economic agents are active in which markets and the flow of goods within the economy.

There are two types of commodities in this economy:

- The first are the SL commodities. In the market labeled $SL_{i,j}$, the agents can buy or sell the right to access SL type j from SCP i, i.e., whether they want to increase or decrease values of x_{ijk}.

- The second is the M_0 commodity. This represents the total cost of buying an allocation of service logic. If an SSP wants to obtain more service logic it must do so at the expense of "borrowing" more from the M_0 market. This is reflected in a lower utility value.

We have assumed that the behavior of the agents is competitive means that each agents tries to maximize its own utility function. Under this behavior M_0 commodity ensures that while the agents maximize their self-interest, in doing so they also maximize the global interest [19]. Other agent behavior models have not been investigated in this paper.

The economic agents labeled SSP_k represent the SSPs in the network. Their objective is to "buy" allocations of service logic from different SCPs given their incoming demand level and the price of the SL commodities. The behavior of these agents is similar to a *consumer* in classical economic theory. Their preferences for a particular set of goods, $\overline{x}$, is determined by a utility function, $U(\overline{x})$. The consumers try to maximize this utility given the prices of all of the commodities. The utility function that is used in our algorithm is based on a reduced load approximation:

$$U_k\left(x_{1k},...,x_{jk}\right)=\sum_{j=1}^{J}R_j\lambda_{jk}\left(1-B\left(\lambda_{jk},x_{jk}\right)\right)+m_{0_k} \qquad (3)$$

Where $x_{ij}=\sum_{i=1}^{I}x_{ijk}$ = The total allocation of SL type j available to SSP k.

λ_{jk} = The incoming rate of SL type j at SSP k.

$B\left(\lambda_{jk},x_{jk}\right)$ = The probability the call is blocked because of insufficient resources.

m_{0_k} = SSP k's allocation of the commodity M_0, that is, the cost of SSP k's allocation of SL.

This utility function represents the expected revenue that will be generated per second given an incoming call rate of λ_{jk} for SL type j and a total allocation of x_{jk}. The formulation of this utility function means that the solution is not unique in x_{ijk}. This allows a higher level decision making unit of the agent to choose which markets to make bids in.

The economic agents labeled SCP_i try to "sell" services in the various markets based on the demand level for different services in the network. The behavior of these agents is similar to that of a *supplier* since this agent initially starts with all of the commodities and then tries to sell them against a cost function. The following utility function is identified that has the required cost function properties:

$$U_i\left(x_{i1},...,x_{ij}\right)=\alpha_i\log\left(\sum_{j=1}^{J}s_jx_{ij}\right)+m_{0_i} \qquad (4)$$

Where $x_{ij} = \sum_{k=1}^{K} x_{ijk}$ = The total number of calls of SL type j still not allocated to

any SSP.

m_{0_i} = SCP i's allocation of the commodity M_0, that is, the cost of SCP i's

allocation of SL.

α_i = A gain parameter that can fine tune the performance of the algorithm.

This utility function reflects the increasing cost of buying SL from SCP i as the SCP utilization approaches its capacity. The price function generated by this utility function has a low price when resources are abundant and rises the amount of remaining capacity is limited. The price of the resources becomes extremely nonlinear and in fact approaches infinity when the SCP capacity approaches saturation. This ensures that demand will never exceed the supply of resources. The gain parameter α_i may be used to fine tune the performance of the algorithm.

In order to trade a particular commodity, an economic agent must send a demand bid to the auctioneer for the commodity in question. For agent l the bid, $d_l(p)$, takes the form of a demand curve which is the quantity of the good demanded as a function of its price, assuming all other prices are fixed. The consumer agents find the demand values by performing a maximisation of their utility function as a function of the price of the commodity. The supplier agents try to buy and sell allocations of SL in order to balance the demand levels for different services. Once the auctioneer has received all the bids from the agents, it finds a market clearing price, p^*, such that:

$$\sum_{l=1}^{L} d_l(p^*) = 0 \tag{5}$$

The new clearing price is fed back to the economic agents in an iterative way until the system has reached an equilibrium. The auction is then said to be complete. Further information about the auction process can be found in [17].

5. Multi-agent Architecture

Given the computational economy defined in the previous section, the task of 'agentifying[1]' the system now comes into focus. This involves a mapping between the algorithmic view and the agent entity view of the system; in other words, viewing the system in terms of *software agents*. This is a loosely constrained process and to a certain degree subject to the designer's goals and objectives.

On one hand, we can have the entire system represented by a single, monolithic agent. However, this method, apart from harboring problems associated with centralizing an intrinsically distributed system, works against the principles we laid forth in Section 3 concerning multi-agent systems.

---

[1] Note that in some texts 'agentification' is used differently to refer to the process of encapsulating an existing legacy software systems in an agent shell.

On the other hand, we can model each component, however simple and minute, by an autonomous agent. This approach complicates most solutions and introduces communication costs beyond what is strictly necessary. Additionally, we must bear in mind that, if the resultant system is to be comprised of *autonomous* and *intelligent* agents in any sense, then too simplistic agents tend to make manifestation of such characteristics difficult.

The agent 'granularity' we have chosen is shown in Fig. 3. This approach gives us the best of both worlds while simultaneously being intuitive with respect to the underlying theoretical market concept.

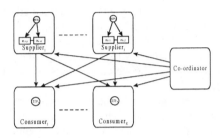

Fig. 3. Multi-agent system architecture

The system incorporates three distinct types of agents: (i) *Supplier* agents, associated with and acting on behalf of SCPs, (ii) *Consumer* agents, associated with and acting on behalf of SSPs and (iii) a *Co-ordinator* agent, which maintains an overview of the market and has global information regarding ongoing auctions and their participants.

It is worth noting here that the Co-ordinator does not function as a centralizing point of the auctions and consequently does not run the danger of becoming a bottleneck. The auctions are performed in a purely distributed fashion. The Co-ordinator simply offers services similar to those found in the Agent Management System (AMS) and the Directory Facilitator (DF) of an agent platform. It is envisaged that in different realizations of this agent system on practical systems, it will be possible for the Co-ordinator to be implicit in the system, with its functionality provided by the underlying layers, be they part of an agent platform or existing network services and facilities.

A single agent is conceptually comprised of two distinct parts: (i) the algorithmic or computational part, which follows the steps dictated by the algorithm and (ii) the agent part, which deals with the participation in the agent society, communication with other agents, maintenance of historical and statistical data and which, in general, performs tasks more characteristic of an agent at a higher level.

The agent's algorithmic part is implemented in C++ to give the benefits of increased speed of execution. The agent part is implemented as an APRIL (Agent PRocess Interaction Language)[2] process, whose symbolic and higher-level messaging features make it better suited to agent applications. Communication between these two parts is via the TCP/IP protocol and is virtually free when, as is typical, both parts are run on the same host.

---

[2] http://quimby.fla.com/Activities/Programming/APRIL/april.html

6. System Realization

Agents participating in auctions behave along the lines given in Section 4. In formalizing this behavior, we have defined a series of *Message Sequence Charts* (MSC), which detail the 'conversations' each agent engages in, as well as the interactions and dependencies between agents.

We have defined MSCs for the following:

(i)   At the platform level, interactions between Suppliers/Consumers and the Co-ordinator. These are concerned more with the logistics of the system rather than its functionality, so they are not described in detail in this paper. However, after these interactions have taken place each agent has its own unique identifier (address) and is tied to a specific network element (SCP or SSP), and each agent has knowledge of every other agent that participates in the same auction and with which it needs to communicate to carry that auction out successfully.

(ii)   At the auction level, interactions between Suppliers and Consumers to carry out the auction algorithm itself. These are shown in Fig. 4.

Agents communicate with one another in a manner compliant with the Foundation for Intelligent Physical Agents (FIPA) standard. Message headers are realized in FIPA *Agent Communication Language* (ACL) [13] format, using the language SL1. For comparison purposes, the content of the messages is also realized in the same language.

There are also several facets of the algorithm in which real-time constraints have to be observed. These particularly relate to (i) The auction interval used by the Suppliers, (ii) The time interval between successive iterations within the same auction at the Suppliers and (iii) The periodic and iterative nature of the bids sent in by the Consumers. More specifically:

- Auctions are periodic and an auction must finish within a specified period of time.
- Having received a bid, Suppliers only accept additional bids for a specified period of time before updating and sending out the new allocations.
- Having sent a bid, the new allocation must be received within a specified period of time before the Consumer engages in pre-emptive behavior and submits a new (not necessarily identical to the old one) bid.

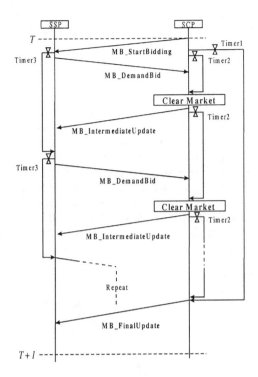

Fig. 4. Auction MSC

These impose real-time constraints on both Suppliers and Consumers. We have limited the inter-agent real-time constraints to the following assumption: Auctions in all Suppliers are assumed to run in lock-step synchronization. In practical terms, this means that all auctions start aligned in time with one another. No further assumptions are made about the time of delivery of a message or, indeed, its delivery itself. Periodic synchronization between Suppliers and Consumers is only implicitly maintained in the *MB_StartBidding* message of the Suppliers, which serves as a temporal checkpoint for every auction participant.

The interaction of the components in the IN with those in the computational economy also needs to be considered. Once a Consumer agent receives an allocation of SL, it can either pass this information to the SSP or for each new call arrival the SSP can ask their corresponding Consumer to determine if it should accept or reject the call. In the former case, the SSP needs to monitor and estimate the incoming arrival rate and pass this information to the Consumer. In the latter case, the Consumer itself can monitor the incoming arrival rates for services and no further interaction is required. Integration of legacy systems that do not inherently support the features of the SSP described above may be achieved by converting the allocations of SL into a set of ACG parameters that can be passed to the SSP. The Consumer would then have to monitor the call arrivals before entering the SSP to be able to estimate its utility function.

The SCP agent can obtain SCP capacity information and service cost information from the SCP IN component, but this could also be calculated off-line. Information concerning new services at the IN level can be propagated to the various agents through the Co-ordinator agent.

7. Results & Evaluation

In evaluating the performance of the algorithm, we compare it to the benchmark ACG [2] algorithm. We investigate a system with 8 SSPs, 4 SCPs and 3 service types. The services used are a Virtual Private Networking service, a Ringback service and a Call Forwarding service which are assigned a revenue of 5, 10 and 3 units respectively. These services can be provided by any of the SCPs. In Fig. 5 we compare the algorithms in terms of the revenue generated for different levels of load. We can see that the systems perform identically while in underload conditions, but as the load increases the ACG performs worse due to oscillation in the ACG algorithm.

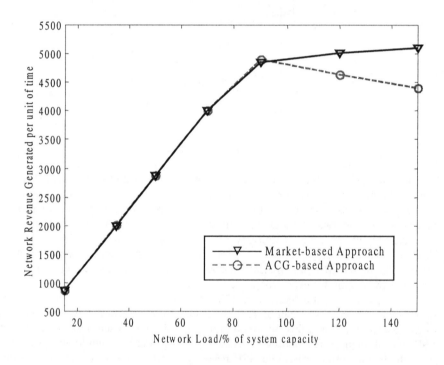

Fig. 5. A comparison of revenue generated versus load level

In the proposed scheme, the results of an auction may not be feasible if a sufficient number of iterations have passed. Moreover, the number of iterations required to find a feasible solution combined with the time required to perform a single iteration is a guideline for the required interval between auctions for effective performance. In Fig. 6 we investigate the expected convergence time, in terms of the number of iterations, versus the load level. A normal load of 35% of system capacity is applied to the network. The load is then increased instantaneously to monitor convergence. The auction interval represents a real-time deadline in which a result must be found. If a feasible solution is not found before the deadline time, a heuristic algorithm [18] is used to provide a feasible solution, but at the cost of optimality. Fig. 6 shows that the number of iterations required to converge is dependent upon the load placed upon the network. Fig. 5 shows that even if the algorithm is stopped before the optimum solution is found the quality of the result in steady state conditions still better the ACG algorithm.

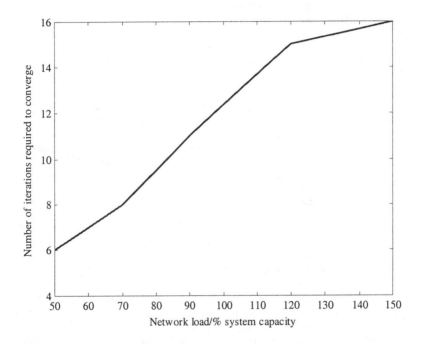

Fig. 6. Convergence time versus load level

Furthermore, we investigate the traffic generated by the algorithm for different numbers of auction participants in Fig. 7. It can be seen that the system performs linearly with increasing number of Suppliers or Consumers.

The process of agentification comes with an associated overhead. There is a minimal amount of start-up overhead (within 8.5% of the generated traffic per auction), the messaging traffic generated (40% increase in the average message size by inclusion of agent headers), and also an increase in the processing time to pack and unpack the messages (an effective 18% increase in the message transmission time).

Part of this 'agent overhead' lies in adopting the FIPA ACL standard for inter-agent communications. FIPA ACL is structured in a flexible and expandable way that is easier to read for humans and is supported by virtually every ubiquitous network transport protocol. However, this flexibility comes at a significant cost when it comes to assembling, transporting and parsing messages.

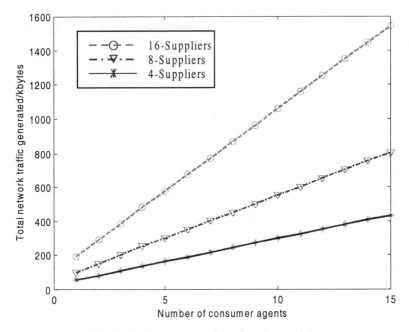

Fig. 7. Traffic versus number of auction participants

It is worth emphasizing here that the agent layer is based on a fully-fledged, general-purpose agent platform, and therefore carries a disproportionate performance overhead. This can be to a great degree recovered by using a customized implementation. Some services provided by the agent platform may be performed much faster directly by the lower-level network layers. Additionally, by using a data representation tailored to the actual network transport used, the traffic overhead can be reduced greatly. At this stage, however, our intention is to provide a fully generalized solution that can be mapped later on to different infrastructures on an independent case-by-case basis.

8. Conclusions

An IN load control algorithm forms an integral part in the provisioning of a high level of QoS to users of IN services. Previous approaches to this problem have been restricted to a single service architecture, where only one SCP provides a service, or they require a centralized controller to solve the global problem.

Recently, market-based schemes for distributed resource allocation have been proposed in the public literature [12][14]. In this paper we have proposed and investigated a market-based model for solving a distributed IN load control problem. The proposed model makes use of agents to form a multi-agent system. We presented simulation results for this algorithm which suggest that using a market-based algorithm can perform better than the benchmark ACG algorithm under overload conditions and therefore this is a viable alternative solution to distributed IN architectures.

Benefits brought into the system by agents (such as decentralized decision making, autonomy and intelligence) came at an added cost. In the current state of the art, the added overhead is significant enough so that, for all practical purposes, the agent system must be considered separately and *superimposed* onto rather than *embedded* into the IN system. Nevertheless, the question of whether the added benefit of an agent system is greater than the overhead is still largely an open issue and depends on many parameters, like the underlying network structure, the particular algorithm used and the specific agent system implementation, among others. What our experience suggests, however, and what we have pointed towards in this paper, is that agents can be a powerful paradigm as well as a valuable tool when applied to the domain of IN load control.

Acknowledgements

This work was partially supported by the EC-funded ACTS project AC333 MARINER. The authors wish to acknowledge the valuable contribution that their colleagues in MARINER have made in this paper.

References

[1] X. H. Pham and R. Betts, "Congestion Control in Intelligent Networks", Computer Networks and ISDN Systems, Vol. 26, No 5, pp. 511-524, 1994.

[2] D. E. Smith, "Ensuring Robust Call Throughput and Fairness for SCP Overload Controls", IEEE/ACM Transactions on Networking, Vol. 3. No. 5, pp. 538-548, 1995.

[3] J. Bedoy et al, "Study of Control Mechanisms for the Effective Delivery of IN Services", IEEE International Communication Conference (ICC'98), pp. 1738-1742, 1998.

[4] A. Hac, and L. Gao, "Congestion Control in Intelligent Network", IEEE International Performance, Computing & Communications Conference, pp. 279-283, 1998.

[5] F. Lodge, D. Botvich, and T. Curran, "A Fair Algorithm for Throttling Combined IN and non-IN Traffic at the SSP of the Intelligent Network", Proceedings of Intelligence in Service & Networks (IS&N'98), Antwerp, Belgium, 1998.

[6] M. Kihl, and C. Nyberg, "Investigation of Overload Control Algorithms for SCPs in the Intelligent Network", IEE Proceedings-Communications, Vol. 144, No. 6, pp.419-424, 1997.

[7] H. Kawamura, and E. Sano, "A Congestion Control System for an Advanced Intelligent Network", Proceedings of NOMS '96 - IEEE Network Operations and Management Symposium, Kyoto, pp. 628-631, 1996.

[8] A. Arvidsson, B. Jennings, and L. Angelin, "On the use of Agent Technology for Load Control, with an example Intelligent Network (IN) 'Market-based' mechanism", Proceedings of ITC-16, 1999.

[9] M. P. Wellman, "A Market-Oriented Programming Environment and Its Application to Distributed Multicommodity Flow Problems", Journal of Artificial Intelligence Research, Vol. 1, No. 1, pp. 1-22, 1993.

[10] N.R. Jennings and M.J. Wooldridge, "Applications Of Intelligent Agents", in Nicholas R. Jennings and Michael J. Wooldridge (Ed.), "Agent Technology Foundations, Applications, and Markets", Springer-Verlag, 1998.

[11] H.S. Nwana & D.T. Ndumu, "A Perspective on Software Agents Research", The Knowledge Engineering Review, Vol. 14, No. 2, pp 1-18, 1999.

[12] B. Jennings et al, "FIPA-compliant Agents for Real-time Control of Intelligent Network Traffic", Computer Networks 31, pp. 2017-2036, 1999.

[13] FIPA'97 specification Part 2: Agent Communication Language, http://www.fipa.org, 1997.

[14] Gibney, M.A. and Jennings, N.R., "Market Based Multi-Agent Systems for ATM network management", in Proceedings of the 4th Communications Network Symposium, Manchester, UK, 1997.

[15] A. Chavez and M. Moukas, "Challenger: A Multi-agent System for Distributed Resource Allocation", Proceedings of the First International Conference on the Practical Application of Intelligent Agents and Multi Agent Technology, London, UK, April 1996.

[16] M. Chatzaki, S. Sartzetakis, N. Papadakis, and C. Courcoubetis, "Resource Allocation in Multiservice MPLS", *Proceedings of the 7th IEEE/IFIP International Workshop on Quality of Service*, London, UK, 1999.

[17] M. P. Wellman, "Market-oriented Programming: Some Early Lessons", in S. Clearwater Ed., *"Market-based Control"*, World Scientific Publishing, 1996.

[18] F. Ygge and H. Akkermans, "Duality in Multi-Commodity Market Computations", in C. Zhang and D. Lukose, eds, *Proceeding of the Third Australian Workshop on Distributed Artificial Intelligence*, Perth, Australia, pp. 65 - 78, 1997.

[19] S. Bikhchandani and J. W. Maner, "Competitive equilibrium in an exchange economy with indivisibilities", *Journal of Economic Theory*, Vol. 74, pp. 385-413, 1997.

Virtual Home Environment

Abarca Chelo

Alcatel France

The Virtual Home Environment (VHE) has been defined by 3GPP[1] as *a concept for personal service environment portability across network boundaries and between terminals*. The idea behind this system concept, which is one of the key features of Third Generation mobile systems, is that users are always provided with their personalised services, no matter the technology of the network they roam to, no matter the type of the terminal they choose to use.

The key objectives of the realisation of the VHE system concept in 3G mobile systems are the support of:

- personalised services,
- a seamless set of services, from the point of view of the user,
- global service availability,
- a common set of services for any kind of access (fixed or mobile),
- a common service control and data which is also access independent.

It would be too limited to realise the VHE concept as a kind of enhanced GSM. The VHE is a concept of integration − fixed/mobile, telecom/data − and globalisation. Third Generation mobile systems will implement the VHE as a vehicle to reach their main aim, which is providing users with new and innovative services.

One of the first steps towards this goal is the shift from service standardisation in GSM to a more flexible standardisation of service capabilities, where only a framework for service creation and provision is given. This extra flexibility will help service providers compete in the service market, but it also introduces some technical difficulties.

Paper [1] raises the problem of how services, if they are not standardised among providers, can be available to roaming users; how the "anywhere, anytime" can be achieved if, at the same time, it is not imposed that the services are the same in every network. After describing the problem they propose a solution based on agent technologies: services are implemented as software components that reside in the home network; when the user roams, these agents may either remotely provide the necessary functionality or migrate to the network the user is visiting. They present the service inter-working architecture developed in the ACTS project CAMELEON, and how it can be used to solve the problems above.

Paper [2] addresses another key issue for the realisation of the VHE: user location. The fact that a user roams poses some problems for service providers, and therefore several methods for positioning the user have been developed; they are presented in

---

[1] Third Generation Partnership Project

the paper, and the advantages and drawbacks of each are analysed. At the same time there are services that are improved where the position of the user is known; these services are presented in the paper as well. Finally an architecture is proposed and described: an hybrid architecture, that gathers the best features of the different positioning systems described, and aims at the maximum positioning accuracy and the minimum cost.

Paper [3] focuses on the demand from Third Generation users for new, sophisticated services that combine the worlds of telecommunications and data, that can be personalised. It describes how current architectures cannot face this challenge, and proposes PANI: an open and dynamic service platform, based on agent technologies, that aims to provide value-added service solutions. PANI allows users to personalise their services by means of composing basic services into service packages; at the same time users are aware of the availability of services, but not of their location. The paper describes the proposed architecture, a trial that developed a set of user services, and the GUIs used for interaction with the user. It also shows how PANI could be integrated with WAP.

Paper [4] addresses the VHE issue of personal mobility: it gives a high level business view of the very competitive telecommunications world of the future, and presents a business case where a roaming user is allowed to select and configure a terminal in a visited domain, and then receive service invitations in that terminal. It then analyses the requirements that arise from that scenario, and presents a candidate architecture that fulfils those requirements. The architecture proposed is an extension of the TINA architecture; extensions are proposed in a modular way, that is, as new components, for minimum impact. These components are proposed to be implemented using mobile intelligent agent technology. The paper is based on results from the ACTS project MONTAGE.

The four papers in this session address different requirements for the realisation of the VHE system concept in Third Generation mobile systems, and proposed solutions for each of them. They represent important steps towards the definition of a system architecture for 3G, currently under discussion in standardisation.

References

[1] "An agent based service inter-working architecture for the virtual home environment", Jaya Shankar P., Chandrasekaran V., Desikan N.
[2] "Architectures for the Provision of Position Location Services in Cellular Networking Environment", Sofoklis Kyriazakos, and George Karetsos.
[3] "PANI: Personal Agent for Network Intelligence", David Kerr, Richard Evans, John Hickie, Donie O'Sullivan, Setanta Mathews, James Kennedy.
[4] "Agent-based Support of Advanced Personal Mobility Features", M. Louta, K. Raatikainen, A. Kaltabani, N. Liossis, E. Tzifa, P.Demestichas, M.Anagnostou.

An Agent Based Service Inter-working Architecture for the Virtual Home Environment[1]

Jaya Shankar P., Chandrasekaran V., Desikan N.

Centre for Wireless Communications, Singapore
jshankar@cwc.nus.edu.sg

Abstract. One of the major efforts for standardising of service capabilities rather than standardising the services in UMTS is to achieve service differentiation and system continuity. While pre-UMTS systems thrived on the ability to access services in a same manner when user roamed, these new deregulation efforts of services will pose some problems in provisioning of services across networks. The concept of the virtual home environment (VHE) addresses problems like this but does not rigidly specify how services can be deployed. This contribution will discuss how agent technology can be applied for designing a service provisioning architecture based on the VHE concept. More importantly, the paper will focus on the importance of service inter-working for supplementing service functionality and discuss how agent communication based on FIPA[2] ACL standard can be used to overcome this issue. Finally, the paper will discuss the design and implementation of a middleware known as a common communication module (CCM) used by service components on the network and the mobile nodes to perform ACL communication.

1   Introduction

One of the key characteristics of GSM system was the standardising of services that led to the global availability and consistent performance of telecommunication services across networks. Although such rigid stadardisation had some benefits in solving network-related issues, in reality it actually limits the operators from introducing new and innovative services to cater to the end-users' needs. The evolution towards the third generation mobile systems such as UMTS will offer greater flexibility in service introduction by providing a framework for service creation and provision rather than a complete set of services as done in the traditional approach. As a consequence of this approach, the same services may not be available to users when they have roamed into another network. To overcome this problem the

---

[1] This work was partly funded by the European Commission as part of the ACTS project AC341 "Cameleon". The views expressed in this paper are those of the authors and do not purport to represent the Cameleon consortium.

[2] The Foundation for Intelligent Physical Agent (FIPA) is a non-profit organisation which is promoting the development of specification for generic agent technologies that maximise interoperability within and across agent based applications.

concept of the Virtual Home Environment (VHE) being proposed by European Telecommunication Standard Institute (ETSI) and the International Telecommunication Union (ITU) was introduced to provide a focus for the issues surrounding support of non-standard services to roaming users.

Due to the inherent nature of agents being autonomous, mobile, collaborative, goal-oriented, adaptive etc., software agents were proposed as one of the means to realise the Virtual Home Environment (VHE)[1]. The ACTS research project, CAMELEON, aims to use agent technology to realise the service roaming feature and an agent based support environment that will support rapid service deployment[2]. This paper will focus on the service inter-working aspect of the VHE service concept and will discuss how agent technology can be exploited for this particular aspect. The paper sets out to discuss the concept of VHE and highlights the notion of providing the same service in a visited network. Following this, the use of agent technology, in particular mobile agent technology for the VHE service roaming concept will be discussed. Several key factors that influence the agent-based service architecture will be discussed. The paper will then present a service inter-working architecture which is primarily based on migrating intelligent agentx communicating using FIPA's[3] Agent Communication Language (ACL). Finally, the design and implementation of an agent communication framework based on ACL/XML, which can be used to realise the inter-working architecture will also be described in the paper.

2   Virtual Home Environment

The Virtual Home Environment (VHE) is defined as a concept for personalised service portability across network boundaries and between terminals. The concept of the VHE is such that users are consistently presented with the same personalised features, user interface customisation and services in whatever network and whatever terminal (within the capabilities of the terminal), wherever the user may be located[1]. The exact configuration available to the user at any instance will be dependent upon the capabilities of the USIM (Universal Subscriber Identity Module), Terminal Equipment and Network currently being used or on the subscription restriction (user roaming being restricted). A user with her USIM in another terminal should receive maximum capability limited only by the terminal characteristics. Additionally, the user may access new services directly from the service network or from Value Added Service Providers attached to the serving network. Services obtained directly from these service providers are not managed by the Home Environment and therefore are not part of the VHE provided by the Home Environment.

Both the provisioning of additional services by the visited network as well as the emulation of the home environment should be possible within the VHE concept. This flexibility will allow service providers and network operators to compete on the basis of proprietary services and provide greater benefits to the user in terms of service choice and service personalisation.

3   Realising Service Roaming with Mobile Agent Technology

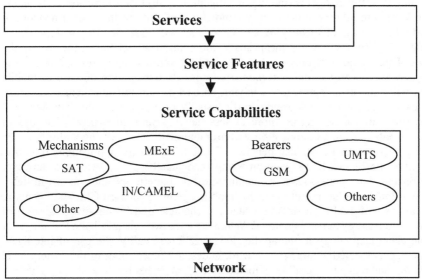

Fig. 1. Service architecture adapted from the VHE concept

The UMTS service concept necessitates that UMTS networks provide a service feature layer to simplify creation and provisioning of services[4]. Service features are building blocks that can be used to create services. Some examples of service features include security, access control, address translation, location information, messaging, service control, etc. The functionality provided by a network's service feature will depend on the underlying service capabilities layer. The relation between the service feature, service capabilities and the network in the UMTS architecture can be envisaged as shown in figure 1. To provide proper adaptation of technology, we view that standardised interfaces should be provided between service capabilities and network and between service feature and service capabilities. The interface between the service feature and the services will depend closely on the technology used to realise these layers.

With the need for a service roaming feature and a rapid application development environment as stated in the VHE concept, we see agent technology as a suitable candidate for deploying the services. With the use of agent technology, especially mobile agent technology, several benefits can be achieved. For example, in some new application or service, communication will span across networks while requiring some degree of QoS. Mobile agents can be helpful in monitoring and managing the communication nodes while ensuring the QoS required. In service provisioning, through migration of agents, the service logic can be moved closer to the user or resources and hence reduce the network traffic load. Providing on-demand service while ensuring personal mobility is another strong point of mobile agents. Mobile agents can be dynamically configured to suit a particular remote end-device and act as service front-ends to bring the functionality to the user[5].

As a solution to the service roaming concept, we view that most services in a UMTS network can be implemented as a collection of software components consisting of software agents (mobile, intelligent or both) that will reside in the home network and provide the functionality remotely or migrate to the visited network to provide the benefits of locality. A key factor that will influence the exploitation of mobile agents in a service is the primary communication schemes required by the service. Three major schemes can be identified for various applications; *background traffic*, *interactive traffic* and *real-time streams*[4]. In our opinion, the use of mobile agents will benefit the applications relying on background traffic schemes the most. Applications based on background traffic usually do not expect the data within a certain time and are therefore more or less delivery time insensitive. The use of mobile agents in such an application will promote off-line processing and reduce the end-user's connection cost. The benefits of mobile agents for applications requiring interactive traffic such as web-browsing and database retrieval are however questionable. The request-response communication pattern and the round trip delay time will influence the requirement for mobile agents. The use of mobile agents in applications with real-time streams which are characterised by one-way transport of real-time data are even less beneficial. While the later two schemes have lesser use for mobile agent technology in the application level functionality, the use of intelligent mobile agents located at strategic network nodes and functioning in the system level to monitor and manage overall QoS for a service is more likely.

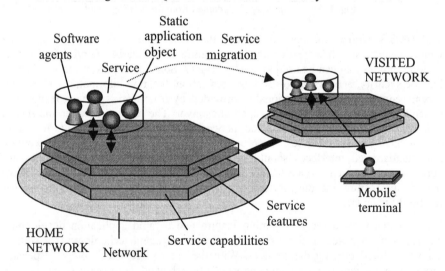

Fig.2. Service migration using mobile agents

Examples of service distribution topologies can be envisaged in figure 2 and figure 3. Whether the components migrate to a visited network and the entire distribution of the components will largely depend on the nature of the service itself. The location of the end-resources (hardware and software) with which agents would need to act on will influence the decisions for agents to migrate to a visited network. Figure 2 depicts the migration of service using mobile agent and static components to a visited network to support a user. As an example, the service that would use the distribution of service components shown in figure 2 is the Virtual Workspace Application. With such a service, users are allowed to create a virtual workspace and move important data such as documents, e-mails, etc. as close as possible to the roaming user. A backup copy of the virtual workspace data is also maintained in the home environment. Users will be able to manipulate and use the data in the virtual workspace while being connected to the visited network. Mobile agents on the user's terminal will be able to keep track of the changes and update the document in the visited virtual workspace. At an appropriate time, synchronisation can be performed between the visited virtual workspace and the home virtual workspace.

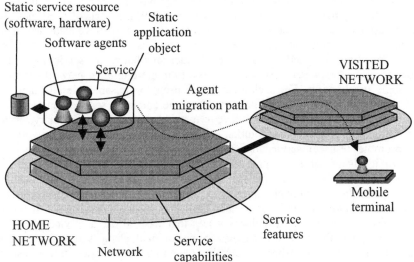

Fig. 3. Remote service execution using mobile agent

Another possible service distribution topology is depicted in figure 3. In such a case the service components need not migrate to the visited network to provide the service. Only a mobile agent with the service front-end (GUI) need to migrate to the user's end device. Users will be able to assign a task to the agent that will finally migrate back to the home network to perform the task. In such a scenario, a possible restriction for moving the end-resource in the home network necessitates that such a topology be applied. A typical application that may use this topology is a home banking application which is implemented using a proprietary banking interface (hardware and software) that is found only in the user's home network.

4   Realising Service inter-working with intelligent agents

In figure 2 and 3, ideally the service components in the visiting network and the home network should inter-work with the service feature in the visited network. Current research trend points to a strong likelihood for Distributed Object Technology (DOT) such as CORBA or object-oriented service architecture such as TINA to be used in third-generation mobile telecommunication systems[6]. Although object-oriented architecture are well suited for implementation of the lower layer service architectures and promoting potential for reusability, they are insufficient for expressing the service layer that will consist of new and diverse applications. With the given flexibility for service vendors to design services, a dynamic behavior to the service's interface and functionality would be expected. This is especially true if service vendors or service providers are likely to constantly improve their service to ensure competitiveness; we see the co-operation interfaces being constantly upgraded to match these improvements. In order for applications to embrace the constant changes and also to cope with the overwhelming complexity of distributed technology, intelligent agents can be used.

The UMTS service concept also has a strong provision for supplementary services which can enhance the capability of a service[4]. While each of these new breed agent-based services can be invoked as a stand-alone service, we also view them as having the capability to supplement other services. When services in a home environment are designed to make use of other supplementary services, it can also complicate matters when these services need to migrate to other networks to provide the same functionality. For such a service to provide the same overall functionality and performance in the visited network, the supplementary service components will have to migrate or the service will have to use other similar supplementary service in the visited network. The later solution is more favorable as it reduces network traffic and promotes re-use of service logic.

Two levels of service inter-working are required, i.e. between an agent-based application and the service feature level and between two or more agent-based applications. In a traditional approach, service inter-working can be solved using a RPC-based paradigm. This would mean that each service exposes a set of functions through a well defined API. Other services that wish to use a service will have to adhere closely to these APIs. In a dynamic environment, changes in APIs would mean that software using the earlier functionality have to be re-compiled. In contrast to the RPC-based interoperability paradigm, services can be designed to communicate using FIPA's ACL[7]. ACL offers a universal message-oriented communication language approach providing a consistent speech-act based interface. Using the intelligent agent approach, services exposing their functionality will describe their entity, attributes, behaviour and the process that bind them together in terms of ontologies. An ontology defines the basic terms and relations comprising the vocabulary of a topic area, as well as the rules for combining terms and relations to define extensions to the vocabulary. This method provides a more flexible approach for communication between agents or entities and promotes reusability at the knowledge level.

The problem of service inter-working is not limited to components on the network only. Mobile end-terminals have to be properly equipped to interact with the service components on the network and handle the basic provisioning of services. The terminals would be capable of connecting to various networks and therefore should be able to handle the features provided by various networks. These basic needs will also impose a service platform inter-operability requirement on the VHE architecture from the terminal's point of view. A flexible service inter-working architecture is needed to ensure that services communicate and inter-operate effectively while coping with the dynamic nature of the service interfaces.

5   Service Inter-working Architecture

Figure 4 shows a logical view of the CAMELEON service inter-working architecture based on ACL communicating agents. The outlined architecture will allow a user to perform the basic task of service discovery, service startup and service interaction.

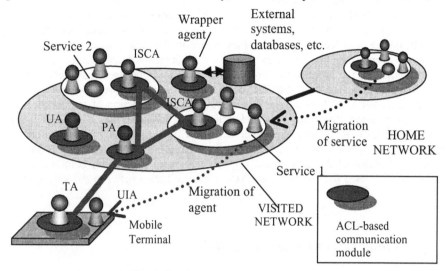

Fig. 4. Service Inter-working Architecture

In figure 4, each CAMELEON service can be made up of several agents and static objects. The agents can be a mobile or a static agent. Each agent will perform a certain function and as such contain a varying degree of intelligence. In contrast, the static objects are non-agent objects such as application data. Each of the CAMELEON services will be designed with an Inter-Service Communication Agent (ISCA) to allow other services to inter-work with this service. The function of an ISCA is also to provide the initial contact point for other agents such as the Terminal Agents (TAs) to start a service. Another key component of the service is the GUI agent which is a mobile agent that migrates to the user's terminal to provide the service to the user. The migration of a service to another network may result in all or part of the service components to migrate to the user's visited network.

The Terminal Agent which resides on the users terminal is a static agent with respect to the user's terminal. It is assumed that this agent is initially obtained from the home service provider and can be viewed as a service-container used to host other service's front-end agent (GUI agent). To the user, the TA can also be viewed as a desktop application which sets up, updates and presents the user with service links. When a user roams into a new domain, the initial contact point of the TA is the Provider Agent (PA). In CAMELEON, we introduce the concept of a provider agent (PA) which acts as a gateway for TAs originating from any domain to query, subscribe and use the available services in a visited domain. Before the above task can be carried out, the user has to be authenticated with the visited domain. This can be achieved by either using a migrating User Agent (UA), USIM or a combination of both UA and USIM to provide the necessary information to the PA to authenticate a user. User Agents are agents that carry the user's subscription information and other related profiles to the user's visited network. The Terminal Agent is also linked to profile information such as the user's personal preferences, service specific profiles, terminal profile and the USIM. Together some of these profiles are used in the service provisioning process which results in the preferred service front-end (GUI Agent) being finally presented to the user.

In the architecture shown above, external applications or legacy systems such as data repositories can be wrapped to appear as agents. Agent wrappers are more sophisticated than object wrappers as they maintain the rich semantic model of the system they represent. The ontology service for the inter-working architecture can be supplemented by the architecture proposed by FIPA[8]. The service inter-working architecture is designed such that the Wrapper agents, ISCA, PA, TA and UA will communicate using ACL at the application level. The abstraction of the architecture above allows alternative standards and delivery technologies to be deployed, e.g. CORBA, DCOM, RMI, etc. to deliver the ACL messages. At a lower layer, the architecture also allows messages to be sent on various bearer services. The following section will highlight the design and implementation of a common communication module (CCM) used for managing the ACL communication between the various core agents (TA, UA, PA, etc.).

6   Common Communication Module

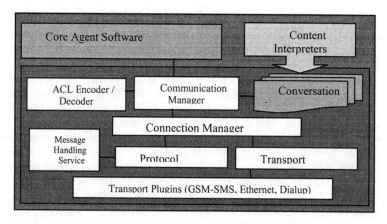

Fig. 5. Common Communication Module architecture

The Common Communication Module is represented in figure 5. The CCM represent a communication middleware that allows agents to communicate over various bearer services. In figure 5, the core agent software differentiates itself from other agent by maintaining a knowledge base consisting of its attributes, capabilities and knowledge of how to perform tasks. Each core agent module will also maintain a set of conversation classes that will determine the conversation schemes to be used in the inter-agent communication. A conversation can be defined as an ongoing sequence of communicative act exchanged between two (or more) agents relating to some ongoing topic. Many conversations can also be performed with the same agent but in the context of different task or subject. The conversation classes will specify the set of states, conversation rules and error rules that are specific to a type of conversation. In designing the CCM architecture, a mix of wireless and wired environment was assumed. In figure 5, the transport plug-ins layer can be replaced by the UMTS adaptation layer[9] when designing for the UMTS network. The following section highlights the CCM's internal architecture.

6.1　Communication Manager

The Communication Manager (CM) is the central process of the CCM which manages all communication needs of the core agent software. It interacts with the other CCM components to provide the communication status to the core agent software. The following are the major functions of CM.

- *Processing of time:* ACL allows a time expression to be used for communication between agents. To indicate a time/date expression by which an agent should reply, the ACL *:reply-by* primitive is used. The *:envelope* primitive can be used to stamp the date/time of message being sent or being received. The CM will manage the synchronisation of time for each agent communication channel. Since there is a possibility that an agent might be located in a different time zone, CM will handle the offset. The time token encoded in the message will be based on ISO 8601 format.

- *Resolution of address:* The agent residing on a platform can be connected to a naming server to perform a mapping of logical address to physical address. In the end-terminals, a light weight naming server which maintains the addresses of the more recent and frequently used agents is stored on a smart card. The Personal Naming and Directory Service (PNDS)[10] used in the current implementation is an example of such software.

- *Establishing a logical connection*: This function is performed after an address of an agent has been resolved. Based on the destination agent's name, which carries the bearer type, CM checks with the connection manager (ConM) to see if a similar bearer service is present in its agent platform. If the bearer service is available, CM issues a request to ConM to make a connection. If an alternative bearer service is present, CM can also request the ConM to route it through another special agent on the network which provides translated messages from one bearer service to another. Provider Agents(PA) are assumed to provide such functions. If ConM succeeds in making a connection, a profile for this new connection will be maintained by CM. The profile includes a timing offset, agent name, connection profile (direct or indirect), type of transport used, etc. The connection here is termed as a logical connection only because it is possible that a conversation to an agent has not ended but the physical transport via the bearer service is not available.

- *Maintaining a logical connection:* It is the duty of CM to maintain a logical connection to a particular agent. The "time" entry in the agent profile will be used to check the status of the logical connection. This function will constantly check its time-out functions associated with a link. If a time-out occurs for a particular connection because the physical connection has been terminated or replies were not received within the specified time, CM will inform the core agent software. The core agent software will use its error recovery techniques/schemes specified in the conversation classes to handle the ongoing conversation.

- *Terminating a logical connection:* This function ends a logical connection. It will be carried out by CM when a request is issued by the core agent software or the destination agent to end the conversation. CM uses ConM to send out messages to end the connection. If the operation is successful, CM performs a cleanup to remove the connection profile.

6.2   Connection Manager

As its name implies, the connection manager handles all connections related issues for incoming and outgoing communication. The ConM receives request from CM to make a connection. The ConM then checks the availability of a particular bearer service and reports to CM if the selected bearer is available. The implementation of this function will be closely dependent on the agent platform used. For an agent platform running on TCP-IP, this function is assumed to open a socket for listening and transmitting. Any broken TCP-IP connection will only be discovered if ConM tries to send/receive some data over the broken socket. For a connectionless link such as GSM SMS, the ConM can only check if the transport is available. The ConM can only ensure that the message is sent out but it cannot ensure whether it has been received by the destination agent. To ensure safe delivery, additional acknowledgement and retransmission handling is performed by an additional module known as the message handling service.

6.3   Message Handling Service

The message handling service (MHS) receives the ACL messages from CM and also from incoming connections. Once the connection is up, CM uses MHS to send out the messages. The message handling service can handle both connection and connectionless links. For a connection-oriented link, the ACL message will simply be passed through the connection. For a connectionless, out-of-order, fragmented messaging link such as SMS, the message delivery service will perform segmentation and re-assembly. Extra tagging fields can also be used for acknowledgement purposes. It can ensure that messages sent out are received by the desired Agent. Additional tags can also be used to signal error conditions such as loss of fragments by the receiving end. In such cases, a retransmission can be performed.

6.4   ACL Encoder

The ACL Encoder converts the internal format data to ACL to be sent out to another agent.

6.5 ACL Decoder

The ACL Decoder decodes ACL messages to the internal format of the core agent software. The decoder can also handle new primitives used specifically by CAMELEON agents. The decoder will detect any irregularities in the message structure and reports them to the core agent software.

6.6 Transport Plugin

The transport plugin has interfaces to the transport mechanisms connected to an agent platform. It will provide information on the availability of the transport mechanism.

7 Conclusion

This paper highlights the importance of service inter-working and service inter-operability in a Virtual Home Environment and discusses the use of software agent technology namely the FIPA ACL as a means to solve this issue. The paper also presents a high-level service inter-working architecture and discusses how a VHE based service provisioning architecture can be designed and implemented. The diverse communication schemes required by various types of application, the network characteristics, the software technology employed by the developer and the diverse end-user devices are some of the issues that pose a challenge to the service inter-working and inter-operability issue in the VHE concept. Within the framework of the CAMELEON research project, two prototype agent-based services and the common communication module were implemented and used to test the service inter-working concepts presented. The service inter-working platform was designed based on FIPA's technology and implemented using JAVA. Using a knowledge-sharing approach based on shared ontologies, a more flexible approach for communication was perceived. A common communication module capable of processing ACL messages carrying XML-based content was implemented to test the communication within the architecture. In the lower layer, the CCM provides access to a diverse bearer service to transport the ACL messages. Residing on distributed nodes, the CCM will facilitate the inter-working and interoperability of services.

References

1. 3G TR 22.970, 3rd Generation Partnership project; Service aspects, Virtual Home Environment, v 3.0.1.
2. Jens Hartmann, Carmelita Georg, Peyman Farjami, "Agent Technology for the UMTS VHE Concept", ACM/IEEE MobiCom'98, Workshop on Wireless Mobile Multimedia, Dallas, United States, October 1998.
3. http://www.fipa.org
4. 3G TR 22.105, 3rd Generation Partnership project; Service aspects, Services and Service Capabilities, v 3.4.0.

5. Hartmann Jens, Song Wei. "Agent Technology for Future Mobile Networks", Second Annual UCSD Conference on Wireless Communications in cooperation with the IEEE Communications Society, San Diego, United States, March 1999.
6. Lars Hagen, Thomas Magendanz, Markus Breugst, "Impacts of Mobile Agent Technology on Mobile Communications System Evolution", IEEE Personal Communication Magazine – August 1998 Vol.5 No.4 pp. 56-69.
7. FIPA 97 Specification, Part 2, Agent Communication Language, v1.1.
8. FIPA 98 Specification, Part 12, Ontology Service, v1.0.
9. Draft UMTS 23.01, "Universal Mobile Telecommunication System (UMTS), General UMTS Architecture", v. 0.1.0, Sept.1997.
10. David Carlier, Alain Macaire, "A Personal Namimg and Directory Service for Mobile Users", 6th International Conference on Intelligence and Services in Networks, IS&N'99, Barcelona, Spain, April 1999.

Architectures for the Provision of Position Location Services in Cellular Networking Environments

Sofoklis A. Kyriazakos and George T. Karetsos

National Technical University of Athens
Department of Electriacal and Computer Engineering
9 Heroon Polytechniou, 157 73 Athens, Greece

skyriazakos@telecom.ntua.gr, karetsos@cs.ntua.gr

Abstract. In this paper we discuss the possible position location architectures in cellular networks, their advantages and disadvantages. Localization of mobile terminals is an essential application for cellular networks. The ability to locate subscribers will give rise to developed networks, that will offer a variety of new services. Standardization groups are already investigating the possible solutions and preparing the networks for the integration of the new elements, related to localization. The recently issued FCC requirements are presented together with the current technological status. Then the candidate location-based systems, based on the generic architecture proposed by ETSI for location based services in GSM, are given. These systems are evaluated and finally a network architecture based on a hybrid positioning method is presented, which appears quite advantageous compared with the other techniques.

1   Introduction

The importance of localization services has boosted the activities of many telecommunication companies and institutions to develop positioning systems. The United States Federal Communications Commission (FCC) has required that cellular operators have to be able to offer precise information concerning subscriber's location. Initially, FCC required 125 m positioning- accuracy in 67% of all trials by 2001 [4]. In September 1999, FCC revised its Phase II rules to deploy Automatic Location Identification (ALI) [9]. Specifically FCC requires that wireless carriers adopting self-positioning systems (GPS), must begin selling and activating ALI-compatible terminals before March 1, 2001 and ensure that at least 50% of all new handsets activated are ALI-capable before October 1, 2001. By October 1, 2002 wireless carriers must ensure that 95% of all digital handsets are ALI-capable. FCC classified the positioning techniques in network- and handset-based. Network-based methods must be accurate to within 100 m for 67% of all calls and 300 m for 95% of all calls. Handset-based methods must be accurate to within 50 m for 67% of calls and 150 m for 95% of all calls. FCC focuses its rules to support emergency calls (E-911).
In Europe on the other hand, ETSI (European Telecommunications Standards Institute) has started from October 1998 working towards the creation of a set of standards for the provision of position location services in GSM. This work has

resulted in the publication of ETSI TS 101723 set of standards on the 30th of August, 1999 [3]. The architecture for location based services proposed by ETSI in TS 101723 is used as the reference system for supporting the various position location techniques described here.

This paper is organized as follows: In the first section we present the methods for position location. Then the foreseen set of services that will take advantage of this technology are given. To realize the positioning algorithms, a suitable architecture must be constructed. Each of the proposed positioning methods will be discussed and an architecture will be proposed. Advantages and disadvantages of each architecture will be taken into account for the evaluation. Finally a hybrid positioning system is proposed, which fulfils all FCC requirements for position location and additionally is a cost effective solution. Its architecture is presented and the functionality of each network element, in addition to overall system, is described.

2   Methods of Position Location

There is a variety of localization methods that can be applied to cellular networks for subscriber-positioning. The localization methods can be classified in self-, remote- and indirect-positioning, according to the place where the measurements and the evaluation of them take place [2]. On the one hand, in the self-positioning systems, the handset makes the measurements, the evaluation and finally estimates its position. On the other hand, in the remote-positioning systems, the fixed part of the network makes the measurements and estimates the mobile terminal's position. Finally, in the indirect-positioning systems the mobile station makes the measurements and transmits them to the fixed network for evaluation[1] [2] [9].

The existing positioning techniques are the following:

- Time Of Arrival (TOA)

TOA is the time that takes for a signal to travel between the base station and the mobile terminal. If one TOA measurement can be performed, the MS is around the BTS in a radius equal to the distance that represent the TOA. According to this measurement the possible user's location is a circle with known radius. Additional measurements result in more circles and finally the MS location can be estimated with an accuracy of less than 125m.

- Time Difference Of Arrival (TDOA)

The mobile station listens to a series of base stations. If the transmitters (BTSs) are synchronized (very accurate clock) and the MS can perform measurements of the time difference of arrival it is possible to locate the subscriber. For this method three BTS are needed and two TDOA-measurements. If we assume that the MS performs one TDOA measurement, then the area of location is a hyperbolic curve as shown in fig.1 A second TDOA-measurement defines another hyperbolic curve and the intersection of them is the area of location.

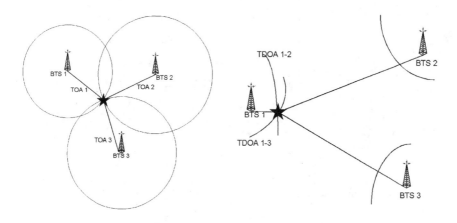

Fig. 1. TOA positioning system   **Fig. 2.** TDOA positioning system

- Angle Of Arrival (AOA)

The angle of arrival method can be applied if one of the MS or BTS is able to measure the angle of the signal arrival. Two AOA measurements are required to locate a user. AOA is also known as Direction of Arrival (DOA). This method requires array type antennas.

- Pattern Recognition Techniques [6]

In all cellular networks the signal strength varies from place to place due to the propagation loss. The propagation loss depends on the distance between MS and BTS and the area-obstacles as well. Therefore, it is possible to classify roads or parts of them according to the signal pattern that appears. Consequently, a pattern matching algorithm can result the location of the subscriber. The pattern matching method is not the most accurate method for positioning, but it is the most cost efficient, since it can be easily integrated to any cellular network. The accuracy varies, but is within the 125m that FCC requires.

- Satellite Positioning Systems (GPS, GLONASS, Galileo)

The most common and accurate positioning methods are the satellite ones. GPS, GLONASS and Galileo are three of the satellite positioning systems. GPS is based on simultaneous propagation measurements that can be performed from the mobile unit. If the object to be positioned, has a sight of at least three satellites it is possible to localize the object. The accuracy of a satellite positioning technique is between 50 and 80 meters (GPS). By means of differential GPS the accuracy can be improved up to 10 meters.

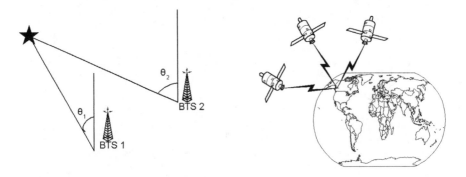

Fig. 3. AOA positioning system   **Fig. 4.** Satellite positioning system

At this point, it has to be mentioned that the cellular networks do not use any of the above described positioning techniques. There are many differences between each of the methods. The most important criteria for choosing one of the above are: installation costs, and accuracy. Time Of Arrival(TOA) has the disadvantage that it is difficult to be measured or approximated. For example the synchronization in GSM can be achieved by means of the Timing Advance (TA) parameter. The value of TA represents the number of bits that the MS has to start transmitting before originally programmed, in order to arrive on time. As a matter of fact, TA cannot provide enough information to locate the terminal. TOA can be measured by means of access bursts generated by the MS. These bursts are received and measured by serving and neighbor BTSs. This method requires additional hardware at the listening BTSs' to accurately measure the TOA of the bursts.

TDOA is a very precise method that requires very well synchronized BTSs. The BTSs have to be either synchronized and transmit always simultaneously, or provide the exact time-information of each transmission. Moreover, the mobile station must listen to at least three base stations and additionally to be able to measure the differences of each incoming package, by taking into account the time of transmission. Moreover the exact morphology area has to be known in order to define the hyperbolic curves of constant absolute difference. The last requirement for this system is the ability to send the TDOA measurements to the BTS. That is possible if the measurement reports are respectively modified. Although the idea of TDOA is based on mathematical fundamentals that ensure the success of the method, there are many technical difficulties for the implementation.

AOA is a method that needs just two BTSs and that is possible most of the time. Either the MS has to detect the area of arrival, or the BTS. The first one is very difficult, since it is difficult to integrate an antenna of this kind to the handset. The latter is more possible, but still special antennas are needed.
Localization based on pattern recognition is an attractive technique, since no extra equipment is required. The focal points of this method are the accurate prediction maps and the pattern recognition algorithms.

Satellite positioning systems are the major candidates for positioning systems. The main disadvantages are the cost, the integration of satellite receivers to the terminals and the non-accurate results in cases of indoor positioning or positioning in urban areas. Satellite systems are able to locate in all other cases the mobile stations with quite good performance.

Finally it has to be mentioned that a combination of more one techniques, providing a hybrid system is also a localization proposal, since it will work in different ways according to the situation. The cost and implementation complexity will be higher in comparison with the other methods.

3 Location based services in cellular networks

Position location in terrestrial networks appears to be a key application for cellular operators, since there is a variety of value added services which are based on the ability to locate mobile terminals. They can classified as follows:

- Safety Services

Mobile phones can be very useful in cases of emergency, if the operator can locate the subscriber anytime there is a need. Emergency roadside service, early warning evacuation and rescue operations are some of the localization-applications that characterize the safety services.

- Billing Services

In this category belong all services related to the differential billing. One possible application is the construction of home zones. Each of the subscribers can provide information to the cellular operator about the location of his house. The operator can define a virtual area around his house with a radius of 300m (DECT maximum coverage). All calls that are initiated from this area belong to home-zone calls and will have a sensitive billing. In this way the subscribers will have the opportunity to select the most concurrent provider (including the fixed network provider) and save money. Modifications of this idea are also known as Wireless Office and Residential Cordless.

- Information Services

This category describes all services that based on the user's location, these become helpful by providing information related to the area where the user is moving. Traffic-related information can be seen as a great asset to drivers. The traffic information can also support a navigation system. By knowing the places of traffic jams it is possible to navigate the vehicles in a way to save time. It has to be mentioned that in cases of movement on highways and other main roads, the localization of several vehicles can provide additional information about the mean velocity, which can also be used for predicting the traffic-progress.

- Tracking Services

Mobile terminal tracking is a valuable application of localization that enables fleet management, children tracking and can be also applied for stolen vehicle service.

There are many thoughts and discussion about these kinds of services. Many think that tracking can violence the policy of the subscribers. Nevertheless, the services based on tracking are definitely useful.

- Multimedia Services

As mentioned above, navigation is one of the major information services. Navigation can also be used in other situations. For example the subscribers in an airport can follow the arrows indicated on the display of their phone and find the terminal they are searching. Information like remaining distance and area conditions (queues) can be a great asset to the subscribers. Beyond GSM and cellular networks of second generation, the next generation mobile networks will provide multimedia services as well. Most of the multimedia services require high data rates. Localization methods can lead to dynamic multimedia services. Subscribers searching for restaurants around their location can receive a list of possible places.

4   Network architectures for supporting location services

According to ETSI TS 101 723 [3], the general GSM network architecture for supporting location services is the one indicated in figure 5.

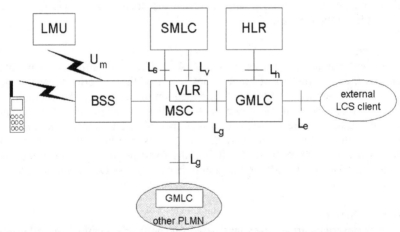

MLC: Mobile Location Center   LMU: Location Measurement Unit GMLC: Gateway MLC
LCS: Location Services   HLR: Home Location Register   SMLC: Serving MLC
MSC: Mobile Switching Center VLR: Visitors Location Register BSS: Base Station Subsystem

Fig. 5. General network architecture

According to the above architecture, whenever localization is needed, GMLC will request routing information from the HLR via the L_h interface. It also sends positioning requests to and receives the terminal's location from MSC, after performing registration and authentication. The BSS receives the measurements performed by LMU and provides them together with the Cell-Id and Timing Advance(TA) information to MSC. According to ETSI, LMU must transmit its measurements over the air interface. VLR is responsible for registering an LMU in its associated SMLC after LMU has performed a location update.

Since many of the positioning methods are remote or indirect, LMU could be integrated in the BSS or another part of the network in order to avoid having a wireless link with the system. Therefore, we propose that the LMU is integrated within the fixed part of the network. Finally, SMLC manages the overall coordination and scheduling of resources required to perform positioning. This network element has a major role, since it calculates the final terminal's location and accuracy.

As described in the previous sections, there is a variety of different kind of methods that can be applied to cellular networks in order to perform position location services. The implementation complexity of each method depends on the measurements that are needed, as well as on the requirements for system and protocol-flow modifications. In the following subsections we will describe some network architectures, according to the positioning method that will be implemented. In order to evaluate the possible positioning methods, the following table indicates the modifications that are needed in each positioning system.

Table 1. Network element functionalities supporting location services for each positioning technique

	MS	LMU	BSS	SMLC GMLC
TOA	LMU integration (possibility 1)	Signal delay measurement	LMU integration (possibility 2)	Triangulation
TDOA	LMU integration	Measurement of signal arrival difference	-	Evaluation of two TDOA measurements
AOA	LMU integration (possibility 1)	Signal angle of arrival	LMU integration (possibility 2)	Evaluation of two AOA measurements
Pattern Recognition	-	-	-	A_{bis} - data evaluation
Satellite Positioning	LMU integration	3 satellite TOA- measurements	-	Evaluation of three TOA measurements

4.1 Network architecture for TOA-positioning

TOA is a positioning method that can lead to localization by triangulation of time delay measurements. Either several TOA measurements have to be made by the MS, or several BSS have to measure the time delay of the signal propagation from the MS to the BSSs. Therefore, the LMU element has to be integrated in the handset or the BSSs. There is a proposal for application of TOA in cellular networks by means of access bursts. GMLC will send the position location requests and the LMU will make the appropriate measurements. SMLC will then calculate the distances that represent the propagation delays and finally will estimate the user's location. For TOA positioning there are new hardware components needed to be integrated in the BSS in order to measure the exact time delay, but the rest of the network infrastructure including the MSs will remain the same. This technique is also supported by ETSI as a candidate positioning method.

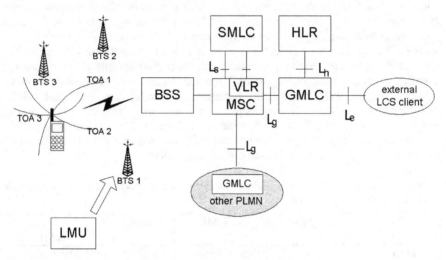

Fig. 6. Network architecture for TOA positioning

4.2   Network architecture for TDOA-positioning

For applying this method the LMU element has to be integrated in the mobile handset. When the GMLC will send the positioning request, the LMU has to measure the time-arrival difference of two signals. Two TDOA measurements are enough for estimating the subscriber's location. The SMLC must define the hyperbolic curves that represent the TDOA measurements and locate the terminal. This method requires well-synchronized BTSs and mobile phones that can perform TDOA measurements. In the SMLC part of the network, there must exist a workstation containing area maps with pre-simulated and defined hyperbolic curves, in order to locate subscribers in real time. The main disadvantage of this method is the need to use new mobile terminals, that will be able to measure the time difference of a pair of signals on arrival.

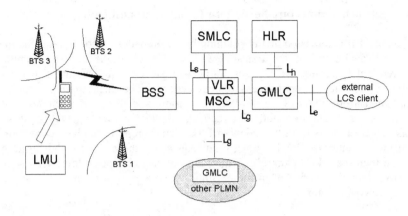

Fig. 7. Network architecture for TDOA positioning

4.3 Network architecture for AOA-positioning

AOA positioning can be used as a self-, remote-, or indirect-positioning system. Since it is very difficult to integrate smart antennas to the terminal, the self-positioning mode will not be taken into account. After GMLC will send the positioning request, SMLC must collect the AOA measurement from the LMUs that are integrated in the BSSs. Two AOA measurements can result in the terminal's location. This positioning technique requires array-type antennas that will be able to determine the angle of the signal's arrival which is leading to increased system's cost.

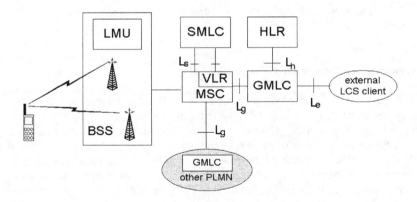

Fig. 8. Network architecture for AOA positioning

4.4   Network architecture for Pattern Recognition-positioning

In table 1 it is indicated that positioning based on pattern recognition techniques is based on evaluation of the transmitted data. Therefore the network architecture required is the simplest. An LMU is not needed, since the measurements that will be analyzed are the common measurement reports. The measurement reports are transmitted from the MS every 480 ms during a call and contain all required information for the operator, like cell-Id, signal strength, power control and measurement reports from the neighbor BTSs. These reports can be recorded or read from the A_{bis} interface. SMLC must send request for these reports and furthermore, analyze them based on pattern matching algorithms. This positioning technique has the advantage that is based on the existing network infrastructure, therefore it is a cost-effective solution.

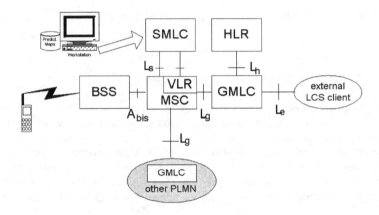

Fig. 9. Network architecture for Pattern Matching positioning

4.5   Network architecture for GPS-positioning

Satellite positioning is based on measurements performed by the object, whose position needs to be estimated. In GPS systems, the object (e.g. mobile terminal) measures the signal time delay of three satellites. The time delay of each satellite describes a sphere about the satellite. If the terminal can measure three TOAs, then it is possible to locate the subscriber by means of triangulation. The receiver's position is the intersection of three spheres. Therefore, the LMU element must be integrated in the terminal. Since measurement-evaluation requires databases and high performance tools, only indirect- and remote-positioning systems have to be taken in account. Therefore the GMLC will send the positioning requests and the MS will respond, by providing the LMU measurements to the BSS. The BSS will then provide the measurements and the additional information to SMLC, where the positioning will take place. SMLC will estimate the position and provide this information to GMLC for location services. The LMU measurements can be transmitted to BSS by means of measurements reports, during a call. The measurements reports provide information for up to six neighbor BTSs. Instead of six BTSs, there can be less information for neighbor BTSs transmitted in order to enable positioning information. In cases of idle mode, localization can be achieved only on-demand. For satellite positioning GPS antennas have to be integrated in all mobile terminals.

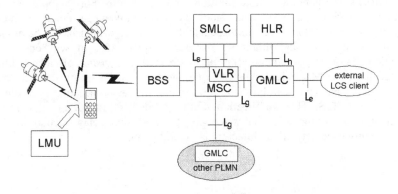

Fig. 10. Network architecture for satellite positioning

5   A hybrid positioning architecture

In the previous sections, the network modifications for each of the positioning systems were described. The two major parameters on which the evaluation is based are: The reliability and accuracy of the positioning and the system's cost. For example, the use of smart antennas for supporting an AOA positioning method could result in a very accurate localization, but that would require the modification of all existing antennas resulting in high system cost.

In order to overcome the difficulties and disadvantages of each architecture on the one hand and to fulfill the increasing requirements for accurate position location on the other hand, a solution would be the implementation of a hybrid positioning technique. We propose to combine the pattern matching method, since it is the simplest technique from architectural point of view and the TOA technique which is commonly accepted from the standardization bodies and more realistic to implement. There will be no network modifications needed for the pattern matching method. The measurement reports are sent every 480 ms during a call and contain all the required information. These reports are available on the A_{bis} interface of the GSM position location architecture. The evaluation can take place locally in each SMLC by means of a workstation where the positioning algorithms, as well as a database containing model-information will be stored. According to trials the computational cost is not high. During the trials we have created models for one cell in a suburban area. The computation time for estimating the location of one subscriber was less than 1 sec. [5]. Moreover, the matching algorithms can be improved in the future to provide even faster positioning. In urban areas where the signal strength strongly varies, this method will have accurate results. Since the MS does not transmit any measurement reports in idle mode (although the measurements are performed), the localization is only possible when the subscriber sets up a call. Nevertheless, a protocol modification could enable the positioning if the network requests a measurement report on demand. Pattern recognition has the great advantage that it is a very cost-effective method, but on the other hand it is not a reliable solution for all occasions. For example, location estimation of pedestrians is difficult, since the user's movements do not influence enough the reported signal level. In these situations, the algorithm has to be assisted from another method, that might be of greater overhead, but will ensure the localization. TOA measurements can be performed by means of access bursts generated from the MS. Silent calls (no ringing tone) can be set up from different BTSs, or forced handover can be made for gaining the TA information according to more than one BTS. The modification that is needed, is an additional hardware component in each of the BTSs that will enable the exact time delay calculation of the bursts. The proposed network architecture is the following:

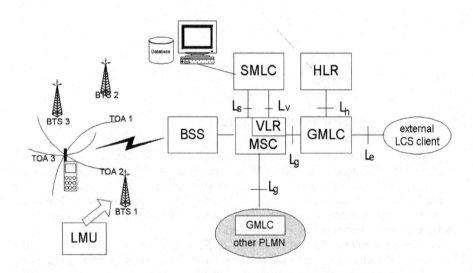

Fig. 11. Hybrid positioning system architecture

Anytime localization is needed, GMLC will send positioning requests via interface L_g. If the MS is in idle mode either a measurement report on demand has to requested from the MS, or the positioning will be made with the TOA technique. If the subscriber has set up a call, the localization can be performed by means of the pattern recognition technique. The measurement reports will be forward to SMLC via L_s interface, where they will be processed and analyzed. By observing the process results for a few seconds it is possible to decide if the accuracy is high enough, in order to base the localization in this method. If the error probability is higher than a specific limit, the localization has to be continued by the TOA technique. SMLC will request access bursts from the MS. These will be received from the BTSs and the performed measurements (LMU) can be linked to SMLC where they will be analyzed. In cases where the MS is in idle mode, silent calls have to be made. These requests will be sent also by SMLC. Especially for all existing cellular networks this is a suitable solution, since the system modifications are limited. The accuracy will be high enough and it is an appropriate solution for all occasions. It can also be applied for indoor positioning, where most of the other methods fail.

In the flowchart diagram (figure 12), the generic localization protocol for the hybrid estimation is shown. The optimal case is the use of the pattern recognition technique that can be applied only if the MS is not in the idle mode. If the subscriber has set up a call and needs to be located, the pattern matching method will be taken into account first. If the results are accurate enough, then the process continues to estimate the user's location by means of the initial method. Otherwise, the TOA method will be used for estimating the position.

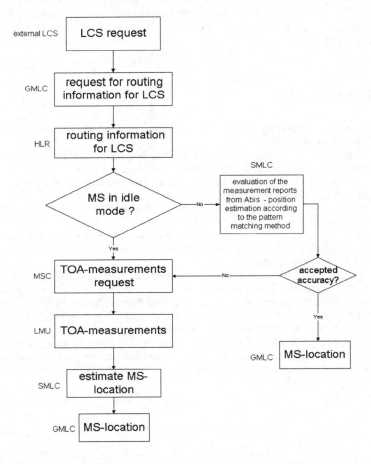

Fig. 12. Flow-chart diagram for the hybrid position location architecture

6   Conclusion and Future Work

In this paper we have presented the candidate systems for supporting position location services based on the architecture proposed by ETSI for GSM. Then the advantages and disadvantages of each system have been pointed out in terms of cost and implementation complexity. Additionally we have presented a hybrid system which also relies on the general ETSI architecture and combines two known techniques. The effectiveness of this architecture is based on the ability to use two totally different methods, according to the situation and the requirements. System modification will be less in comparison to other techniques.

Finally this method can be upgraded anytime with new features for enhanced positioning capabilities. New pattern matching algorithms, precise signal prediction maps and more accurate LMUs can be integrated in the future, in order to enable the system's improvement. The architectures described in the previous sections refer to the GSM network. One of the major goals of our research is to develop a positioning method that could be efficiently applied in all kinds of cellular networks like IS-95, AMPS and UMTS.

References

1. C. Drane, "Positioning Systems in Intelligent Transportation Systems", Artech Hause, Boston, London, 1998
2. C. Drane, M. Macnaughtan, C. Scott, Computer Systems Engineering, Sydney, "Positioning GSM Telephones", IEEE Communications Magazine, April 1998
3. ETSI TS 101 723 V7.0.0, "Digital Cellular Telecommunications Systems (Phase 2+), Location Services(LCS) Functional Description", Aug. 1999
4. FCC, "Revision of the Comission Rules to Ensure Compatibility with Enhanced 911 Emergency Calling Systems", CC Cocket No. 94-102, Oct. 1994
5. S. Kyriazakos, Chair of Communication Networks, RWTH Aachen, "Position Location of Mobile Terminals in GSM and UTRA, based on Pattern Recognition Techniques", Master's Thesis, Apr. 1999, Aachen
6. S. Mangold, S. Kyriazakos, Chair of Communication Networks, RWTH Aachen, "Applying Pattern Recognition Techniques based on Hidden Markov Modells for Vehicular Position Location in Cellular Networks", IEEE VTC 1999-Fall, September 1999, Amsterdam
7. B. Ludden, E. Villier, L. Lopes, B. Saleh, Motorola, UK, "Mobile Subscriber Location Unsing Handset-assisted Positioning Techniques", EMPC 1999
8. T. Rappaport, J. Reed, B. Woerner, "Position Location Using Wireless Communications on Highways of the Future", IEEE Communications Magazine, pp.33-41, Oct. 1996
9. Rural Spectrum Scanner, " http://www.bennetlaw.com/rss/538-0920.htm#2", Vol. 5, No. 38, September 20, 1999

PANI: Personal Agent for Network Intelligence

David Kerr, Richard Evans, John Hickie, Donie O'Sullivan, Setanta Mathews,
James Kennedy

Broadcom Éireann Research Ltd.,
Kestrel House, Clanwilliam Place,
Dublin, Ireland.
re@broadcom.ie
jh@broadcom.ie
do@broadcom.ie
sms@broadcom.ie
jky@broadcom.ie

Abstract. Current IN services are no longer sufficient to feed the increasingly sophisticated requirements of the customer for feature-rich, multimedia, internet services. The development and deployment of a platform to support these services, their composition into tailored user services, and support for user interaction, were the primary objectives of this last phase of the PANI project. Intelligent Agent technology provides mechanisms to tailor the user's profiles, rules and services to their requirements, and act on their behalf even when they are not online. Furthermore agent technology enables the development of an open service environment into which new services can enter and leave dynamically without the need for manual integration or development of customised User Interfaces. At the core of the PANI service platform is the Agent Services Layer (ASL), an open, distributed, language and operating system independent Agent Platform developed by Broadcom[1,2]. On this foundation, the goal of flexibility and openness in service delivery is achieved through the Service Definition Meta-Language (SDML) and Service Composition Meta-Language (SCML). SDML allows both event and notification services to advertise their presence and capabilities to the PANI system and therefore enter and leave the service-space with no requirement for changes to the user interface or the user's Personal Agents. SCML allows the user to compose their individually tailored service(s). Access to and configuration of these services is enabled through dynamic User Interfaces which render themselves according to the characteristics of the available services.

1   Introduction

PANI is an open, dynamic, service platform which deploys elements of agent technology in the provision of value-added service solutions. PANI provides the foundation for the dynamic delivery of feature-rich, multi-media, Internet services which are user customisable and adaptable to the user's changing environment [3].

Previous work in the area of Intelligent Agent Technology (IAT) [1] has identified IN as a target domain where the deployment of IAT is beneficial. This work specifically looked at a number of facets to the delivery of advanced user services, from an underlying architecture to the services themselves and their ability to be self-describing to the service platform thus enabling their immediate availability to end users. A key ability of the PANI architecture is in the composition of basic services into service packages. As an example, consider the situation where a user wishes to be notified of important emails while travelling. If a service is registered which can deliver information about the user's emails and another service registers stating that it can make telephone calls and synthesise speech, then the user can themselves compose a service whereby emails are delivered to the telephony system under their required conditions, e.g. high priority emails, or those from a specific person. This approach can be taken further with services composed from a combination of information sources. Basic weather and timer services could easily be combined to create a weekday wakeup call service that triggers 30 minutes early if the weather is wet (as commute times are likely to be longer).

This type of flexibility in service composition and the ability for external services to register themselves and their capabilities makes PANI a unique system. The use of Intelligent Agents allows for powerful rule analysis to be performed and for rules to be fired producing messages to gateway agents such as those responsible for the delivery of voicemails or fax messages.

2   PANI Services

A PANI service can be described as being an event service (providing sensor data) or an action service (providing activator data). Sensor data provides the central rules agent with the data which is entered to it's fact base for processing. Sensor data can be further subdivided into two classes: those which are specific to a particular user (e.g. incoming email messages) and those which are generic to the user population (e.g. stock ticker information). The fact base is in a proprietary format [4] and therefore to allow for portability across rule systems, a fact meta-language was defined. The services themselves are atomic in that they provide a single function (stock information, GSM SMS message transmission).   The user provides the conditions under which the services are combined.

2.1   What is the user's perspective?

The user's perspective is that there are services available to the service platform, distributed across heterogeneous networks, which are available for interaction through a client. It is not important to the user which location is serving them with information.

2.2   Components of the PANI Service Architecture

Services are one of two basic types. *Event* (source) services provide data into the system and *Action* (sink) services provide mechanisms of notification for the user. The events from one service are joined to another service by the RulesAgent, which contains user-specified rules for their connection. One simple configuration of a PANI platform would have a single event service, e.g. stock quote provider, a single RulesAgent and a single action service, e.g. an SMS sending service. The user would configure a rule to send stock quotes pertaining to the Ericsson B share on the Stockholm stock exchange to the SMS service when a condition is met, e.g. the price exceeds SEK300. The service is made available through its registration with the RulesAgent in SDML, the user's GUI renders itself based on available event and action services and the user composes the rule with the RulesWizard client which pushes SCML into the RulesAgent for verification and activation. The next section will go through this process in more detail.

2.3   Dynamic Service Provision – Service Registration

In order for services to be made available to users dynamically, a process of service registration must be made with the rules agent of the system. This is achieved using the meta-language developed – Service Description Meta-Language (SDML). The service registering on the service platform interrogates the platform for an agent that has registered itself as having rules processing capabilities. After a name has been obtained from the platform the agent sends it's registration message in SDML. This has the general format and specific example for a news service as shown below.

```
:tell

:ontology info_service_registration

:language pani_sdl

:content(<service_name>(<param1_name>

(<param1_type>), ... , <paramn_name>  (<paramn_type>))
```

```
:tell

:ontology info_service_registration

:language pani_sdl

:content(Deliver_News_Article(Headline(string),
Article(istring))
```

Once the RulesAgent has processed this message, the service can be made available to the users. The information in an SDML registration allows the RulesAgent to dynamically build executable behaviour which will subsequently be used to manage information provided by the service (in the case that it is an event service) or to build executable behaviour which will allow that service to be invoked (in the case that it is an action service). Therefore the RuleAgent is continuously modifying its own capabilities and behaviour on the basis of SDML registrations which it receives.

3   Service Description and Composition Languages: SDML & SCML

SDML serves two purposes in the PANI system. Firstly it enables services to be registered on the service network and secondly it allows the user interface render itself according to the specification of the service and to constrain the user in his generation of SCML statements. The dynamic and open nature of the PANI platform place additional requirements on the generation of SCML statements as they are immediately published in the RulesAgent and activated. The client must generate correct SCML statements and the user must not be able to modify the generated statements as they are published in a live system. A comprehensive RulesWizard was developed in order to structure the way which users create rules and publish these rules to the RulesAgent. The core of the RulesWizard is an SDML/SCML parser/generator which analyses the registrations of the active services and renders GUI widgets appropriate for them 'on-the-fly'. In this manner, the RulesWizard is itself a dynamic component in that it has not got a hard form, but rather a soft one which changes according to the state of the system at the time of client connection.

The range of services that can be implemented is wide and includes all those services that can be described by SDML. Tests have been done to implement varied services both in the datacomm and telecom domains, ranging from delivery of news information to the delivery of alarms in a telecommunications network. Action services in both domains have also been implemented, ranging from sending of emails to speech-synthesis over a PSTN network. The core of the service discovery is thus in the SDML basic types supported. To date several basic types are supported (as shown in Table 1) with a view to building more complex types as required.

Table 1. Basic types offered through SDML

Type	Range
integer	-maxinteger to +maxinteger
real	-maxreal to +maxreal
atom[]	list of symbolic terms explicitly enumerated
string	any character string
sstring	any user-searchable character string

At present these basic types are the only ones supported by SDML but it is possible to extend the language to support complex types if required by a specific service. These basic types offer a number of different operators to the user, and accordingly to the GUI, which the user manipulates in order to build a service. For example, if a service offers a parameter of type `real`, the GUI will first render the available operators for that type in a dropdown box (i.e. <, =, >, <=, >=) and then a textbox into which the user will type the operator value.   In turn, the RulesAgent will dynamically offer search capabilities to the instances of the service, e.g. a stock value.

4   Services Offered in the PANI Prototype

In order to demonstrate the technology developed, a trial system was chosen which would enable users of the trial to configure their own services from the basic services available. The architecture of the trial system is shown in Figure 1.

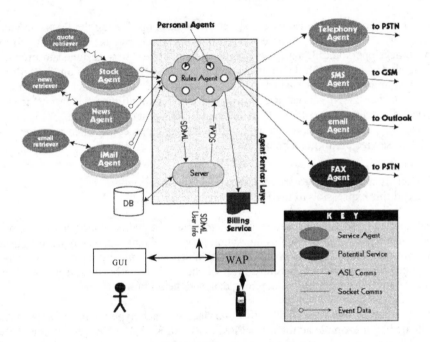

Fig. 1. PANI Trial System

The services developed for the trial were:

Event Services:

News Story – slashdot.org, Dagens Nyheter (with kind permission)
Stock Quote – NASDAQ, AMEX, NYSE, Stockholm Stock Exchange
iMail – incoming mail event service (restricted users only) [5]

Action Services:

Telephony – including dynamic text-to-speech synthesis [6][7]
SMS sending
Email sending

A comprehensive GUI was developed in three forms:

- Standalone java application (fat client approach)
- Servlet based browser application (thin-client approach)
- A WAP interface for mobile devices such as WAP enabled phones

A trial was conducted from March 10th 1999 until March 31st 1999 with the trial participants drawn from both Broadcom and Ericsson P-VAS.

4.1 PANI Services – Implementation

Each of the services provided by the system were developed using the PANI Service API. This is an OO API (implemented in Java) which can be used by platform service developers and 3rd party service developers alike. It is designed to provide non-agent developers with the tools necessary to deploy their services on the ASL in the PANI architecture. The API consists of a number of objects which represent the service in the PANI service platform and a number of utilities which enable the services discover and register with the relevant RulesAgent on the system.
   Broadly speaking, each service will follow the same procedure when deploying in the PANI service platform. These are:

- Discover a RulesAgent
- Construct an SDML expression describing their service
- Send this expression to the RulesAgent

In addition to these three steps, one more step is required and is dependent on whether the service is an Event service or an Action service. Event services typically inject data into the system for processing and Action services typically have data injected into them for processing. Therefore most of the work done by these services is in dealing with either incoming or outgoing data and the API reflects this.

Objects exist in the API which allow services of each type to have these tasks automated. For example action services will have all KQML messages automatically parsed for them before the parameters to the services are presented to the service. Likewise, Event services will have their data packaged into appropriate messages which are dispatched to the RulesAgent for processing. This automation of service functions enables services to be deployed in the PANI service architecture very rapidly, as illustrated in Figure 2. Once this service is deployed, the service will update the RulesAgent with the current time every minute.

```
public class TimeProcessor extends EventServiceProcessor implements TimeConsumer
{
    private final String CURRENT_TIME = new String("Current_Time");

    public TimeProcessor() {
        super();
        SERVICE_NAME = "Time";
        serviceDescription = new ServiceDescription(SERVICE_NAME);
        serviceDescription.addServiceComponent(CURRENT_TIME,
                                    ServicePrimitive.STRING);
        TimeScheduler scheduler = new TimeScheduler(this);
        scheduler.start();
    }
    public void updateTime()  {
        sendEventUpdateNotify();
        String[] timeInstance = new String[1];
        java.text.SimpleDateFormat sdf =
                new java.text.SimpleDateFormat("HHmm", java.util.Locale.UK);
        java.util.Date current_date = new java.util.Date();
        String current_time = sdf.format(current_date);
        timeInstance[0] = current_time;

        try {
                sendEventUpdate(timeInstance);
            }
        catch (EventUpdateException e)   {
                System.out.println("[ EventUpdateException: "
                                    + e + " ]");
            }
    }
}
interface TimeConsumer
{
    public void updateTime();
}


class TimeScheduler extends Thread
{
    private TimeConsumer consumer = null;
    public TimeScheduler(TimeConsumer con)  {
        this.consumer = con;
    }

    public void run() {
        while (true)  {
                try  {
                        this.sleep(60000);
                        consumer.updateTime();
                } catch (InterruptedException e)  {
                    System.out.println("[ TimeScheduler:
                    Interrupted - ignoring ]\n" + e);
    }       }       }       }
```

Fig. 2. Complete Implementation of a time service on the PANI Platform

5   Client Interaction

5.1   Fat Client

The GUI incorporates an SDML parser and an SCML generator. These are used in order to generate widgets required for the user to interact with the services which are currently available on the PANI service platform. The GUI connects to the server using TCP/IP and obtains a list of all available services. Then the RulesWizard is ready to start processing user rules.

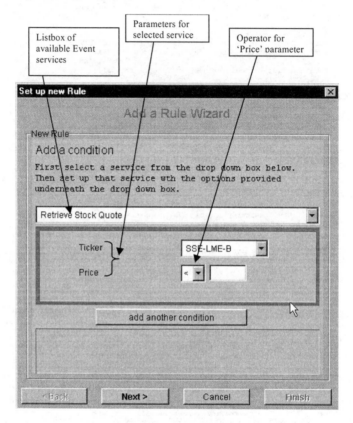

Fig. 3. Selecting Event Condition for a new Rule

The first step in setting up a new rule is to select the event service which provides the data to which the user attaches a filter (see Figure 3). The user can select a number of heads (conditions) to the rule and can combine events from a number of different services. For example, the user may wish to be notified of a stock event to their mobile telephone after working hours and to their desk telephone during working hours, and therefore would combine the stock service with the time service to achieve this.

After selecting the event and applying a filter, the user must select a notification method (or methods). Each notification method has a number of different parameters that need to be filled in. In the case where the event service supplies a hidden field (e.g. a news article to which the user cannot apply a filter) then this hidden field is automatically filled into a parameter of the action service, as depicted in Figure 4. In the case where the event service does not supply a hidden field to the action service, the user must supply all parameters.

Once the rule is submitted to the RulesAgent in SCML, it is guaranteed by the RulesWizard to be syntactically correct and so can be activated immediately in the RulesAgent. The SCML is converted into a programmatic representation and is asserted into the RulesAgent's rulebase and activated. The first instance of a match to the rule will trigger the firing of the required notification method (SMS, email, telephony). Therefore the user is required to exercise restraint when entering rules so that they have some actual meaning (e.g. a rule to notify the user if the share price of the Ericsson B share exceeds SKr10 would not be as useful as one where notification occurs if the Ericsson B share exceeds SKr300).

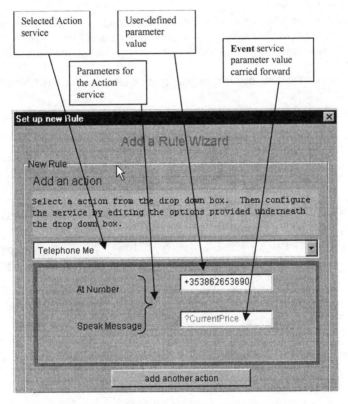

Fig. 4 Selecting Action Condition for a new rule

5.2   PANI WAP Client

The Wireless Application Protocol (WAP) is an application environment and set of communication protocols designed to allow the development of applications and services that operate over wireless networks[8,9]. Interoperability is an important aspect of the bearer independent WAP standard. Wireless devices from different manufacturers can all avail of the services offered by wireless network operators. WAP makes use of existing Internet standards such as XML, URLs, scripting, CGI etc. It also utilises mobile networking standards such as GSM, IS-136, etc.

One of the most prominent applications of WAP is to facilitate web browsing through hand held devices. Figure 5 illustrates an example WAP network that would allow the realisation of this application. While a discussion of this WAP application will not be exhaustive it will introduce some aspects of WAP development that are relevant to this document.

Some Internet standards are too inefficient over mobile networks to be integrated into the WAP standard. Examples of such standards are HTML, HTTP, TCP, etc. These standards require large amounts of text based data to be sent. This is not appropriate
for WAP due to the low bandwidth and high latency characteristics of mobile networks.

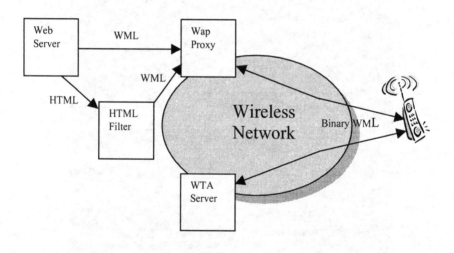

Fig. 5. An Example WAP Network

5.2.1   PANI & WAP

The integration of WAP with PANI greatly increases the versatility, availability, value and commercial viability of the PANI project. The user's ability to avail of PANI services without having to be sitting at, or to even own, a computer adds to the attractiveness of the PANI concept. The mobile device is an adequate interface to PANI.

5.2.2   The WAP Application

The features of WAP necessary to implement the application to be integrated with PANI will now be discussed.

WML data is structured as a collection of *cards*. A single collection of cards is referred to as a WML *deck*. Each card contains structured content and navigation specifications. Logically, a user navigates through a series of cards, reviews the contents (both text and images can be supported) of each card, enters requested information, makes choices and navigates to another card or deck or returns to a previously visited card or deck. The mobile device maintains a certain amount of state information. This information is the *browser context* and includes navigational history for efficient backward navigation and a number of variables and their values used within WML cards. The browser context can be initialised by the *newcontext* attribute of the card element.

Input from the user can be entered as text or the user can choose from a presented list of options. Navigation is event driven. For example, when the user presses the OK button on a mobile phone an ACCEPT event is generated. This event is caught and triggers navigation to a new card or deck. Decks and cards are identified by URLs.

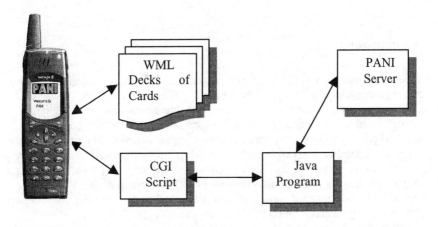

Fig. 6. WAP Application Architecture

The Common Gateway Interface (CGI) is a method of allowing a URL to refer to a program rather than a static WML deck or card. If a URL in WML points to a CGI then the server hosting the CGI program will execute it. It is possible to pass variables contained in the browser context to the CGI program as part of the URL. The CGI program must create decks containing cards in order to display its output on the mobile device.

This is all the information that is needed to collect user input, process it and display results on the user's mobile phone. The CGI program was implemented as a perl script. This was because there were modules available for the simple and efficient creation and display of WML decks and cards. It was decided the perl script would pass the information to a java application. This application would do the following.

- Take input from the CGI program.
- Formulate and send a request to the PANI server using this input, e.g. adding or deleting a rule.
- Extract the relevant information from replies from the PANI server.
- Return this information to CGI program.
- Dynamically generate WML decks of cards displaying the services offered by PANI, the existing rules for the user, etc.

The perl script then creates the necessary deck of WML cards and displays these on the user's mobile phone. Sockets were used as the communication mechanism between the CGI and java program. Figure 6 details a basic overview of the WAP application.

With the incorporation of the java program it is possible to remove much of the complexity from the mobile phone. It would be possible to maintain the java program on a WTA (where the WML cards and decks and the CGI programs would more than likely be hosted). The java application could update the WML decks as services offered by PANI are added or deleted.

6   Conclusions

The appetite of users for specialised and personlised information services being is catered for increasingly by services based both on IN architectures and on the World Wide Web. This trend is set to accelerate, with important growth areas being in services delivered over mobile computing platforms and digital set top boxes.

The PANI project addresses the issues of allowing users and third party developers the maximum flexibility in creating and customising information services for delivery using a range of telephony media. The key component in the PANI architecture is the use of meta-languages to support creation of service components that are self-describing. This enables naïve users to plug together and customise their own services using service components supplied by third party developers.

Success in the creation of this dynamic service environment is achieved through the use of Intelligent Agent technology. The Rules Agent is considered intelligent as it modifies both its capabilities and behaviour dynamically on the basis of messages received in SDML and SCML. In the case of SDML registrations the Rules Agent dynamically builds code to manage data from information services or to invoke action services. It also plans on the basis these registrations, for example building plans to clean up data from information services at regular intervals. In the case of SCML messages the RulesAgent translates these rules into its own executable format, that is the rule-based language JESS and adds them to its rule base. Firing of these rules is then handled automatically by the JESS engine. A consequence of the use of the

JESS engine is that the approach to rule conflicts is very simple. If a rules conditions are met then that rule will fire, it is the responsibility of the user to ensure that the effect of a rule firing is beneficial. For example, a poorly conceived rule could result in the user receiving tens of telephone calls per day indicating that a particular share price is greater than 0.

The authors believe that the use of Intelligent technology and meta-languages presents an elegant solution to the development of dynamic service environments. While it is true that there are more conventional technologies available which approach a type of dynamic behaviour, for example the CORBA Dynamic Invocation Interface (DII) there are drawbacks to such approaches. Firstly the PANI meta-languages capture not only structural information (the types and values of parameters) they also capture semantic information (although it is true that this information is only of benefit to the end user), a feature not present in the CORBA DII. Secondly the use of an interpreted language and the adoption of the agent paradigm allows for an elegant and realisable method of adapting behaviour in a dynamic environment, an aspect difficult to realise through traditional programming approaches.

More information on the PANI project can be found at the PANI website[10].

Acknowledgements

This work was kindly sponsored by ERICSSON P-VAS.

References

1. Kerr D., PANI: Personal Agents for Network Intelligence™, Communicate, Volume 4 - Issue 2, pp4-15, January 1999
2. Kerr D, O'Sullivan D, Evans R, Richardson R and Somers F, Experiences using Intelligent Agent Technologies as a Unifying Approach to Network Management, Service Management and Service Delivery, Proceedings of IS&N98 (http://www.alcatel.be/isn98/)
3. Busioc, M., Winter C., and Titmuss, R., Distributed Intelligent Agents for Service Management, In IJCAIU'95 Workshop, Montreal, Canada
4. Friedman-Hill E. J., Jess, The Java Expert System Shell (Version 5.0a6), Internal report SAND98-8206 (revised), Distributed Computing Systems, Sandia National Laboratories, Livermore, CA, 30 June, 1999
5. Microsoft Outlook '98, http://www.microsoft.com/outlook/
6. Rhetorex home page (Lucent Technologies) http://www.rhetorex.com/
7. Microsoft Research Speech SDK, http://research.microsoft.com/stg/
8. Wireless Markup Language Specification
9. [http://www1.wapforum.org/tech/terms.asp?doc=SPEC-WML-19990616.pdf]
10. Wireless Application Protocol (WAP) Tutorial
11. [http://www.webproforum.com/wap/index.html]
12. The Pani Website at http://pani.broadcom.ie/

Agent Based Support of Advanced Personal Mobility Features

M.Louta[1], K.Raatikainen[2], A.Kaltabani[1], N. Liossis[1], E.Tzifa[1],P.Demestichas[1], M.Anagnostou[1]

[1]National Technical University of Athens, 9 Heroon Polytechneiou Str, 15773 Athens, GREECE
email: etzifa@cc.ece.ntua.gr
[2]University of Helsinki, Department of Computer Science, P.O. Box 26 (Teollisuuskatu 23), FIN-00014

Abstract. Support for personal mobility will be among the key factors for success in the competitive communications market of the future. This paper proposes enhancements to the personal mobility support capabilities of existing service architectures. The TINA service architecture is considered to be our starting point. The key issues in this paper are the following. First, the definition of the required functionality and its design by means of appropriate service components. Second, the use of mobile intelligent agent technology for the implementation of the new components. Third, the integration of the agent based personal mobility support components to the "standard" TINA.

1 Introduction

It is often stated that the ongoing liberalisation and deregulation of telecommunications will introduce new actors in the respective market of the future [1,2,3,4]. In principle, the main role of all the players in such a competitive environment will be to constantly monitor the user demand, and in response to create, promote and provide the desired services and service features. Some key factors for success are the efficiency with which services will be developed, the quality levels, in relation with the corresponding cost, of the new services and the provision of universal service access to the widest possible set of users, regardless of physical locations, terminals used and the inherent heterogeneity of communication systems.

In the light of the challenges outlined above, the aim of this paper is to propose enhancements to the personal mobility support capabilities of existing service architectures [3,5,6,7]. Personal mobility may be defined as the ability of a user to access services on the basis of personal data, thus being provided with the sense and feeling of home. More specifically, our aim is to introduce some advanced personal mobility capabilities, to realise them in a flexible manner, and consequently, to show how they can be readily integrated with existing service architectures [8,9]. Our reference service architecture will be the one specified by the Telecommunications Information Networking Architecture Consortium (TINA-C) [2,10].

Our approach in this paper is the following. The starting point is the presentation of our assumptions regarding the entities and the roles at the business level in the telecommunications world of the future. Then, target *business cases*, i.e., important service provision scenarios that should be supported in open competitive communication environments, are specified. Through the business cases presentation, the importance of the personal mobility and service architecture concepts are highlighted. Finally, the use and expansion of existing service architectures, so as to support advanced personal mobility features, is discussed.

A typical view of the competitive telecommunications world of the future may be the one depicted in Fig. 1. Without being exhaustive (e.g., see [11]) four main different entities can be identified, namely, the *user, service provider, retailer*, and *connectivity provider*. The role of the service provider is to design and implement services. The role of the retailer is to promote, sell, customise, deliver and operate these services. Limited by techno-economical or administrative reasons each retailer offers services only inside a *domain*. Users may have subscription contracts for specific services with one or more retailers (*home retailers*). Through the subscription, the user acquires the right to use the services in the retailers' domains (*home domains*). Finally, the role of connectivity providers is to offer the network connections necessary for supporting the services.

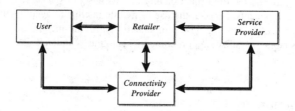

Fig. 1. A view of the future competitive telecommunications environment

Such highly competitive and open environments should encompass mechanisms that will enable users to register on a terminal for future service use in a target domain as well as to receive incoming calls or invitations to join a service session from another party, directed to the pre-registered terminal. Such mechanisms should provide means for establishing association with the most appropriate retailer for the delivery of service, i.e., those offering at the given period of time, adequate quality of service in a cost efficient manner. This offer will constitute a key factor for success.

Hence, the following business case, composed of two phases, may be pursued. As a first phase, to enable users to be remotely registered to a terminal outside the domain they are currently located, in the context of which the most appropriate retailer (*visited retailer*) is found. As a second phase, to receive an incoming invitation for a scheduled delivery of service (i.e., scheduled delivery of pieces of news from a service provider). This business case falls into the realm of personal mobility, as it comprises both Remote Registration and Incoming Call features. The relative study has been made by the ACTS project MONTAGE [3,5,6,7].

The realisation of the examined business case requires a significant amount of co-operative computing among the involved business level entities. These interactions are described in more detail in Section 2. Section 3 introduces the extensions necessary for the support of advanced personal mobility concepts, and explains how they may be integrated with the TINA framework. The extensions will be specified in terms of the new functionality needed, the corresponding service components, and the software technologies applied for the realisation of the new components. In particular, the use of *intelligent mobile agent* [12,13] technology is considered for the implementation of the new personal mobility support components. Hence, it will be shown how agent-based personal mobility support components may inter-work with standard (non-agent based) service components of the TINA service architecture. Section 4 includes future plans and some concluding remarks.

2   Business Case

A number of target business cases may be found of interest in the competitive communications world of the future. In this paper we focus on a "noble" scenario that may be decomposed in two phases. The first phase is targeted to the selection of the best retailer as well as the selection and configuration of the most appropriate terminal through which a user, that is found in a foreign domain, may access and use a service. As a second phase, we consider the reception of an incoming call for a scheduled delivery of service (i.e., pieces of news from a service provider) or invitation to join a service session directed to the pre-registered terminal.

Remote registration constitutes an important personal mobility feature that should be supported in future service architectures. The main idea underlying the overall remote registration concept is to settle everything for the user for service access and usage before he/she is actually located in the target domain. Thus, the best retailer in terms of service offerings and corresponding cost and terminal in terms of availability, capabilities supported and cost imposed should be found. After the remote registration procedure has ended, both counterparts (user and retailer) are informed about the results of the registration, that is the terminal identifier and selected retailer identity. In this way, the user is given accurate information about where to "seek" and request access to the service and the retailer reserves the resources necessitated for the provision of the service. Additionally, in case the logic of the service requested is not available in the target domain, the whole mechanism supporting the required service can be downloaded in the selected retailer's domain. Considering a service involving data retrieval, matching the user requirements, preferences and constraints encoded in his personal profile, hasty users, demanding immediate access to the results could be satisfied by applying the overall concept of remote registration. Thus, after a successful authentication process, the user could have immediate access to the requested service or its results.

Let us now consider a user who has already established an access session with a retailer. The user issues a request to be remotely registered to a terminal located at a target domain for future service usage through an appropriate interface. A fundamental assumption in the scenario is that a number of terminals, maintained by retailers, could be utilized for rendering services. A retailer may have acquired the list with the potential terminals through contracts with the terminal owners. This issue is out of the scope of our scenario and will not be further examined. Terminals are differentiated in terms of the capabilities they support (special services, high quality levels) and the cost imposed by the retailers as a return for service rendering. Each terminal may be supported by a number of retailers who in turn offer services with different characteristics both in terms of the service features and the corresponding quality levels as well as the cost for delivering the service to the user. Thus, since the terminals may be served by various candidate retailers, an interesting option that is offered to the user is the selection of the best candidate retailer and terminal selection and configuration. The basis of the selection comprises the relevant service data (e.g., desired quality levels), the list of eligible candidate retailers and the service offerings (e.g., cost at which the desired quality levels are provided, as well as the cost imposed by the retailers for the service rendering.

Incoming call or invitation to join a service session is a challenging functionality in the personal mobility context. Examples include receiving an ordinary telephone call (B Party) as well as a scheduled delivery of pieces of news (pre-ordered by the user) from a service provider. In the personal mobility context, the binding between a user and a terminal is temporal. In addition, the user may have different bindings for different services. Therefore, user registration at a specific terminal is a prerequisite for receiving an invocation to join a service session. Thus, as a second phase of this business case we consider the reception of an incoming call (i.e., invitation to join a session) directed to the pre-registered terminal.

3   Introduction of Advanced Personal Mobility Support Features

This section introduces the extensions in the service architecture that are necessary for supporting the set of personal mobility support features required by the business case of section 2. Initially, the new functionality required will be identified in sub-section 3.1. This functionality will be attributed to service components in sub-section 3.2. Finally, integration issues among the new components and the TINA service architecture will be provided in sub-section 3.3.

3.1   Requirements: Identification of additional functionality

TINA access session offers the framework for user authentication, basic personal mobility support, service invocation and user invitation for joining a service session. The novel issue that is not (at least straightforwardly) supported is the overall task of terminal selection in conjunction with the retailer selection. As a first step, this aspect requires an entity that will act on behalf of the user. Its role will be to capture the user preferences, to deliver them in an appropriate form to the appropriate retailer entity, to acquire and evaluate the corresponding retailer offerings, and ultimately, to select the most appropriate terminal-retailer tuple.

As a second step, retailer selection requires an entity that will act on behalf of the retailer. Its role will be to collect the user preferences, and to make a corresponding offer after contacting the underlying connectivity providers. The overall object model is depicted in Fig. 2.

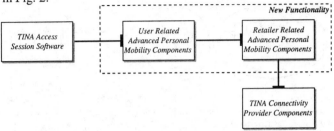

Fig. 2. High level representation of the required personal mobility support functionality

3.2   Analysis: Definition of new components

In this sub-section we define the components that will undertake the functionality identified above. New components are defined, in order to comply with the requirement to introduce the functionality by minimally impacting the service architecture. Furthermore, the use of mobile intelligent agent technology is considered, due to performance reasons (as explained in detail in [14,15]), and due to the capabilities offered for computing platform independence and user specific customisation. Consequently, the *Remote Registration Agent* (RRA) is defined as whose role is to select on behalf of the user the most appropriate terminal-retailer for the service at hand. Moreover, the *Retailer Agent* (RA) is defined and designated with the role to promote the services offered by the retailer. More specifically, the RRA knows the user requirements and constraints, may interact and negotiate with the RAs, and may select the most appropriate terminal-retailer for the desired service. Moreover, it may be extended to record past experience (e.g., terminal-retailer selections, service usage characteristics, etc.). The RA promotes the service offerings of the retailer, maintains a list of terminals utilised for rendering services, negotiates with RRAs, and the underlying connectivity provider mechanisms.

3.3   Design: Service component interactions

This section describes the integration of the new components with the TINA service architecture. This is accomplished by providing a formal description of the business case and the interactions among the involved service components. For simplicity, the whole analysis is decomposed in two phases. Thus, the dynamic interactions among the service components in the context of Remote Registration and Incoming Call features are presented, in respective subsections.

Remote Registration. This section elaborates on the first phase of the Business Case. In particular, it shows the associations among the service components (Fig. 3) that appear in the context of *Remote Registration* and describes the dynamic interactions among them. The textual description is accompanied by Message Sequence Charts depicted in Fig. 5, Fig. 6 and Fig. 7.

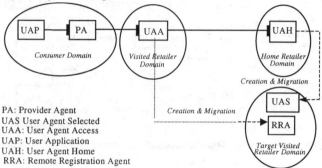

PA: Provider Agent
UAS User Agent Selected
UAA: User Agent Access
UAP: User Application
UAH: User Agent Home
 RRA: Remote Registration Agent

Fig. 3. Computational model for remote registration

We assume that the end-user has logged into a retailer as a known user. The following steps should be performed:

1. The end-user (userId) requests from the PA, through the access session related UAP, to perform remote registration for a given service type (serviceId). The remoteRegistrationRequest() of the end-user provides as input parameters in addition to the userId, the desired service (serviceId), the target domain area (targetDomainId) and a time interval defining the starting and ending time of registration. The input parameters are not depicted in the MSCs for simplicity.
2. The PA forwards the remoteRegistrationRequest() to the UAA. The request includes the arguments received from the UAP.
3. The UAA checks the request against the end-user's subscription data. We assume that the UAA already has the required subscription information, so it performs the check.
4. Assuming that the user is allowed to be remotely registered, the RRA-related mechanism is instantiated. In particular, UAA creates and instantiates the RRA with the user identifier (userId), user's service requirements (serviceProfile), the Domain identifier (targetDomainId) and the time interval.
5. In the sequel, the RRA identifies by means of a Brokerage service the eligible candidate retailers supporting the requested service in the target visited domain.
6. As a next step the RRA migrates to the identified node of one of the candidate retailers constituting the list (randomly selected).
7. Thereafter, the "parent" RRA is replicated and each replica migrates to the identified nodes of the retailers that potentially could deliver the service to the user.
8. Next the RRA attempts to find the most appropriate terminal-retailer (terminal-retailer selection phase) for the user to be remotely registered. The overall model of the RRA-based mechanism is depicted in Fig. 4. This phase involves the following subtasks:
   ➢ Each RRA replica interacts with the respective RA in order to retrieve the list of terminals that could be used for remote registration. This list is returned to

each RRA replica along with the terminal properties (i.e., availability, capabilities supported, etc.). RA is the component that would be responsible for maintaining terminal specific information (that is terminal availability, capabilities supported, list of end-users registered and the respective registration-period, charge for remote registration, etc.)

➢ As a next step, this list is filtered in terms of terminal availability and capabilities supported. Only the terminals that support the requirements imposed by the user and are available at the time-period the user has specified at his request will be included in the filtered list.

➢ In the sequel, each RRA replica would pose to the respective RAs questions regarding the charge of a fully specified service combination (that is combination of media and their respective QoS) for all the terminals contained in the list acquired during the previous step. Each RA will in turn contact the connectivity providers providing connections in his domain, estimate the charge for the service combination requested and calculate the total charge. The total charge is the estimated charge for the service usage plus the charge imposed by the retailer to the user for being remotely registered to the terminal.

➢ Each RRA replica forwards the results of the negotiation to the "parent" RRA which in turn makes a decision on the most appropriate terminal, retailer and connectivity provider in terms of QoS offered and total estimated charge.

9. After the terminal-retailer selection has finished, the "parent" RRA requests the selected RA to register the user to the selected terminal for the service combination offered and for the time interval specified by the user at the initial request. The arguments specify the end-user (userId), the service finally offered (serviceOffer), the time interval and the selected terminal identification (terminalId), needed to address the terminal in the retailer domain. As a response, the RA updates the necessary data structures.

10. After the "parent" RRA has completed its operations, it invokes the UAH component in order to inform it about the result of the remote registration procedure. The UAH, in turn, cancels a possible obsolete registration and stores the updated registration information comprising the selected terminal identifier, the selected retailer identity, the service offer and the time interval (starting and ending time of registration).

11. After the new registration is completed, the UAH returns a registration reply to the UAA at the Default Retailer Domain.

12. The UAA, in turn, returns a reply including the termination condition to the PA.

13. The PA via the UAP informs the end-user. The reply comprises the selected terminal identifier and the selected retailer's identity required, so as to address the terminal in the target domain.

14. Finally, when the registration period starts, the UAH creates a new UAS instance which in turn migrates to the selected visited retailer domain and represents the user in the target-domain. The new UAS instance will be initialized with information needed to serve the end-user for the requested service and the initialization of the subordinated objects. The existence of the UAS is only for the

time interval the user has specified at his request for remote registration. The UAH stores the reference of the UAS component in the selected retailer domain. At this point, the network knows the user and the service session could start.

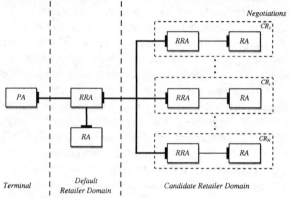

Fig. 4. Negotiation between the RRAs and the RAs

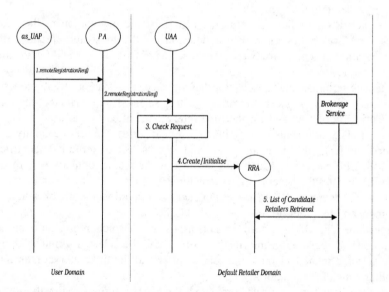

Fig. 5. Dynamic Model for Remote Registration (Steps 1-5)

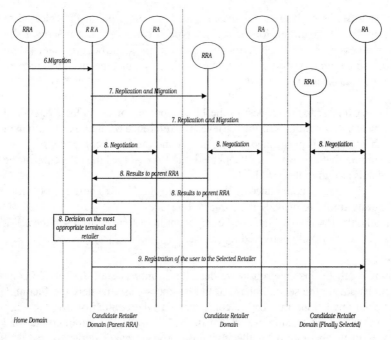

Fig. 6. Dynamic Model for Remote Registration (Steps 6-9)

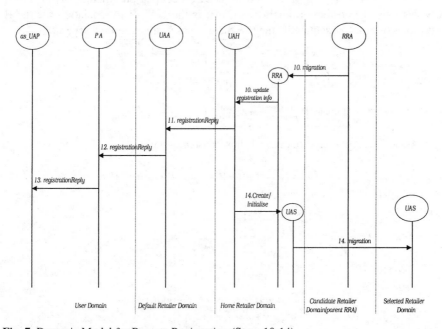

Fig. 7. Dynamic Model for Remote Registration (Steps 10-14)

The end-user is registered, for the indicated service type, at the selected terminal and an obsolete registration for the service type is cancelled. Let us now assume that the user accesses the selected terminal, provides his identification data (login name and password) and wishes to start the service he has been registered for. The following subtasks are performed:

15. The PA forwards the request to the IA component of the default retailer. The IA by means of a naming service retrieves the reference of the UAH component.
16. Authentication is performed among the IA and the UAH component.
17. Assuming that the user is authenticated, the PA resolves the reference of the UAS, which is stored in the UAH.
18. The PA requests commence of the service to the UAS. The UAS notifies the PA about its ability to provide the requested service.
19. Finally, the PA invokes the UAS component in order to start a service session and the service is finally delivered to the user.

Incoming Call. In this subsection we describe in more detail the service component interactions during the second phase of the Business Case, namely the Incoming Call. The associations among the service components in the context of inviting a user (a B Party) into a service session are shown in Fig. 8, while the minimum set of known references in the beginning of the invitation procedure are depicted in Fig. 9. The course of actions during the invitation procedure depends on whether or not the B Party has logged into the terminal or not, that is the UAS-PA association for the B Party exists or not. Consequently, the two distinct cases are examined separately.

Fig. 8. Associations between entities and their establishment times

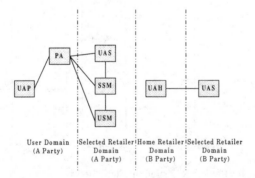

Fig. 9. Minimum Set of Known References in the Beginning of Invitation to join a Service Session

Case 1: The B-Party has logged into the system

If the PA-UAS association for the B Party exists, that is the user has logged into the system implying that UAS and PA have acquired the references for each other, then the procedure is straightforward:

1. The A-party requests the SSM in the selected retailer domain through the access session related UAP to invite B Party to join in a service session. The *invite()* operation provides as input parameters the inviting party and the service identifier.

2. The SSM, as a result, resolves the reference of the UAH of the invitee party (B-Party), and in the sequel, sends to it an incoming call invitation.

3. The UAH of the invitee (B Party) receives an invitation for joining a session through the SSM of the inviting party (A Party), and forwards it to the UAS of the B Party.

4. The B Party's UAS has the reference of the B Party's PA and, therefore, it invokes the PA.

5. As a next step, the PA starts the incoming call application that shows the invitation details and asks the user either to accept or reject the invitation.

6. The user could either reply with "rejection" or "acceptance" of the invitation. If the user rejects the invitation, then the PA replies a "rejection" to the UAS which in turn is forwarded to the requesting SSM (reference obtained in the "join request" from the UAH). If the user accepts the invitation, then the PA replies an "accept" to the UAS which, in turn is forwarded to the requesting SSM.

7. The SSM invokes at the A Party's UAP either the *notifyInvitationRejected()* or the *notifyInvitationAccepted()* method.

8. The UAS of the B Party invokes the local SF in order to create the USM for the B Party.

9. The USM of the B Party activates the UAP for the B Party and invokes the *usmRequestResponse()* operation of the UAS (of the B Party).

10. The USM of the B Party contacts the SSM of the A Party which, in turn, notifies the both UAPs through the USMs that the end-to-end Service Session has been established.

Fig. 10 and Fig. 11 depict the conceptual Message Sequence Charts of the beginning (steps 1-7) and the end (steps 8-10) of the invitation procedure.

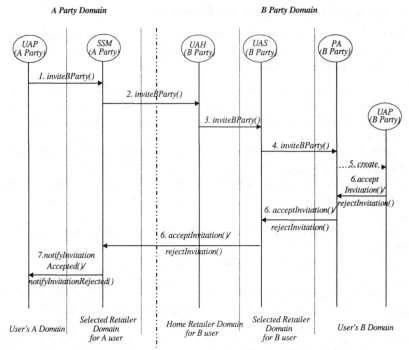

Fig. 10. Beginning Phase of Inviting a User to a Service Session (Steps 1-7)

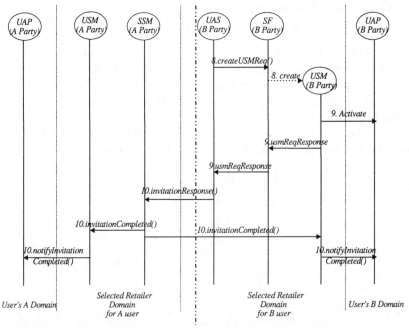

Fig. 11. Ending Phase of Inviting a User to a Service Session (Steps 8-10)

Case 2: The B Party has not logged into the system

If the PA-UAS association for the B Party does not exist, then a paging procedure is needed in order for the UAS to retrieve the reference of the PA and establish association an with the terminal to which the B-Party has been registered to. The first three steps are the same as in the previous case and hence they are not depicted in the relative MSC.

1. The A-party requests the SSM in the selected retailer domain through the access session related UAP to invite B Party to join a service session. The *invite()* operation provides as input parameters the inviting party and the service identifier.
2. The SSM, as a result, resolves the reference of the UAH of the invitee party (B-Party), and in the sequel, sends to it an incoming call invitation.
3. The UAH of the invitee (B Party) receives an invitation for joining a session through the SSM of the inviting party (A Party), and forwards it to the UAS of the B Party.
4. The UAS does not have the reference of the PA, thus contacts the RA in the selected retailer domain in order to get the reference of the Connectivity Provider that provides connectivity to the terminal the User (B Party) has been registered to.
5. The Selected RA invokes the Connectivity Provider that, in turn, invokes the terminal to show a login request.
6. When the User (B Party) responds to the login request, then the User is authenticated in the Home Retailer domain.
7. After authentication, the association between the PA and the UAS is established. Therefore, the next steps are the same as described in the previous case.

Fig. 12 depicts the conceptual message sequence chart of the paging procedure.

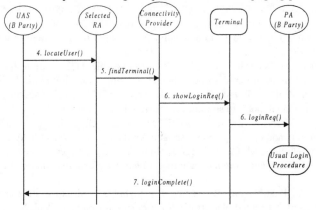

Fig. 12. Paging Procedure

4 Discussion and Conclusions

This paper introduced necessary extensions for the support of advanced personal mobility concepts and explained how they may be integrated with the TINA framework. These extensions were realised as new components, so as to minimally impact the existing service architecture model. This last requirement, in conjunction with performance reasons, designated the intelligent mobile agent technology as an appropriate alternative for the implementation of the new components.

The presented approach, integrated with various versions of the logic of the RRA and RA, will be validated in demos in the context of the MONTAGE project. These demos will serve as a demonstration of TINA compliant systems. Feedback will be collected and used for fine-tuning the overall personal mobility architecture.

5 Acknowledgements

This work was supported by the Commission of the European Communities within the ACTS Programme MONTAGE (AC325). The authors would like to warmly thank their MONTAGE colleagues.

References

1. H. Ungerer, N.P. Costello, *"Telecommunications in Europe: Free choice for the user in Europe's 1992 market"*, 2nd ed., Office for Official Publications of the European Communities, Brussels, 1990
2. TINA-C Deliverable, *"Definition of Service Architecture"*, Version 5.0, June 1997
3. ACTS project MONTAGE (AC325), CEC deliverable D21, "Context setting for agent-based accounting and charging and personal mobility support", June 1998
4. A. Lazar, "Programming telecommunication networks", *IEEE Network*, Vol. 11, No. 5, Sept./Oct. 1997
5. ACTS Project MONTAGE (AC325), CEC Deliverable D23, "Algorithms, design and implementation of agent-based accounting and charging for the initial trial", Dec. 1998
6. ACTS Project MONTAGE (AC325), CEC Deliverable D24, "Design and implementation of agent-based personal mobility support for the initial trial", Dec. 1998
7. ACTS Project MONTAGE (AC325), CEC Deliverable D25, "Design of advanced agent-based accounting and charging and personal mobility support services", April. 1999
8. S. Trigila, A. Mullery, M. Campolargo, J. Hunt, "Service architectures and service creation for integrated broadband communications", *Computer Communications*, Vol. 18, No. 11, 1995
9. S. Trigila, K. Raatikainen, B. Wind, P. Reynolds, "Mobility in long-term service architectures and distributed platforms", *IEEE Personal Commun.*, Vol. 5 No. 4, Aug. 1998
10. T. Magedanz, "TINA - Architectural basis for future telecommunications services", *Computer Communications*, Vol. 20, No. 4, 1997

11. B. Wind, M. Samarotto, P. Nicosia, M. Lambrou, E. Tzifa, "Enhancing the TINA architectural framework for personal mobility support", In Proc. *5th Conf. On Intelligence in Services and Networks 1998 (IS&N'98)*, Antwerb, Belgium, May 1998

12. M. Breugst, T. Magedanz: "Mobile Agents—Enabling Technology for Active Intelligent Network Implementation", *IEEE Network Magazine*, Vol. 12, No. 3, pp. 53—60, Aug. 1998

13. D. Chess, C. G. Harrison, A. Kershenbaum, *"Mobile Agents: Are they a good idea?"*, In Mobile Object Systems—Towards the Programmable Internet, LNCS, pp. 25—47, 1997

14. A. Kind, M.E.Anagnostou, M.D.Louta, J.Nicklisch, J.Tous, "Towards the Seamless Integration of Mobile Agents into Service Creation Practice", In Proc. *Sixth International Conference on Intelligence in Services and Networks "Paving the way for an Open Service Market" (IS&N 99)*, Barcelona, Spain, April 1999

15. ACTS Project MONTAGE (AC325), CEC Deliverable D22, *"Review of applicable agent and service creation technologies"*, Aug. 1998

Integrated and Scalable Solutions for Telecommunications Management

Sebastiano Trigila

Fondazione Ugo Bordoni, Italy
trigila@fub.it

The provision of telecommunications services over an integrated infrastructure including different networks (POTS, Internet, ISDN, B-ISDN, GSM, and UMTS) calls for the integration of a number of management system standards developed by several bodies, for different technologies. For example, the ITU-T TMN series of recommendations and the IETF SNMP are the most prominent types of targeted standards ruling the development and practice of management systems in the areas of Telecommunications and Information Technologies, respectively. The main requirements for integrated management of communications are easy to express, although very challenging to materialize in one single architectural framework:

- Respect and reflect the hierarchical nature of communication networks (for instance, in the direction set by IETF SNMPv2), by aligning the structure of management systems with the tree structure of networks.

- Accommodate the unpredictable growth of the number of network devices and systems to be managed.

- Accommodate major changes in network and domain topology.

- Allow for (limited, compatibly with security) management of resources of any administrative domain from within any other administrative domain.

- Allow for distributed management (see for instance the IETF Distributed Management Charter, http://www.ietf.org/html.charters/disman-charter.html, where it states: "a distributed network manager is an application that acts in a manager role to perform management functions and in an agent role so that it can be remotely controlled and observed").

- Treat both manag-*ed* and manag-*ing* parts of telecommunications systems, with equal levels of focus and the same conceptual paradigm.

Needless to say, SNMP – largely applied in today's corporate Intranets – and TMN – nowadays quite mature as a non proprietary management standard for use by worldwide telecommunications operators – only partially satisfy the above requirements. These could be summarized with a short motto: "hierarchical, scalable, secure, distributed and component-oriented telecommunications management".

The currently prevailing vision is to give up dreaming of a super-framework where all the above requirements could be accommodated and adopt a component reuse approach, so that management solutions developed in different contexts can be suitably inherited, wrapped and applied in a specific management context.

In particular, component-oriented management models, whereby system modules may have both an interface to be managed by other system parts and an interface to manage other system parts, are very promising for open evolution of telecommunications. The long established OSCA/INA by Bellcore, the global architecture proposal known as TINA framework, and the current work of the TeleManagement Forum are very much oriented towards component-based management platforms.

Component-based management platforms are the essential theme of the two papers constituting this Section.

The paper by V. Wade *et al.* provides an excellent and broad overview of the trends towards the use of off-the-shelf *componentware* to cope with the complexity of building telecommunications management services. The main focus is on the ability to integrate components from different sources and different technologies, including CORBA components, Enterprise JavaBeans, Workflow Engines, CMIP-CORBA gateways, and C++ TMN API. A business model for the development of Open Management Systems and the TeleManagement Forum Technology Integration Map are presented. The paper also provides operational hints for integration of management solutions, by presenting two case studies, on CORBA-component integration and on integrating Telecommunications Management by using Workflow Engines and JIDM compliant CORBA/CMIP gateways.

The paper by D. Gavalas *et al.* proposes a solution to overcome the scalability and flexibility limitations of Mobile Agent based approach, which – as applied so far – considers "flat" network models, rather than hierarchical. The key element of the solution is the so-called Mobile Distributed Manager (MDM), a new concept to reduce bandwidth consumption and communications cost induced by heavy and inefficient transfer of management traffic data. The MDM operates at an intermediate level between the (centralized) SNMP manager and the stationary agents. MDMs are mobile agents that may be dynamically delegated responsibility over an assigned domain where they are physically sent "on a mission". The innovative value of the idea is in the possibility to extend its applicability beyond an SNMP context. The paper also discusses, to a remarkable extent, implementation, testing, deployment and performance evaluation issues related to the proposed solution.

Approaches to Integrating Telecoms Management Systems

V.Wade[1], D.Lewis[2], C.Malbon[2], T.Richardson[3], L.Sorensen[4] , C Stathopoulos[5],

[1]Trinity College Dublin, Dublin 2, Ireland
Vincent.Wade@cs.tcd.ie
[2]University College London, London, UK
{D.Lewis,CMalbon}@cs.ucl.ac.uk
[3]UH Communications, Denmark
lbs@uhc.dk
[4]Systems Technology Solutions UK
sts@anglianet.co.uk
[5]Algosystems, Greece
stathop@algo.com.gr

Abstract. Because of the cost and complexity of building bespoke service management systems, telecommunications management developers are moving toward the use of off-the-shelf componentware to satisfy their management requirements. However, a crucial problem with such an approach is the ability to integrate components to realise integrated management solutions. This paper identifies the technology requirements for the development of systems which support telecoms management business processes which are constructed from reusable components. It sets out requirements on component integration technologies that are expected to be key to the development of future operational support systems. The current status of relevant component integration technologies is reviewed in the context of these requirements. The paper outlines two telecommunications management case studies which are being conducted to evaluate these integration technologies. The paper draws some conclusions about the relative merits of the different technologies examined and make some suggestions for further work.

1 Introduction

The liberalisation of telecoms markets across Europe and the world has exposed service providers to a high level of competition. This competition is forcing them to reduce costs, improve customer service and rapidly introduce new services. One key way in which these pressures can be addressed is through the improved integration of the many software systems operated by a service provider. This includes amongst others, the integration of different operation support systems.

Component based reuse is seen as an increasingly important software development aid, both within the telecoms industry and in the wider IT community. Building systems from components that interact through well defined interfaces offer a route to reusing software across projects within a telecom system developer and to integrating commodity third party software into the system. Both of these offer development cost savings and improvements in reliability and maintainability. Emerging standards such as Enterprise JavaBeans and CORBA Components are encouraging the development of platforms that directly support component integration. This is prompting the telecoms industry to move towards the widespread adoption of component-based architectures. For example BT and MCI/Worldcom have already published architectural and requirement documents that encourage the migration of their OSS architectures to component-based platforms. This movement is now also being supported within the TeleManagement Forum. Bellcore has long established its Information Networking Architecture (OSCA/INA) with a similar aim. This has been evolved and validated by the TINA-Consortium, resulting in the standardisation of ODL in the ITU where it is likely to become the basis for the further standardisation of IN Capability Sets. However, many existing architectures, such as TMN, do not directly support component-based systems, and not all the notations and tools currently used in telecoms can fully represent components, e.g. GDMO, UML. The alignment of notations and tools with the modeling constructs and development activities associated with component-based software development must therefore be addressed.

This paper examines the current status of component integration technology and assesses it against the requirements of OSS developers. The technologies addressed are:

- CORBA Components: based on the joint revised submission to the corresponding RFP, which is currently under consideration by the OMG.
- Enterprise Java Beans.
- Workflow technologies based on both OMG and WfMC specifications.
- Contemporary TMN technology, specifically gateways between CMIP and other distributed technologies such as CORBA and SNMP, and the C++ TMN API.

Other relevant technologies such as DCOM, Distributed Databases and Transaction Processing have not been covered in this paper but are worthy of further examination in this area. Also identified in this paper is the current TeleManagement Forum's approach to component integration as specified in their Technology Integration Map. The paper concludes by illustrating some of the technology choices and integration achieved in the ACTS FlowThru project which are performing trials in Service provisioning, Accounting and Assurance Management.

2 Management System Development Requirements

The telecoms industry needs to build solutions to specific management problems from the wide range of architectural and technological approaches e.g. ITU-T, ISO, TM Forum, TINA-C, OMG, ETSI and EURESCOM, among others. In particular, practitioners need to create operational support system solutions from reusable telecommunications management components that may be drawn from multiple origins. Developers of management systems need to be able to make reasoned selections from existing solutions (standardised or otherwise) while ensuring the integrity of the information flows required to satisfy business requirements. Service providers & system developers need an *open market* for reusable management components, which allow the building software systems from reusable, multi-threaded components are reduced. Key requirements include

- The ability to integrate with legacy systems in a cost effective way.
- The ability for components to interoperate even if they have been implemented using different distribution or programming technologies.
- The ability for components to interoperate even when they offer interfaces that have been defined in different languages, e.g. IDL, ODL, GDMO or SMI.
- An integration mechanism that minimises the knowledge needed of other potential interoperating components when a new component is developed.
- The need for an integration mechanism that clearly supports the needs of specific business processes in a clearly observable manner.

3 TeleManagement Forum Technology Integration Map

The TeleManagemet Forum undertook an examination of several different integration technologies for telecoms management called the Technology Integration Map (TIM) [1]. The overall structure of the TIM is represented in Figure. 1.

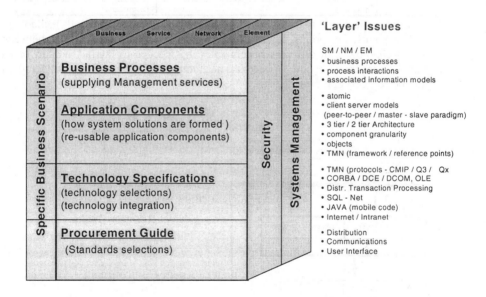

Fig. 1. TM Forum Technology Integration Map Structure

The TIM suggests a possible mapping of technologies and associated application components to particular telecom business needs. The needs of individual service providers are driven by specific business scenarios. The intention of the TMF is to use this mapping for several of its activities, which include:

- Telecoms Operations Map which gives a provider independent view of key business processes.

- Application Component activity, which is at an early stage, but which aims to provide the basis for defining reusable management components.
- Technology Direction specifying the relative merits of different management-related technologies such as CMIP, CORBA, DCE, DCOM, SQL, Java etc.
- Procurement Guide, which would aim to provide guidance to industry practitioners when making purchasing, decisions for components and platforms of management systems.

Currently the TIM focuses on the comparison of the technology specification and does not provide any real guidance on how this might map to *specific business process areas*. It has however recommended where different technologies might be applied to different types of system role. The system type and technology choices are summarised below in table 1.

Table 1. Technology selections

Interface	Type	Technology
1	Customer/Operational Staff access	Web browser / JAVA
2	Business process interaction / backbone distribution	CORBA (+ Workflow?)
3	Business process control of network resources	CMIP/GDMO SNMP/MIBs
4	Business process access to operational data (not discussed)	SQL, SQL-Net, ODBC, Data Distribution?

This categorisation led to the identification in the TIM of several technology integration points, i.e. points where effective interworking were required between different technologies in different system roles that needed to interact as part of an integrated management business process. These integration points (IPs) are identified and numbered (1-5) in Figure. 2.

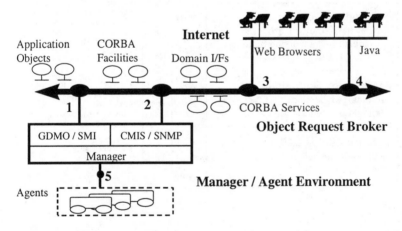

Fig. 2. Technology Integration Points

Essentially, this figure indicates that from the technologies selected, three technology areas will need to be integrated. These are:
- Internet/Web based services.
- Object Request Broker (CORBA) based services.
- Telecom based Manager/Agent services (i.e. CMIP/GDMO and SNMP/SMI).

In order to provide adequate points of integration between these areas of technology, the five IPs have are identified below:

IP1: Provides mapping of objects defined in CORBA/IDL to managed objects defined in GDMO or SMI.

IP2: Provides mapping of appropriates CORBA Services to CMIS and SNMP services.

IP3: Provides a mapping of Web Browser technology access to CORBA objects (for situations where this may be needed as an addition to/replacement of Browser access to a database).

IP4: Provides a mapping between Java based objects and CORBA objects.

IP5: Provides a high level convenient programming interface for the rapid development of TMN based manager/agent interactions. It also provides a convenient point of integration if it is necessary to separate out the two sides of the manager/agent interface from the point of view of technology selection. For example, allowing the manager role to perhaps be supported in a Web-based environment, but giving a good point of integration with a TMN based agent.

While some of these reference points have been addressed by existing gateway standards, e.g. the JIDM CORBA-CMIP specifications, key technologies such as Java, CORBA and Workflow have been rapidly evolving towards direct support for reusable components in a way not currently addressed by the TIM. The following sections review some of the relevant interworking technologies and the component oriented developments that have emerged recently.

4 Management Component Integration Technologies

This section will discuss three integration approaches, namely TMN API, CORBA / TMN interworking, Component Technology (e.g. CORBA Component Model and Enterprise Java Beans), and Workflow.

4.1 TMN Integration

In the context of this paper TMN means the OSI management framework as standardised in the ITU-T X.700 series of recommendations. The definition of the relationship between managers and agents is modelled in GDMO/ASN.1, and the actual exchange of management information between systems is via the CMIP protocol.

The interoperability between TMN based systems is good. Although CMIP is fairly complex, the experience shows that it is possible to configure systems to interoperate at the protocol level (as opposed to some CORBA based systems)[20]. TMN based management systems have reasonable potential in many aspects. If this is true then why is TMN based systems then not deployed more widely? Probably because early TMN systems got the reputation of being expensive and difficult to implement. However, this does not imply that TMN has no future. The difficulties in implementation are reducing, due to new development toolkits. It can be argued that because of TMN's maturity, its technology has reached a stage where development of a TMN based management system is less demanding in terms of implementation effort compared to any other technology. This is largely due to built-in generic and standardised features that often are needed when building management systems. A modern TMN development platform offers event handling, distribution, naming, persistency, logging, design tools, debugging, test facilities and a number of implemented standardised information models even before the implementation work is started.

The TeleManagement Forum has, with the TMN/C++ Application Programming Interface (API) product set, defined a standard programming interface for use with these international standards for network and systems management. The objective of the TMN/C++ API is to provide a straightforward mechanism to write portable and interoperable management application programs using C++.

To support this objective the TMN API has been designed to:

- Allow widespread portability, with a minimum amount of effort and code modification.
- Be simple, intuitive, and easy to use for experienced programmers.
- Be sufficient to support anticipated applications.
- Be flexible to allow for different implementation strategies, application designs, and operating environments.
- Be useable by a wide variety of application problem domains (i.e. it should be neutral with respect to the actual GDMO content.)

Much of the use of such gateway technology in Network Management is expected at the element management or lower layers of network management, particularly where integration with legacy CMIP based element managers or agents are required.

CORBA / TMN Interworking

The Joint Interdomain Management (JIDM) task force was set up by TeleManagement Forum TMF and X/Open to provide specifications of how CORBA and OSI System Management (SM) could co-exist in the same management environment. Their specifications involve translating information exchange between TMN- and CORBA-based components. There are two streams in this work – mapping between GDMO/ASN.1 and CORBA IDL, and translation between CMIP/SNMP and CORBA operations invocations. The former consists of facilitating information models in either domain to be viewed by its counterpart. The latter is concerned with component interactions in each domain – CORBA and TMN – providing the capability for components in one domain to act like components in its counterpart's domain. Many descriptions of the work and results of JIDM are available e.g. [2]

It also should be noted that, at the time this version of the Technology Direction Statement was produced, the OMG had partially completed the process of adopting a technology for CORBA/TMN Interworking. It also should be noted that the principal submissions to this process were centered on the JIDM and TMN C++ API specifications outlined above. Hence the outcome of the OMG's selection process could well result in a CORBA/TMN interworking specification which is agreed by TM Forum/Open Group and OMG. This would result in greater clarity in the defined roles for both CORBA and CMIP/GDMO, and would see both of these technologies being more fully exploited in future telecom systems.

4.2 Component Technology

The advent of component-based development promises to reduce system and application development to a process of wiring together software "integrated circuit" [3]. To achieve this a component must be able to offer both structural and promotional interfaces. Promotional interfaces allow system developers and third parties to inspect a component in order to determine its usefulness. Structural interfaces allow application developers, integrators and frameworks the means with which they can establish communication.

One key benefit that component technologies promise is software reuse. But why should this be any more attractive than with existing OO development methodologies? The answer lies on the shift of the software development process from a craft-based activity (i.e. building from scratch) to an industrial process. Here, component-based systems reduce the application and system development process to one more familiar in traditional industrial manufacturing. Here, large cost savings are achieved by using standard procedures to assemble complex products from sets of well designed, pre-fabricated parts without a loss of build quality. This is sometimes referred to as the "industrialisation" of the software development process. Such a shift requires a big change in software developer mentality. Existing component-based development techniques are related to well-known OO development practices (such as, business analysis and use case modelling). These will not disappear, but, the deployment of component built systems will shift the emphasis from fabrication to integration, and to this end a component's self-description, or "introspection", will become as important as its ability to implement some set task.

A number of emerging heavyweight technologies are addressing component design and development including Microsoft's Active X Controls/DCOM, Javasoft's Enterprise JavaBeans [4] and the OMG's CORBA Component Model [5]. The following sections outline the last two of these technologies.

CORBA Components

As a component can be viewed from several different standpoints depending on how the component is being used and who is viewing the component the CORBA component specification defines a number of separate models through which a component must be described.

The Abstract Model describes a component from a top-level, design perspective. Also known as the Meta-Model, it contains information on the objects, links, and data values of the component, as well as details on how to this information should be viewed using Document Type Definitions (DTDs) based on the XML Metadata Interchange (XMI).

At a lower level, the specification defines five further models that describe how the component will both function and integrate. These are the Component, Persistency, Packaging, Deployment and Container models.

The Component Model expresses the component as a type. The type definition provides both compile-time and run-time information on its external interfaces. This includes new IDL syntax to provide:

- Unique component identification
- Identification of interfaces that the components both provides and uses
- Details on the events that a component both emits and consumes

- Navigation interfaces that allow the above to examined
- Interfaces that allow runtime attribute and property configuration
- Interfaces for managing multiple component instances

The new syntax is essential to allow an IDL compiler to turn a component definition into an executable implementation in a target language.

The Persistency model defines how a component's state may survive its physical representation in memory, extending a component's lifecycle.

The Packaging Model provides details to enable a component to be assembled with other components within the scope of a distributed application.

The Deployment Model defines how components are configured and readied at run-time. The main aim here is to install the component into a physical computing environment.

Finally, the Container Model deals with a component's runtime environment. In general, components execute in containers and must include details of the system services they require in order to operate effectively. Its container, itself a component, provides access to these services and controls external access to potentially multiple instantiations of the component.

Enterprise Java Beans

SUN Microsystems's Enterprise JavaBeans (EJB) is designed to run within the Java Runtime Environment and therefore requires a suitably engineered Java Virtual Machine. In order that an EJB may be queried by frameworks and other EJBs and connected to those that wish to cooperate, it must provide a number of standard, external entry points. The specification defines a set of naming conventions, "design patterns", that enable a component's entry points to be identified and which builds on the introspection inherent all Java classes. These entry points include: a set of methods or interfaces via which the component can communicate; run-time configurable properties that promote customisation; and events with which a component notifies interested EJBs of changes in state.

Crucially, a JavaBean executes within a container, in itself a JavaBean. A container provides an environment within which one or more beans run and acts as a bean coordinator, controlling access to the bean and creating new instances on request. As the EJB spec, "A container is where an enterprise Bean object lives, just as a record lives in a database, and a file or directory lives in a file system". This enables beans to be managed in a locally scalable way. It can also control bean access to supporting services and scarce resources, and provides a secure environment for each bean to exist within the wider distributed environment.

To facilitate this an EJB exists in one of two different guises, as either a *session* or *entity* bean. Session EJBs are transient and have a limited lifecycle. They are created by clients and usually only exist for the duration of a client/server session. Once the bean is destroyed its data is lost, which, crucially applies in the event of a system crash. The bean must manage any persistency that might exist. On the other hand, an entity EJB is persistent and typically lives as long as the data that describes it. As the data may reside in a database this could be for a much longer time than the components initial representation in system memory. Furthermore, EJBs can survive a server crash and are designed to be transactional.

To complement the EJB distributed component model SUN has defined a comprehensive set of APIs. There are 8 in total and consist of the: Java Naming and

Directory Services (JNDI), Remote Method Invocation (RMI), Java Interface Definition Language (Java IDL), Servlets and Java Server Pages (JSP), Java Messaging Service (JMS), Java Transaction API (JTA), Java Transaction Service (JTS) and Java Database Connectivity (JDBC). Together these form the Java Platform for the Enterprise (JPE) and this is intended to enable any Java application to access an enterprise-class infrastructure, often termed middleware.

CORBA Components and EJBs

CORBA Component and EJB specification are alternative component technologies. The EJB has already presented one of the first working implementations, backed by the Java market. The CORBA Components present a more general modelling framework, advocating the use of CORBA as the underlying technology. The forthcoming CORBA Components standard will constitute an integral part of the third version of the CORBA suite. It will contain both the models and the meta-modelling constructs required in order to deploy a full-scale component technology. Additionally, it will cover the mapping from Enterprise JavaBeans (EJB) components to CORBA components and vice versa. In fact, one of the mandatory requirements set by OMG, in its Request for Proposals for a CORBA Component Model [6], was the compatibility with JavaBeans, while the latest proposal has considered compatibility with Enterprise JavaBeans. It is expected that an Enterprise JavaBean may be inserted into a CORBA container, while a CORBA Component, written in Java, may run in an EJB Container. An EJB interface can be mapped to its equivalent CORBA IDL and an EJB container can use either RMI or IIOP as its distributed communication protocol. More specifically, an EJB may be inserted into a CORBA container and a CORBA component, written in Java, may run in an EJB container. Furthermore, certain JPE services have equivalent CORBAservice counterparts from which they borrow much of their design. JTS is an implementation of the CORBA Object Transaction Service, and the JNDI extends the CORBA Naming Service to give general directory access support that includes the Lightweight Directory Access Protocol (LDAP). The CORBA Component model is currently being aligned to the EJB Model so that JAVA based components are interoperable with (non JAVA) based CORBA components.

4.3 Workflow

There is a growing awareness that workflow management tools and techniques could provide a vital element in the co-ordination of distributed components within different provider domains whilst allowing greater flexibility and the necessary degree of autonomy [7]. Because of the introduction of new services, new relationships with other service providers or new functionality or equipment, the management components (form service provider and network provider systems) must be capable of adapting rapidly to changes in the way the business process is executed.

A workflow management system is a system that defines, manages and executes workflow processes through the execution of software whose order of execution is driven by a computer representation of the workflow process logic. Workflow technology incorporates the benefits of co-operative information systems, computer-supported co-operative work, GroupWare systems, and active databases. Workflow management technology addresses the following requirements:

- Improved efficiency, leading to lower costs or higher workload capacity
- Improved control, resulting from standardisation of processes
- Improved ability to manage processes; identification and analysis of problems
- Cost reductions, where cost can be a euphemism for staff
- Increased quality or capacity while controlling costs
- Construction of unique customised business processes to deal with specialised management work practices
- Improved information distribution, and elimination of the delays caused by the need to move hard copy information around the organisation
- Reduced bureaucracy improved the quality of work, decreased cycle times, and acquisition of better management information about business processes.

Thus workflow management can be considered a very attractive technology for integration and interrelation of telecommunication management components. The purpose of applying workflow management technologies in the service management problem domain is to integrate and re-purpose management components that resolve telecommunication business problems, and to automate telecommunication management services. Such enhancements, i.e. the interrelation of service management components, reduce the business process complexity, improve resilience, and improve the overall performance of the network operators' business process.

There are many differences between the architectures, which are used by workflow systems. However most of the workflow systems fall into one of two broad categories [8]:

- Forms and messages based workflow systems which performs electronic routing of forms to user's e-mail in-boxes.
- Engine based workflow system, which communicates with humans or components via specialised client software.

It is the workflow engine based approach, which we will focus on to achieve management component integration.

Workflow Engines

A Workflow Management System (WFMS), as defined by the Workflow Management Coalition, is a system that defines, creates, and manages the execution of workflows through the use of software, running on one or more workflow engines, which is able to interpret the process definition, interact with workflow participants and, where required, invoke applications (or components) [9]. A workflow engine is the basic workflow management control software. This software is often distributed across a number of computer platforms to cope with processes, which operate over a wide geographic basis

The workflow engine controls the flow of work (sequences of management activities which form a management business process) through the system by interpreting the management process rules to determine the scheduling of required activities, and invoking the relevant management components. The engine is responsible for:

- Business process creation, deletion, and management of process execution from instantiation through completion

- Control of the activity scheduling within an operational (business) process.
- Interaction with management components and/or human resources (which execute the required management activities).
- Monitoring and control of the management processes in execution.

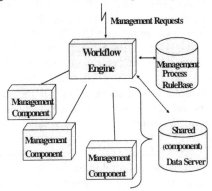

Figure. 3. General Workflow Architecture

Figure 3 illustrates a generalised workflow engine which accepts a (management service) request and based on its process rulebase, invokes the correct sequence of components and store application specific values in a shared data server.

There are a great many products and research projects which offer support for workflow management. In April 1996, 250 products claimed to support workflow features and/or workflow management[10]. This constituted a market size of more than one billion dollars. The number of products has risen since then.

Many of the products simply provided a means of graphically representing a business process using techniques such as dataflow, digraph, flowchart, network, orgcharts, pertcharts etc. e.g. Zippen [11]. Others are data management systems, which use e-mail, imaging, databases, electronic forms, engineering drawings etc. to collaboratively process documents or data. Groupware also forms part of this group, Lotus Notes being a good example.

All of these systems have an emphasis on office processes, e.g. imaging, document routing, enhanced mail. However a number of limitations are evident with these types of workflow systems [12].

1. Lack of support for heterogeneous computer systems
2. Incompatibility between workflow products
3. Failure to capture distributed/true nature of infrastructure in business model
4. Scalability not achieved
5. Very little support given towards fault-tolerance and reliability.

Most of these products were designed for small collaborative projects with small loads. As such, they are unsuitable for large-scale workflow management, potentially involving several thousand users, hundreds of thousands of concurrently running processes and several thousand sites distributed over wide area networks [13]. Research projects, however, have confronted many of these issues. Many current research projects are drawing from technologies such as objects, the World Wide Web, CORBA, transaction processing, Java and others in order to help solve some of the problems mentioned above.

CORBA and Workflow

CORBA is used in varying degrees within many workflow research prototypes. In the simplest case, CORBA is used for database access or as a wrapper around legacy applications. The Mentor project uses CORBA to provide a uniform interface to heterogeneous invocable applications [14]. The WebFlow project has a CORBA based CLF co-ordinator which communicates with legacy applications, a relational database and a document management system through CORBA [15].

The WorkWeb project uses a network of CORBA-based agents, where each agent represents a resource or a participant [16]. These agents collaborate and vie for resources. In OrbWork, again based on METEOR2, transactional concepts are implemented using CORBA to achieve fault tolerance[17]. It includes a layer of CORBA-based system components and failure detection mechanisms, which increase availability and allow recovery. Persistence and scalability are other key requirements, which led to the use of CORBA.

The OrbWork project is a CORBA-based workflow engine which contains a 'Workflow Model Repository', a task manager (combining the duties of the scheduler and dispatcher), a monitor (holding state for the system as a whole) and tasks (wrappers around legacy applications).

Workflow Standardisation

The standardisation of workflow systems has been on-going since 1993 with the formation of the Workflow Management Coalition WfMC (an industrial consortium which set about standardising an architecture for workflow engine based systems, and several interfaces for application invocation, process definition, process management and system interoperability. In 1998, the OMG ratified the definition of a workflow facility, which was based on the WfMC standards.

5 Case Studies – Integration of TINA based Telecommunications Management Architecture Components

The ACTS FlowThru project is currently evaluating the component integration technologies mentioned above, across a range of TMF related business processes, namely Service Fulfilment, Service Accounting and Service Assurance. Two such case studies are sketched below. The first, looking at Service Fulfilment uses CORBA component technology can be equally applied to Enterprise JavaBeans due to their close similarities. It is the subject of a yet to be published paper which presents the analysis, architecture and design of a system that integrates service and network management building blocks in order to satisfy business process requirements for service order fulfilment [18]. The second looks at Service Accounting and uses a workflow engine to integrate a range of accounting components.

5.1 Case Study 1 –CORBA Component Integration

Briefly, this takes a number of existing TINA compliant service and network management components, themselves the products of previous projects and uses proposals set out in the CORBA Components specification [5] to produce components that satisfy the TMF FulFilment business process. The scenario relates to the fulfilment of customer orders for a switched ATM service.

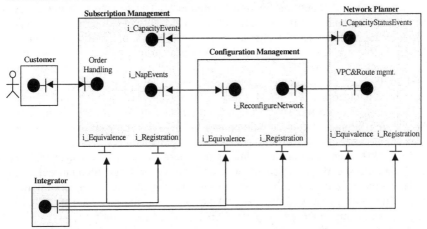

Figure. 4. TINA based Service and Network Architecture integration using CORBA Components

The three principle components: Subscription Management, Network Planning, and Network Provisioning are used as off-the-shelf components. An 'publish-subscribe' eventing model is constructed around each component to provide inter-component asynchronous event communication. The eventing model draws heavily on the Observer design pattern espoused in [19] and the EJB Delegation Event Model. Each component provides standardised introspection and registration interfaces through which a "framework" integrator object can extract their consumed and produced event interfaces. A component desire to receive, or consume, certain events is matched by the integrator object and the relevant publisher informed through a registration interface (Figure 4). Once registered events are 'pushed' from publisher to consumer, responses where needed are managed similarly establishing an 'event protocol'. Other components can be introduced seamlessly to the system and are similarly integrated so long as they can be matched with a cooperating component. Decoupling components in this way encourages reuse as existing components can be adapted to take advantage of the Fulfilment event protocol. New protocols can be developed to extend functionality, as the integrator object will match cooperating components.

5.2 Case Study 2 – Integrating Telecommunications Management using Workflow Engine, and JIDM compliant CORBA/CMIP gateway

The ACTS FlowThru project is also evaluating the workflow based component integration technologies, across a set of TMF related Service Accounting business processes.   In this area, FlowThru is integrating a TINA compliant service management platform with a workflow engine and re-usable accounting service components. The workflow engine is used to integrate a range of accounting components e.g. account manager, Tariff Control, Bill Control, Charge Control, User Metering Management etc. Figure 5 outlines the TINA architecture and the workflow engine. The scenario depicted is that for a customer accessing a TINA based Service Management system (Accounting Session Component). The service is being offered by a service operator how is independent of the ATM network operator. The ATM network is managed using TMN technology and ATM usage information is compiled into Call Data Records and emitted as charging notifications. They are received by a JIDM compliant CORBA/CMIP gateway (in the service operators domain) and delivered to the service operators accounting system in CORBA format, where charges for each customer for ATM usage is extracted. The service management system has interfaces into the Subscription Component and Accounting Component. The Accounting Component is actually made up of a WorkflowEntry Agent, the workflow engine and several accounting specific components.

It is the workflow engine that co-ordinates the interactions and sequences of these service accounting specific components to ensure the desired management activities are carried out e.g. generate a bill for the customer, generate tariff information, provide account management controls etc. All component interfaces are specified in CORBA IDL and components execute of a CORBA based infrastructure.

The use of the engine for the accounting component allows the easy introduction of new accounting components without disruption or changes to other accounting components. On introducing a new component into the accounting system, the operator needs to either compose new or amend existing (accounting management) business process(es) and data flow rules. This composition/amendments are required as the engine uses these business rules to decide which accounting components to invoke to satisfy a management request.

In order to execute an accounting business process, an invocation is made to the 'work entry agent' component of the workflow engine. This agent contains a mapping of external accounting service offerings to accounting business processes. On determining the appropriate accounting business process, the work entry agent creates an instance of that business process and invokes the scheduler to start enacting the accounting (business) process instance. The work entry agent also stores any parameters (which have been given when invoking the accounting service) in the Shared Data Exchange (SDE). The SDE is a distributed information server, which is interrogated on behalf of the accounting components during the execution of accounting activities. Once invoked, the scheduler logs the business process instance (workflow) information into the workflow information server. This workflow information server maintains the state of execution of the business process throughout its execution. However, all accounting (component or application) data is not passed through the engine but is maintained in the Shared Data Exchange. The scheduler then asks the Knowledge Server to determine the next activitie(s) for the process instance. The knowledge server contains several strategy objects which are associated with different business process knowledge. These strategy objects interrogate, at runtime, a knowledge base of process rules to determine the next activitie(s) to schedule. The separation of knowledge base, knowledge server and scheduler allows the knowledge to be potentially represented in different techniques/technologies.

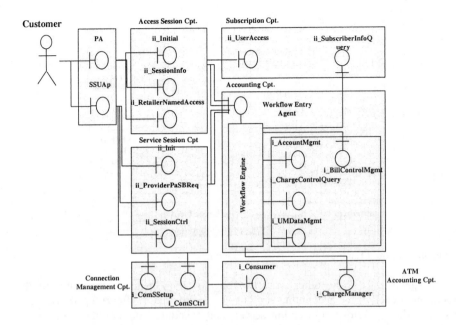

Figure 5. TINA based Service architecture using Workflow engine based Accounting Component Integration.

For the trial, a Java Expert System Shell based knowledge-base was used to specify the accounting business processes. Figure 6 illustrates the workflow engine architecture and shows the interactions between the work entry agent, knowledge server, accounting process rulebase and workflow information server. When the scheduler has identified the next activitie(s), it invokes the dispatcher. The dispatcher is responsible for mapping the activitie(s) to component(s). The dispatcher then invokes agent(s) on the component(s) to ensure the activitie(s) are carried out. The agents are responsible for retrieving the necessary input parameters for the accounting components from the Shared Data Exchange. This retrieval is made simple for the agent as the Shared Data Exchange offers a high level query interface and contains meta-data concerning the information flows between activities (and components) required to execute the accounting business processes. The agents invoke the accounting components in whatever means those components support. For simplicity of diagramming Figure 6 only depicts three account components. However in the trial seven accounting application components were used. Most of these accounting components were originally developed in other ACTS research project e.g. Prospect, Vital.

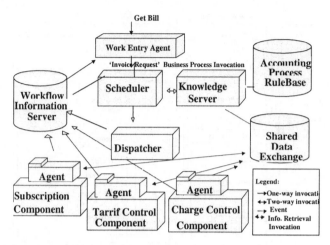

Fig. 6. Workflow Engine driven Service Accounting

The agents inform the other workflow engine components of the status of the execution of each of the accounting activities by emitting (workflow) typed events. In this way, the workflow information server, and scheduler monitor the state of the business process instance and can schedule further activities in the accounting business processes. The workflow entry agent, dispatcher, scheduler and component agents are multi threaded to allow concurrent execution of business processes. The agent code is 80% pre-written for any application component. The only new code needed is to actually invoke the component itself (i.e. all interactions with the workflow engine, shared data exchange, and use of the CORBA naming service). However even this code could be pre-written if the application component is a CORBA object and a dynamic interface type is used to invoke it.

On completion of the last activity the scheduler sends a 'process instance completed' event. The workflow entry agent retrieves the results of the process instance from the shared data exchange and passes it back to the original accounting service requestor.

In the FlowThru trial, the workflow engine and accounting components were seamlessly integrated with a pre-existing TINA SA compliant platform. All workings of the engine were transparent to the TINA platform.

6 Conclusion

This paper has outlined the context in which telecommunication management components would be developed and operated. The irresistible move towards componentisation of the management systems has been discussed and the current TMF position on component technology has been described. The paper has also discussed four key integration technologies, namely TMN API, TMN/CORBA interworking, Component Technologies from OMG and SUN, and Workflow.

The CORBA Component case study illustrates how legacy components can be seamlessly integrated using introspection interfaces and a 'publish-subscribe' event model. Asynchronous event communication, although more complex from a design point of view, decouples components so that additional componentry can be introduced, for example event logging purposes, without impacting current functionality. Similarly existing functionality can be adapted and extended by introducing new event protocols. Introspection foresees the introduction of component frameworks responsible for connecting cooperating components, here simulated by our integrator object. The model can be easily extended to provide property interfaces that would allow components to be configured 'on-the-fly'.

With regard to the Workflow case study, the use of the engine had several positive effects on the overall design and implementation. Firstly, it was possible to directly map the accounting business processes into the workflow engine rulebase. This provided a means of validating the system behaviour at a high level. The engine also facilitated the ease of introduction of new components as well as the composition of new business processes based on the existing accounting components. Research into the workflow engine is ongoing to provide improved (management) business process management. Research is also ongoing to provide tools for management service creation based on the workflow engine integration.

Acknowledgement

The authors would like to gratefully acknowledge that this research was partly funded under the EU ACTS programme.

References

[1] NMF Technology Map, Draft NMF GB909, July 1998

[2] http:\\WWW.opengroup.org

[3] Cox, B., Novobilski, A., Object-Oriented Programming : An Evolutionary Approach, Addison-Wesley, May 1991

[4] Matena, V., Hapner, M., Sun Microsystems: Enterprise JavaBeans, Version 1.0, 21 March 1998

[5] CORBA Components: Joint Revised Submission. OMG TC Document orbos/99-02-05, Draft, March 1999.

[6] RFP for a CORBA Component Model. OMG TC Document orbos/97-06-12, July 1997.

[7] M Rusinkiewicz, A Helal "Workflow Management Systems", published in Journal of Intelligent Information Systems, Vol 10, Number 2, March/April 1998

[8] "Ovum Evaluates: Workflow" - Heather Stark, Laurent Lachal, ISBN 1898972605, Ovum, 95.

[9] Workflow Management Coalition Homepage http://www.wfmc.org/

[10] Amit Sheth, "State of the Art of Commercial Technology and Research in Workflow Management" DARPA/ISO Workshop on collective Action Tools - April 10, 1996,

[11] Zippin Project Homepages, Stefania Castellani, Xerox Grenoble, http://www.rxrc.xerox.com/research/ct/prototypes/aippin/home.html

[12] Amit Sheth, "From Contemporary Workflow Process Automation to Adaptive and Dynamic Work Activity Coordination and Collaboration", University of Georgia SIGGROUP Bulletin, Vol. 18, No 3 (December 1997)

[13] G Alonso, D Agrawal, A El Abbadi, C Mohan, "Functionality and Limitations of Current Workflow Management Systems", IEEE Expert, Vol 12, No. 5, Spet/Oct 1997

[14]J Weissenfels, D Wodtke, G Weikum, A Kotz Dittrich, http://paris.cs.uni-sb.de/public_html/papers/mentor.html

[15] Antonietta Grasso, Jean-Luc Meunier, Xerox, WebFlow Homepage, http://www.xrce.com/research/ct/prototypes/webflow/home.html

[16] H Tarumi, K Kida, Y Ishiguro, K Yoshifu, T Asakura, "WorkWeb Systems - Multi Workflow Management in a Multi Agent System", SIGGROUP 1997, Pheonix, Arixona, USA

[17] S Das, K Kochut, J Miller, A Sheth, D Worah, "ORBWork: A Reliable Distributed CORBA-Based Workflow Enactment System for METEOR2", Technical Report #UGA-CA-TR-97-001, Department of Computer Science, University of Georgia, Feb. 1997

[18] Lewis, D., Pavlou, G., Malbon, C., Stathopoulos, C., Jaen Villoldo, E., "Component-Based Integration of Customer Subscription Management with Network Planning and Network Provisioning"

[19.] Gamma, E., Helm, R., Johnson, R., Vlissides, J., Design Patterns : Elements of Reusable Object-Oriented Software, Addison-Wesley, October 1995

[20] L Bjerring, C Grabowski, S Dittman, R Lund, L Bo Sorensen, S Rasmussen, "Experience in developing multi-technology TMN Systems", NOMS '98, New Orleans, USA

Deploying a Hierarchical Management Framework Using Mobile Agent Technology

Damianos Gavalas[†], Dominic Greenwood[*], Mohammed Ghanbari[†],
Mike O'Mahony[†]

[†]Communication Networks Research Group,
Electronic Systems Engineering Department,
University of Essex, Colchester, CO4 3SQ, U.K.
E-mail: {dgaval, ghan, mikej}@essex.ac.uk

[*]Distributed Network Management and Agent Technology Research Group
Fujitsu Telecommunications Europe Ltd.,
Northgate House, St. Peters Street, CO1 1HH, Colchester, U.K.
E-mail: D.Greenwood@ftel.co.uk

Abstract. The use of Mobile Agent (MA) paradigm has been proposed by many researchers as an answer to the scalability and flexibility limitations of centralised Network Management (NM). Nevertheless, while large enterprise networks are already hierarchically structured, MA-based management has not yet moved from 'flat' to hierarchical structures. That results in non-scalable flat architectures, particularly when the management of remote subnetworks is considered. In this context, the deployment of a hierarchically structured MA-based management framework is a reasonable approach. The migration to hierarchical structures is achieved with an additional management entity, the Mobile Distributed Manager (MDM), which takes the full control of managing a given network segment. This architecture exploits the mobility features of MDMs to dynamically adapt to mutable networking conditions. Empirical results indicate a substantial reduction of the overall management cost compared to both centralised and MA-based flat management approaches.

1.   Introduction

Contemporary large enterprise networks span applications, organisational and geographical boundaries. In order to cope sufficiently with the unpredictable growth of the number of network devices, structuring networks in logical hierarchies is being employed as a design and deployment principle. Accordingly, the design of such networks' management system needs to be aligned with the corresponding managed network hierarchical structures. Hence, several hierarchical/distributed management solutions have been proposed both by researchers [1][2] and standardisation forums [3][4].

Despite these efforts towards management distribution, the majority of traditional systems still rely on centralised architectures, wherein operational data is collected by stationary *agents* embedded in network devices and subsequently gathered by a central platform (*manager*) using a management protocol, such is the IETF Simple

Network Management Protocol (SNMP) [5]. The centralised nature of NM protocols results in massive transfers of NM data that stress the network resources to their limits, while causing processing bottlenecks at the manager host. Another serious disadvantage of centralised paradigm is its intrinsic architecture inflexibility as the functionality of both managing and managed parties is rigidly defined at design time.

Hierarchical NM models help to overcome the scalability limitations of their centralised counterparts through delegating part of the management functionality to dual-role entities; these are responsible for a set of devices, which play the role of the agent when managers request information, while acting as managers for the agents located within their domain. Yet, these models cannot sufficiently meet the flexibility requirements of today's networks, with the fluctuating traffic patterns and topology structures. This is due to the static definition of the incorporated NM components and their corresponding roles in the management hierarchy. Such static configurations would possibly provide a feasible solution for moderately small and/or not very dynamic networks. However, it is not in step with the evolution of large-scale enterprise networks.

In search of more flexible management solutions, the Mobile Agents (MA) paradigm has recently attracted considerable attention in the field of distributed NM. MAs introduce a new software communication paradigm that allows code migration between hosts for remote execution. However, there is a notable inconsistency between the currently hierarchically structured networks and the emerging Mobile Agent Frameworks (MAF) that insist on 'flat' models. These models bring about scalability issues when the management of large networks is considered, whilst resulting in heavy use of low-bandwidth WAN links connecting remote subnetworks to the backbone network, due to the frequent MA transfers.

This work attempts to address these issues by coupling the concepts of hierarchical management and Mobile Agents. Therefore, we propose the deployment of a hierarchically structured, dynamic MA-based management infrastructure. To attain this objective, we introduce a novel management entity termed the "*Mobile Distributed Manager*" (MDM), which operates at an intermediary level between the manager and the stationary agents. MDMs are essentially MAs that may be dynamically assigned to / removed from a network domain in response to a change in traffic distribution or network's topology. Upon their migration, MDMs take full control of managing the assigned domain, localising the associated management traffic. In addition, MDMs may transparently move within their domain when their hosting processor becomes overloaded, thereby optimising the usage of local resources.

The rest of the paper is organised as follows: Section 2 comprises an overview of several MAFs employed to distributed NM applications, whilst Section 3 explains the rationale and introduces our proposed hierarchical approach. Section 4 provides an overview of the MAF used to support this work. Section 5 discusses the implementation details of the introduced architecture with a performance evaluation given in Section 6 and conclusions drawn in Section 7.

2.   Mobile Agents-based Distributed Management

The use of mobile code has attracted tremendous attention during the last few years. MAs offer a new powerful abstraction for distributed computing, answering many of the flexibility and scalability problems of traditional centralised archetypes. Regarding distributed NM area, MAs can provide all the functionality offered by static delegation agents, having the additional benefit of mobility. In that sense, MAs can be regarded as a 'superset' of MbD agents, leading to more efficient use of computing resources on the managed entities, as management functions are executed only so long as the MAs reside and are active on the NEs [6][17]. An exhaustive review of all the MAFs proposed for distributed NM in the last few years is beyond the scope of this paper, so we have chosen to concentrate on the most representative.

Thus, several works [6]-[12] deal with implementations of MAFs employed for NM, while others [13][14] are confined to describing the general concepts behind the introduced frameworks and discussing the arising theoretical aspects. Another set of research papers [15][16] include performance evaluations of MA-based management.

Focusing on the works supplemented by implementations, many researchers [6]-[11] have deployed new MAFs from scratch whereas some others [12] have chosen general-purpose, commercial platforms to comprise the core of their architectures. In the majority of the former [7]-[10], MAs can interact with standard SNMP agents providing access to legacy systems. Currently, only the frameworks described in [10] and [11] comply with the OMG *Mobile Agent System Interoperability Facility* (MASIF) [18] emerging standard that defines agent mobility and management policies. In addition, Java programming language [19] has dominated, being the implementation platform in all cases.

Regarding their practical use within NM, MAs have been utilised to address a broad spectrum of applications, among others: network monitoring [11][13]; traffic analysis [12]; collecting atomic Management Information Base (MIB) object values from multiple hosts [8]; calculating aggregation functions combining several MIB values [9][10]; obtaining snapshots and filtering the contents of SNMP tables [17]; automated fault management of SDH networks [6], etc.

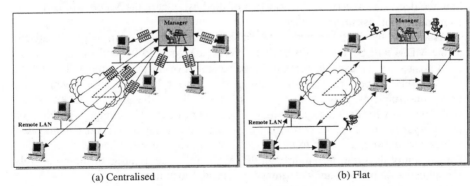

| | |
| (a) Centralised | (b) Flat |

Fig. 1. Centralised vs. 'flat', MA-based Network Management

The common denominator below the aforementioned MA-based management architectures is the assumption of a 'flat' network architecture, i.e. a single MA is launched from the manager platform and sequentially visits all the managed NEs, regardless from the underlying topology (see Figure 1b). However, although relaxing the network from a flood of request/response SNMP messages (see Figure 1a), such an approach brings about scalability issues, especially when frequent polling is required. That is, for large networks the round-trip delay for the MA will greatly increase, whilst the network overhead may overcome that of centralised paradigm (the MA size will grow after visiting each of the nodes included into its itinerary [15]). The situation seriously deteriorates when considering management of remote LANs, connected to the main site through low-bandwidth, expensive WAN links. In this case, frequent MA transfers may potentially create bottlenecks and considerably increase the management cost. Hence, the deployment of a hierarchically structured MA-based management framework seems a rational approach to overcome this problem.

3.   Mobile Agent-based Approach to Hierarchical Management

Surprisingly, MA-based architectures proposed for distributed NM have not yet moved from flat to hierarchical structures. This represents an inconsistency though, as these management systems are not directly mapped to the managed networks. In this work we address this issue through combining the concepts of Hierarchical/Distributed Management and Mobile Agents for NM. Specifically, we introduce the concept of Mobile Distributed Manager (MDM), referring to a management component that operates at an intermediary level between the manager and management agent end points. MDM entities are essentially MAs that undertake the full responsibility of managing a network domain, when certain criteria (determined by the administrator) are satisfied. Upon being assigned to a domain, the MDM migrates to a host running in that domain (see Figure 2) and takes over the management of local NEs from the central manager. In a later section, we discuss how the decision concerning the selection of the host, where the MDM will carry out its management tasks from, is made.

As a result, the traffic related to the management of that domain will be localised, as the MDM will be able to dispatch and receive MAs to collect NM data from the local hosts, or even execute centralised management operations on them. The MDM will continue to perform its tasks without the manager's intervention, even if the interconnecting link fails. A first-line response will also be given to tackle trivial faults/alarms, with the manager being notified only in case of a complex problem or an emergency situation. In performance management applications, only aggregated values and statistics are sent to the manager at regular intervals, thereby diminishing the amount of data transferred through the WAN link. The duration of these intervals is application-dependent and determined by the administrator.

The mobility feature of MDMs allows the management system to adapt dynamically to a mutable environment, optimising the use of network resources. Apart from the fact that management functionality may be added/configured at runtime, this architecture can also dynamically adapt to changing networking conditions. Namely, an MDM entity can be deployed to / removed from a network segment to reflect a change on traffic patterns, or move to the least loaded host to minimise the usage of local resources.

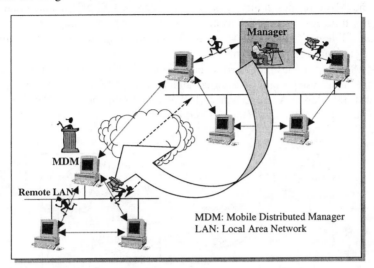

Fig. 2. Hierarchical MA-based management

Similar work has been reported in two research papers. Liotta et al. [13] have conducted an interesting study of an MA-based management architecture adopting a hierarchical, multi-level approach. Interesting cost functions corresponding to various MA configurations are also discussed. However, there is no implementation supplementing this work, while the authors have not considered providing "*Middle Managers*" mobility features, so as to dynamically change their location. In addition, the criteria according to which the managed network is segmented in domains and the way that these domains are assigned to Middle Managers are not explicitly mentioned. In [14], Oliveira and Lopes proposed the integration of the IETF's *Disman* framework [4] into their MA-based NM infrastructure. Again, this work lacks implementation details while the "*Mobile Disman*" architecture they propose is heavyweight (it consists of many resource-demanding components) and would certainly have increased requirements on system resources.

4. Overview of the Mobile Agent Framework

The MAF that comprises the core of the hierarchical infrastructure has been entirely developed in Java chosen due to its inherent portability, rich class hierarchy and dynamic class loading capability.

Our framework consists of four major components [9], illustrated in Figure 3:

(I) Manager Application: The manager application, equipped with a browser style Graphical User Interface (GUI), co-ordinates monitoring and control policies relating to the NEs. Active agent processes are automatically *discovered* by the manager, which maintains and dynamically updates a 'discovered list'.

(II) Mobile Agent Server (MAS): The interface between visiting MAs and legacy management systems is achieved through MAS modules, installed on every managed device. The MAS resides logically above the standard SNMP agent, creating an efficient run-time environment for receiving, instantiating, executing, and dispatching MA objects. Integration with SNMP was vitally important to maintain compliance with the legacy management systems.

The MAS also provides requested management information to incoming MAs and protects the host system against external attack. The MAS composes four primary components: (a) Mobile Agent Listener, (b) Security Component, (c) Service Facility Component, and (d) Migration Facility Component.

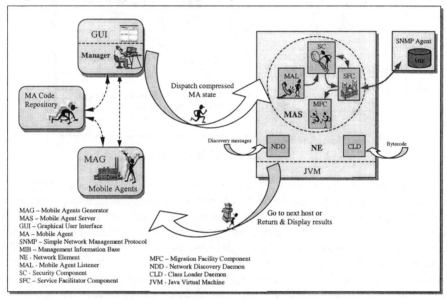

Fig. 3. The Mobile Agents-based NM Infrastructure

(III) Mobile Agent Generator (MAG): The MAG is essentially a tool for automatic MA code generation allowing the construction of customised MAs in response to service requirements. Generated MA code is stored into an MA code repository (see Figure 3). Such MAs may dynamically extend the infrastructure's functionality, post MAS initialisation, to accomplish management tasks tailored to the needs of a changing network environment.

The MAG's operation is described in detail in [9]. However, its functionality has been extended so as to allow the operator (through a dedicated GUI) to specify: the polling frequency (i.e. the polling interval's duration); the transmission protocol to be used for the MA transfers (either TCP or UDP); the security policies.

(IV) Mobile Agents (MAs): From our perspective, MAs are Java objects with a unique ID, capable of migrating between hosts where they execute as separate threads and perform their specific management tasks. MAs are supplied with an *itinerary* *table*, a *data folder* where collected management information is stored and several methods to control interaction with polled devices.

As described in [17], the multi-node movement of MAs can be exploited in a variety of data filtering applications. In particular, MAs may: (i) aggregate several MIB values into more meaningful values, (ii) efficiently acquire atomic snapshots of SNMP tables, and (iii) filter tables' contents by applying complex filtering expressions thereby keeping only the values that meet pre-specified criteria.

5. Implementation Details

5.1. Topology Map

An important element of our framework is the *topology map*, a graphical component of the manager application, used to view the devices with currently active MAS servers. This component not only presents the discovered active devices, but also the underlying network topology, namely the subnetworks where these devices are physically connected as well as how these subnetworks are interconnected.

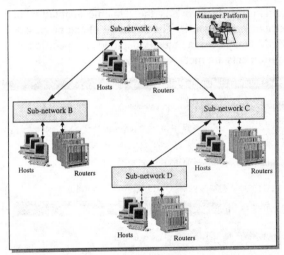

Fig. 4. The topology tree structure

In terms of implementation, the topology map is internally represented by a tree structure (termed *"topology tree"*), where each of the tree nodes corresponds to a specific subnetwork. The node representing the manager's location is the root of the topology tree (see Fig. 4). Each of the tree nodes consist of the following attributes:

- subnetwork's name;
- the names of hosts and routers connected to this subnetwork;
- a flag indicating the presence of an active MDM on this subnetwork;
- the number n_l of local active hosts on this subnetwork;

- the number n_s of active hosts on the subnetwork's *"subtree"* (the term subtree here denotes the set of subnetworks located in hierarchically lower levels in the topology tree, including the present subnetwork itself), hence $n_s \geq n_l$;
- a pointer to the upper level tree node;
- pointers to the next level nodes;
- a list of graphical elements, each corresponding to a specific host, that will be made visible upon discovering an active MAS entity on that host.

For instance, the number of active hosts in the subtree of Subnetwork A (in Figure 4) will be:

$$n_{s,subA} = n_{l,subA} + n_{l,subB} + n_{l,subC} + n_{l,subD} \tag{1}$$

As described in a following section, the topology tree plays a crucial role when the manager application needs to make a decision on which subnetworks require the deployment of an MDM entity.

5.2. Policies on MDMs Deployment

A key characteristic of this work is the dynamic adaptation of our architecture to changes in the managed network. The structure of the proposed model is not rigidly designed, as MDMs may be dynamically deployed to specific network domains, given that certain requirements are met.

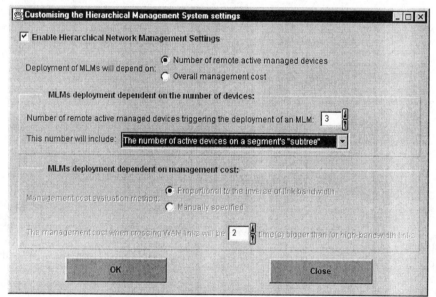

Fig. 5. Graphical User Interface for customising the hierarchical NM system policies

Specifically, the administrator may explicitly set (through the GUI shown in Figure 5) the policies that define the hierarchical NM system operation, i.e. specify the criteria that should be satisfied for deploying an MDM to a network segment. In general, the deployment of MDMs may conform to either of the two following policies:

- **Policy 1**: the population of remotely active managed devices.
- **Policy 2**: the overall management cost.

In the former case (*Policy 1*), the administrator specifies the number of remote managed NEs that will justify the deployment of an MDM to a particular network segment. This number may either denote n_l or n_s. If, for instance, the specified number N denotes the population of the examined subnetwork's local devices n_l, an MDM will be deployed to every network segment S with $n_{l,S} \geq N$, otherwise to every segment with $n_{s,S} \geq N$. In the latter case (*Policy 2*), the management cost may either be: (a) proportional to the inverse of link bandwidth, or (b) manually specified.

5.3. Implementing MDMs Deployment

Upon discovering an active MAS module, the corresponding host is located through scanning the topology tree and finding the subnetwork where the host belongs, whilst the host icon is instantly made visible on the topology map. Then, the number n_l of active hosts on that subnetwork is increased by one and subsequently, through following the pointer to the upper-level nodes, all the topology tree nodes up to the root are traversed and their number n_s of subtree nodes is also updated. A similar procedure is followed when a MAS server is being shut down.

The discovery or termination of a MAS server triggers an event at the manager host. The topology tree is then scanned with the subnetworks that meet specific requirements added to a list. In case that 'Policy 1' is employed, referring to the policies listed in the preceding section, that list will include the subnetworks with n_l or n_s (depending on whether the MDMs deployment is a function of the active devices running locally or in the whole subtree) greater than the specified constant N. If 'Policy 2' is employed, the cost corresponding to the management of each subnetwork is evaluated and the list of subnetworks created accordingly. Ultimately, an MDM will be deployed to each of the subnetworks included in the list.

Certainly, the set of management tasks already performed by the manager on these subnetworks will need to be conveyed to the MDM deployed therein. This is achieved through sending the *Polling Threads* (PT) configurations along with the MDM. PTs are originally started and controlled by the manager application with each of them corresponding to a single monitoring task. Upon its arrival at the remote subnetwork, the MDM instantiates the PTs, which thereafter start performing their tasks without any further disruption of the management process.

5.4. MDM's Migration Within its Domain

Although MDMs have been designed to be as lightweight as possible, they cannot avoid consuming memory and processing resources on the NE where they execute.

The framework should therefore be sufficiently flexible to allow MDMs to autonomously move to another host, when their current hosting device is overloaded.

This is accomplished through the regular inspection of the other domain's NEs, in terms of their memory and CPU utilisation: an MA object is periodically dispatched and visits all the local devices obtaining these figures before delivering the results to the MDM. If the hosting processor is seriously overloaded, compared to the neighboring devices, the MDM will transparently move to the least loaded node. In our early prototype, MAs may only extract memory usage values, but will soon be able to acquire CPU load information also.

5.5. Communication Between Manager and MDMs

One of the key advantages of our framework is that it greatly reduces the amount of information exchanged between the manager platform and the managed devices. This is due to the introduction of the intermediate management level (MDMs). However, that does not obviate the necessity for bi-directional communication between MDMs and the manager host. In particular, MDMs often need to send the manager the statistics obtained through filtering raw data collected from the local devices, inform the manager when migrating to another host, etc. In the opposite direction, the manager may require an MDM to terminate its execution or move to another domain, to download in runtime an additional management service, i.e. a new MA object along with its corresponding PT configuration, etc.

We have chosen Java RMI [20] for implementing the communication bus between the distributed MDMs and the manager host, due to its inherent simplicity and the rapid prototype development that it offers.

6.   Performance Evaluation

Although mobility can often be beneficial for NM, overheads induced by MAs and MDMs in particular, e.g. due to their deployment and management should be accounted for very carefully. Slightly different configurations for a set of MDMs may result in dramatically variant network loads [13]. Hence, it is crucial to define concrete cost functions estimating the corresponding overheads.

In this context, let the "*cost coefficients*" k_{S_i, S_j} denote the cost of sending a byte of information between subnetworks S_i and S_j, where S_0 is the manager host location. For multi-hop connections, the cost coefficients will be equal to the summation of the individual links coefficients. In the following investigation, we make the simplifying assumption that an MDM may manage only the hosts included in a single subnetwork and not a wider set of devices.

Examining a simple performance management application, let us first evaluate the management cost imposed from SNMP-based management. If S_{req} is the average request size (at MAC layer), and polling of N devices, each for v operational variables, is applied, the wasted bandwidth for p polling intervals (PI) would be:

$$C_{SNMP} = \sum_{i=0}^{N} k_{S_0,S^i} * \left[\left(2 * S_{req} \right) + (v-1) * \Delta S_{req} \right] * p \qquad (2)$$

where every extra value included in the SNMP response packet's *varbind* list represents an additional overhead of ΔS_{req} bytes, on average. The index S^i represents the subnetwork including the host *i*.

A simple function characterising the bandwidth consumption for our hierarchical architecture, is the following:

$$C_{hier} = C_{distr} + C_{depl} + C_{pol} + C_{deliv} \qquad (3)$$

where the four terms represent the cost for distributing to the MAS servers the bytecode of the MA that will undertake the monitoring task, the MDMs deployment cost, the bandwidth used for the actual monitoring operation (polling) and the cost for delivering to the manager host the collected data, respectively.

Concerning bytecode distribution, we adopt a lightweight scheme: in contrast with all known MAFs proposed for management applications [6]-[11], wherein both the MA's code and state are transferred on each migration, we have chosen to distribute the bytecode at the MA's construction time and thereafter transfer only the state information, resulting in a much lower demand on network resources (bytecode size is typically much larger than state size [15]). The code distribution scheme proposed in [6]-[11] offers a better starting point in terms of the associated network overhead, since the bootstrapping procedure described above is not required. However, it is outperformed by the scheme adopted by our MAF, after a small number of polling intervals.

The introduction of MDMs reduces the code distribution cost even further: the MAs bytecode is no longer broadcasted to all managed devices, as in flat management [9], but instead it is distributed to the active MDMs, which in turn multicast it to the local NEs. The code distribution cost is therefore given by:

$$C_{distr} = \left(k_{S_0,S_0} * N_0 + \sum_{i=0}^{M} \left[k_{S_0,S^i} + k_{S^i,S^i} * N_i \right] \right) * C \qquad (4)$$

where *M* is the total number of active MDMs, *C* the compressed bytecode size and N_i the number of hosts included in subnetwork S_i.

Likewise, C_{depl} is equal to the cost of broadcasting *M* MDM objects to their corresponding remote domains:

$$C_{depl} = \sum_{i=0}^{M} k_{S_0,S^i} * ST_0 \qquad (5)$$

where ST_i represents the compressed state size of an MA when migrating from the *i*[th] host.

C_{pol} is defined as the summation of the cost induced for polling the NEs being directly managed by the manager host and the cost associated with polling the NEs that operate under the MDMs control, multiplied with the number of PIs:

$$C_{pol} = \left(\sum_{i=0}^{m} k_{S^i,S^{i+1}} * ST_i + \sum_{i=0}^{M} \sum_{j=0}^{N_i} k_{S^i,S^i} * ST_j \right) * p \qquad (6)$$

Clearly, the first term of the summation dominates on the overall polling cost if the m devices managed by the central manager platform are spread among several subnetworks. Specifically, cost coefficients $k_{S^i,S^{i+1}}$ are typically larger when an MA migrates from subnetwork S^i to another subnetwork S^{i+1} ($S^i \neq S^{i+1}$) rather than when it moves within the same subnetwork ($S^i = S^{i+1}$). It is emphasised that MAs state size ST_i does not remain constant, but increases for each visited node. Thus, the polling cost highly depends on the increment rate of the MAs state size, which in turn is a function of "*selectivity*" σ, a metric defined in [13] as the ratio of the amount of data ultimately delivered to that acquired from each host. It is apparent that for small selectivity values (the major part of the obtained data being filtered at the source) the MAs state size will practically remain constant, otherwise the state will rapidly grow. Thus, if b bytes of information are obtained at each host, an MA's state size at its ith hop is given by:

$$ST_i = ST_0 + (\sigma * b) * i \qquad (7)$$

The last term appearing in Eq. (6-2) represents the cost associated with the delivery of the gathered data from the MDMs to the manager host:

$$C_{deliv} = \left(\sum_{i=0}^{M} k_{S^i,S_0} * D/t \right) * p \qquad (8)$$

where t indicates (in number of PIs) how often MDMs package the computed statistics of size D and deliver them to the manager.

The quantitative model introduced in this section has been applied to the test network shown in Figure 6, where the network domain margins are depicted by the dotted curved lines.

Referring to this particular topology, we assign the cost coefficients the following values: $k_{S_0,S_0} = k_{S_1,S_1} = k_{S_2,S_2} = 1$, $k_{S_0,S_1} = 5$ and $k_{S_0,S_2} = 50$. These values are chosen in accordance to the bandwidth of the links they correspond to. We have also measured the set variables: $S_{req} = 90$ bytes, $\Delta S_{req} = 17$ bytes, $C = 1.95$ Kb, and $ST_0 = 447$ bytes. These values have been measured after testing our framework in a real network comprising Solaris and WinNT devices.

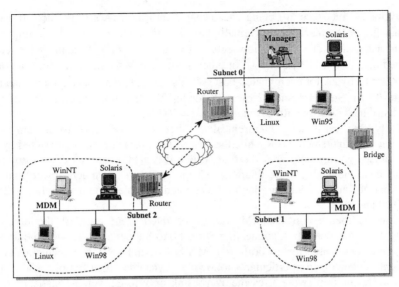

Fig. 6. The test network

Equations (6-1)-(6-7) are applied to compare the performance of SNMP polling against that of MA-based flat and hierarchical NM in terms of the overall management cost, as shown in Figure 7a, drawn on a logarithmic scale. The functions defining the cost of MA-based flat management represent special cases of those developed for hierarchical management.

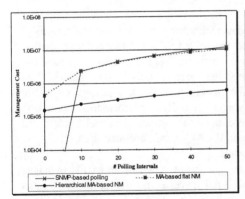
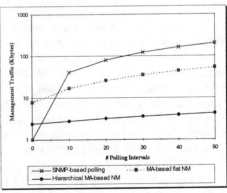

Fig. 7. Comparison between SNMP-based polling, flat MA-based polling and the proposed hierarchical framework in terms of: (a) Overall management cost, (b) Bandwidth usage of the WAN link

We consider a data intensive application, namely polling every host for the contents of the MIB-II *interfaces* table [21]. We assume the minimum of *two* interfaces per host, i.e. 2×21 collected values per host, since each table row includes *21* columns. As described in [17], the MA objects are capable of performing local filtering of the obtained data, so that only the values corresponding to the more heavily loaded interface are being encapsulated into the MA's state and returned to the manager or the MDM that originally launched the MA. That results in improving system scalability, due to the low selectivity ratio $\sigma \approx 1.82\%$ achieved over the obtained data ($b = 714$ bytes/host). In other words, the MAs state size increases only by $\sigma \times b = 13$ bytes, for each visited host. We also assume that management data ($D = 39$ bytes/PI for Subnet 1 and 52 bytes/PI for Subnet 2) are delivered to the manager from distributed MDMs with a frequency of $t = 10$ PIs.

Clearly, the introduced hierarchical architecture gives rise to a remarkable reduction of management cost, while the cost of flat management is surpassed by that of centralised polling only after the first *16* PIs. It is also noted that the starting point for the cost induced by the hierarchical infrastructure is much lower than the equivalent of flat management, due to the adopted scalable code distribution scheme, described above.

Figure 7b focuses on the NM traffic generated from each of the compared paradigms on the WAN link connecting Subnet 0 to Subnet 2. Again, the hierarchical NM framework outperforms both flat MA-based and SNMP management with sufficient distinct. In particular, following bytecode distribution and MDM deployment, our framework uses the WAN link only to deliver the statistics to the manager host, every *10* PIs. In contrast, SNMP heavily utilises the link to broadcast request messages and receive back the associated responses, while in flat management an MA object traverses the link at least twice in every PI, provided that MAs itineraries are optimised so as to poll the remote LAN hosts in sequence.

7.   Conclusions – Future Work

This paper has proposed the use of MA technology for dynamic hierarchical management. In this context, we introduced the MDM, a novel management entity, which is dynamically assigned to a given network segment and localises the associated management traffic. MDMs mobility feature allows the management system to adapt to potential changes of the managed network topology or traffic distribution and optimise the use of local resources. Finally, a performance evaluation in terms of the overall management cost confirms the proposed model's improved scalability over both traditional centralised and flat MA-based architectures.

Future work will address the following issues:

- Optimisation of MAs itinerary, so that in case the manager or an MDM manages more than one subnetwork, the MAs will not traverse the interconnecting links more than twice.
- Investigate the use of CORBA [22] instead of RMI to implement the communication bus between distributed MDMs and the manager host.

List of Acronyms

| CORBA: | Common Object Request Broker Architecture | CS: | Client/Server |
| | | DM: | Distributed Manager |

GUI: Graphical User Interface
IETF: Internet Engineering Task Force
LAN: Local Area Network
MA: Mobile Agent
MAC: Medium Access Control
MAF: Mobile Agent Framework
MAG: Mobile Agent Generator
MAS: Mobile Agent Server
MASIF: Mobile Agent System
         Interoperability Facility
MbD: Management by Delegation
MDM: Mobile Distributed Manager

MIB: Management Information Base
NE: Network Element
NM: Network Management
OSI: Open Systems Interconnection
PI: Polling Interval
PT: Polling Thread
RMI: Remote Method Invocation
RMON: Remote Monitoring
SNMP: Simple Network Management
         Protocol
WAN: Wide Area Network

References

[1]   Goldszmidt G., Yemini Y., Yemini S., "Network Management by Delegation", Proceedings of the 2nd International Symposium on Integrated Network Management (ISINM'91), April 1991.

[2]   Siegl M. R., and Trausmuth G., "Hierarchical Network Management: A Concept and its Prototype in SNMPv2", Computer Networks and ISDN Systems, Vol. 28, No. 4, pp. 441-452, February 1996.

[3]   S. Waldbusser, "Remote Network Monitoring Management Information Base", RFC 1757, February 1995.

[4]   Distributed   Management   (disman)   Charter, http://www.ietf.org/html.charters/disman-charter.html.

[5]   Case J., Fedor M., Schoffstall M., Davin J., "A Simple Network Management Protocol (SNMP)", RFC 1157, May 1990.

[6]   Sugauchi K., Miyazaki S., Covaci S., Zhang T., "Efficiency Evaluation of a Mobile Agent Based Network Management System", 6th International Conference on Intelligence and Services in Networks (IS&N'99), LNCS vol. 1597, pp. 527-535, April 1999.

[7]   Susilo G., Bieszczad A., Pagurek B., "Infrastructure for Advanced Network Management based on Mobile Code", Proceedings of the IEEE/IFIP Network Operations and Management Symposium (NOMS'98), pp. 322-333, February 1998.

[8]   Sahai A., Morin C., "Enabling a Mobile Network Manager through Mobile Agents", Proceedings of the In Proceedings of the 2nd International Workshop on Mobile Agents (MA'98), LNCS vol. 1477, pp. 249-260, September 1998.

[9]   Gavalas D., Greenwood D., Ghanbari M., O'Mahony M., "An Infrastructure for Distributed and Dynamic Network Management based on Mobile Agent Technology", Proceedings of the IEEE International Conference on Communications (ICC'99), pp. 1362-1366, June 1999.

[10] Pualiafito A., Tomarchio O., Vita L., "MAP: Design and Implementation of a Mobile Agents Platform", to appear in Journal of System Architecture.

[11] Bellavista P., Corradi A., Stefanelli C., "An Open Secure Mobile Agent Framework for Systems Management", Journal of Network and Systems Management (JNSM), Special issue on Mobile Agent-based Network and System Management, Vol. 7, No 3, September 1999.

[12] Feridun M., Kasteleijn W., Krause J., "Distributed Management with Mobile Components", Proceedings of the 6th IFIP/IEEE International Symposium on Integrated Network Management (IM'99), pp. 857-870, May 1999.

[13] Liotta A., Knight G., Pavlou G., Modelling Network and System Monitoring Over the Internet with Mobile Agents", Proceedings of the IEEE/IFIP Network Operations and Management Symposium (NOMS'98), pp. 303-312, February 1998.

[14] Oliveira J.L., Lopes R.P., "Distributed Management Based on Mobile Agents", Proceedings of the 1st International Workshop on Mobile Agents For Telecommunication Applications (MATA'99), October 1999.

[15] Fuggetta A., Picco G.P., Vigna G., "Understanding Code Mobility", IEEE Transactions on Software Engineering, Vol. 24, No. 5, pp. 342-361, 1998.

[16] Rubinstein M., Duarte O.C., "Evaluating Tradeoffs of Mobile Agents in Network Management", Networking and Information Systems Journal, Vol. 2, No. 2, July 1999.

[17] Gavalas D., Greenwood D., Ghanbari M., O'Mahony M., "Advanced Network Monitoring Applications Based on Mobile/Intelligent Agent Technology", to appear in Computer Communications Journal.

[18] GMD   Fokus,   IBM   Corp.,   "The   OMG   MASIF   Standard", http://www.fokus.gmd.de/research/cc/ima/masif/.

[19] Sun Microsystems: "Java Language Overview – White Paper" [On-line] (1999), URL: http://www.javasoft.com/docs/white/index.html.

[20] Java Remote Method Invocation (RMI), http://java.sun.com/products/jdk/rmi/index.html.

[21] McCloghrie K., Rose M., "Management Information Base for Network Management of TCP/IP-based internets: MIB-II", RFC 1213, March 1991.

[22] CORBA/IIOP 2.2 Specification, http://www.omg.org/library/corbaiiop.html.

List of Authors

Lecture Notes in Computer Science

For information about Vols. 1–1688
please contact your bookseller or Springer-Verlag

Vol. 1723: R. France, B. Rumpe (Eds.), UML'99 – The Unified Modeling Language. XVII, 724 pages. 1999.

Vol. 1724: H. I. Christensen, H. Bunke, H. Noltemeier (Eds.), Sensor Based Intelligent Robots. Proceedings, 1998. VIII, 327 pages. 1999 (Subseries LNAI).

Vol. 1725: J. Pavelka, G. Tel, M. Bartošek (Eds.), SOFSEM'99: Theory and Practice of Informatics. Proceedings, 1999. XIII, 498 pages. 1999.

Vol. 1726: V. Varadharajan, Y. Mu (Eds.), Information and Communication Security. Proceedings, 1999. XI, 325 pages. 1999.

Vol. 1727: P.P. Chen, D.W. Embley, J. Kouloumdjian, S.W. Liddle, J.F. Roddick (Eds.), Advances in Conceptual Modeling. Proceedings, 1999. XI, 389 pages. 1999.

Vol. 1728: J. Akoka, M. Bouzeghoub, I. Comyn-Wattiau, E. Métais (Eds.), Conceptual Modeling – ER '99. Proceedings, 1999. XIV, 540 pages. 1999.

Vol. 1729: M. Mambo, Y. Zheng (Eds.), Information Security. Proceedings, 1999. IX, 277 pages. 1999.

Vol. 1730: M. Gelfond, N. Leone, G. Pfeifer (Eds.), Logic Programming and Nonmonotonic Reasoning. Proceedings, 1999. XI, 391 pages. 1999. (Subseries LNAI).

Vol. 1731: J. Kratochvíl (Ed.), Graph Drawing. Proceedings, 1999. XIII, 422 pages. 1999.

Vol. 1732: S. Matsuoka, R.R. Oldehoeft, M. Tholburn (Eds.), Computing in Object-Oriented Parallel Environments. Proceedings, 1999. VIII, 205 pages. 1999.

Vol. 1733: H. Nakashima, C. Zhang (Eds.), Approaches to Intelligent Agents. Proceedings, 1999. XII, 241 pages. 1999. (Subseries LNAI).

Vol. 1734: H. Hellwagner, A. Reinefeld (Eds.), SCI: Scalable Coherent Interface. XXI, 490 pages. 1999.

Vol. 1564: M. Vazirgiannis, Interactive Multimedia Documents. XIII, 161 pages. 1999.

Vol. 1591: D.J. Duke, I. Herman, M.S. Marshall, PREMO: A Framework for Multimedia Middleware. XII, 254 pages. 1999.

Vol. 1624: J. A. Padget (Ed.), Collaboration between Human and Artificial Societies. XIV, 301 pages. 1999. (Subseries LNAI).

Vol. 1635: X. Tu, Artificial Animals for Computer Animation. XIV, 172 pages. 1999.

Vol. 1646: B. Westfechtel, Models and Tools for Managing Development Processes. XIV, 418 pages. 1999.

Vol. 1735: J.W. Amtrup, Incremental Speech Translation. XV, 200 pages. 1999. (Subseries LNAI).

Vol. 1736: L. Rizzo, S. Fdida (Eds.): Networked Group Communication. Proceedings, 1999. XIII, 339 pages. 1999.

Vol. 1737: P. Agouris, A. Stefanidis (Eds.), Integrated Spatial Databases. Proceedings, 1999. X, 317 pages. 1999.

Vol. 1738: C. Pandu Rangan, V. Raman, R. Ramanujam (Eds.), Foundations of Software Technology and Theoretical Computer Science. Proceedings, 1999. XII, 452 pages. 1999.

Vol. 1739: A. Braffort, R. Gherbi, S. Gibet, J. Richardson, D. Teil (Eds.), Gesture-Based Communication in Human-Computer Interaction. Proceedings, 1999. XI, 333 pages. 1999. (Subseries LNAI).

Vol. 1740: R. Baumgart (Ed.): Secure Networking – CQRE [Secure] '99. Proceedings, 1999. IX, 261 pages. 1999.

Vol. 1741: A. Aggarwal, C. Pandu Rangan (Eds.), Algorithms and Computation. Proceedings, 1999. XIII, 448 pages. 1999.

Vol. 1742: P.S. Thiagarajan, R. Yap (Eds.), Advances in Computing Science – ASIAN'99. Proceedings, 1999. XI, 397 pages. 1999.

Vol. 1743: A. Moreira, S. Demeyer (Eds.), Object-Oriented Technology. Proceedings, 1999. XVII, 389 pages. 1999.

Vol. 1744: S. Staab, Extracting Degree Information from Texts. X; 187 pages. 1999. (Subseries LNAI).

Vol. 1745: P. Banerjee, V.K. Prasanna, B.P. Sinha (Eds.), High Performance Computing – HiPC'99. Proceedings, 1999. XXII, 412 pages. 1999.

Vol. 1746: M. Walker (Ed.), Cryptography and Coding. Proceedings, 1999. IX, 313 pages. 1999.

Vol. 1747: N. Foo (Ed.), Adavanced Topics in Artificial Intelligence. Proceedings, 1999. XV, 500 pages. 1999. (Subseries LNAI).

Vol. 1748: H.V. Leong, W.-C. Lee, B. Li, L. Yin (Eds.), Mobile Data Access. Proceedings, 1999. X, 245 pages. 1999.

Vol. 1749: L. C.-K. Hui, D.L. Lee (Eds.), Internet Applications. Proceedings, 1999. XX, 518 pages. 1999.

Vol. 1750: D.E. Knuth, MMIXware. VIII, 550 pages. 1999.

Vol. 1751: H. Imai, Y. Zheng (Eds.), Public Key Cryptography. Proceedings, 2000. XI, 485 pages. 2000.

Vol. 1753: E. Pontelli, V. Santos Costa (Eds.), Practical Aspects of Declarative Languages. Proceedings, 2000. X, 327 pages. 2000.

Vol. 1754: J. Väänänen (Ed.), Generalized Quantifiers and Computation. Proceedings, 1997. VII, 139 pages. 1999.

Vol. 1755: D. Bjørner, M. Broy, A.V. Zamulin (Eds.), Perspectives of System Informatics. Proceedings, 1999. XII, 540 pages. 2000.

Vol. 1760: J.-J. Ch. Meyer, P.-Y. Schobbens (Eds.), Formal Models of Agents. Poceedings. VIII, 253 pages. 1999. (Subseries LNAI).

Vol. 1762: K.-D. Schewe, B. Thalheim (Eds.), Foundations of Information and Knowledge Systems. Proceedings, 2000. X, 305 pages. 2000.

Vol. 1767: G. Bongiovanni, G. Gambosi, R. Petreschi (Eds.), Algorithms and Complexity. Proceedings, 2000. VIII, 317 pages. 2000.

Vol. 1770: H. Reichel, S. Tison (Eds.), STACS 2000. Proceedings, 2000. XIV, 662 pages. 2000.

Vol. 1774: J. Delgado, G.D. Stamoulis, A. Mullery, D. Prevedourou, K. Start (Eds.), Telecommunications and IT Convergence Towards Service E-volution. Proceedings, 2000. XIII, 350 pages. 2000.

Vol. 1780: R. Conradi (Ed.), Software Process Technology. Proceedings, 2000. IX, 249 pages. 2000.